ELEMENTS OF
MARINE ECOLOGY

An Introductory Course

R. V. TAIT

Department of Life Sciences,
The Polytechnic of Central London

This edition revised by

R. S. De SANTO, Ph.D.

Chief Ecologist,
C. E. Maguire, Inc.,
Hartford, Connecticut

SPRINGER-VERLAG

NEW YORK

Second English edition published in 1972 by
Butterworth & Co (Publishers) Ltd
88 Kingsway, London WC2B 6AB

First North American edition published by
Springer–Verlag New York Inc,
175 Fifth Avenue, New York, N.Y. 10010

Library of Congress Catalog Card Number 73–20920
International Standard Book Number 0–387–91113–8

Printed in the United States of America

PREFACE

The widening interest in marine biology has led to the establishment of an increasing number of school and undergraduate courses in the subject. There are many books on various aspects of marine biology which students can read with advantage, but few that are suitable as introductory reading at the commencement of studies. This book has been compiled primarily as an aid for zoology students at the start of a special course on marine biology. The text is an introduction to the author's annual course for undergraduates. The aim has been a concise presentation of information and ideas over the general field of marine ecology, with guidance on the selection of more advanced reading. The sources of further information given at the end of each chapter have been chosen as far as possible from books and journals to which students should have reasonably easy access. These lists provide a selection of additional reading which starts at an elementary level and becomes more advanced as the course proceeds.

Students entering the author's course are usually in their third undergraduate year, and a general knowledge of the phyla is therefore assumed. They have previously attended at least two field courses involving work on the sea-shore, directed mainly to the identification of littoral organisms, and have learnt to recognize the commoner species. Although there is further work to be done on identification, especially on plankton, this aspect of the course is not included in this volume. There are already several students' books which are useful guides to identification of marine species, and these are included in the book lists.

Some questions for further study and essays or class discussions, and summaries of some laboratory exercises and fieldwork suitable for undergraduate classes, have been included as appendices.

PREFACE TO SECOND EDITION

The changes in the Second Edition have not altered the broad pattern and approach of the book. They are mostly clarifications and expansions of selected topics with additional illustrations, introduced partly in response to the helpful comments of students and partly to take account of new work, without allowing the book to grow unduly in length. A short chapter has been added which discusses energy relationships in a shallow marine ecosystem. The bibliographies have been revised and updated.

The use of this book in biology courses in schools has been kept in mind by adhering to simple terminology which should present no obstacles to science pupils.

PREFACE TO THE FIRST NORTH AMERICAN EDITION

The aim of this North American Edition is to serve as the core text of a Marine Ecology Course which relates to both the North American and European marine environment. Emphasis has been made as broad as possible and the student is led from general concepts of world wide significance to an understanding of certain specific habitats which are accessible to those interested in class field work.

The text is supported by reading lists which include elementary and advanced material, and there are appendices which indicate topics for class discussion and further study, laboratory projects and field work. Because field work may not be practicable for all classes, the course may be designed around laboratory exercises using materials shipped from various supply houses, some of which are listed in Appendix 3. Since the Metric System has been used in the text, Appendix 5 has been provided which contains a general definition of the system with factors to permit conversion to the U.S. Customary System when necessary. Now that the Metric System is internationally accepted as the system of weights and measures for the sciences, students should be encouraged to use it exclusively.

We wish to thank Jared T. Wibberley for assistance in collecting and interpreting fisheries information in the preparation of this revised edition. We also wish to thank Behrie Knauth for her work on layout of the photographs in this edition.

Robert S. De Santo
R. V. Tait

CONTENTS

CONTENTS

CHAPTER 1

THE OCEANS

EXTENT AND DEPTH

Sea-water covers approximately 71 per cent of the earth's surface, an area of about 361 million square kilometres (139 million square miles) comprising the major ocean areas shown in *Figure 1.1* (*see* page 2). In the deepest parts the bottom lies more than 10,000 m from the surface, and the average depth is about 3,700 m. Although marine organisms are unevenly distributed, they occur throughout this vast extent of water and have been brought up from the deepest places.

Close to land the sea is mostly shallow, the bottom shelving gradually from the shore to a depth of about 200 m. This coastal ledge of shallow sea bottom is the *continental shelf* (*Figure 1.2, see* page 3). About 8 per cent of the total sea area lies above it. Its seaward margin is termed the *continental edge*, beyond which the water becomes much deeper. The width of the continental shelf varies very much in different parts of the world (*Figure 1.3, see* page 4). It is relatively irregular around the North American continent where the continental edge extends as much as about 400 kilometres off of Newfoundland, to virtually no continental shelf along the southern Pacific Coast. The shelf is also broad beneath the China Sea, along the Arctic coast of Siberia, under Hudson Bay, and along the Atlantic coast of Patagonia where the shelf extends out to the Falkland Islands. Many of the shelf areas are of special economic importance because here the major commercial fisheries take place. Over 80 per cent of the landings of world fisheries are taken from the continental shelf, and to a rapidly increasing extent the shelf is also being exploited for sources of oil and gas.

Several processes contribute to the formation of the continental shelf. It is formed partly by wave erosion cutting back the coastline. It may be extended seawards by accumulations of material eroded from the coast, or by river-borne silt deposited on the continental slope. Parts of the shelf appear to consist largely of material held against the continents by underwater barriers formed by reef-building organisms or by tectonic folding. In other places the shelf has been formed chiefly by sinking and inundation of the land; for example, under the North Sea. It is possible that in some regions the shelf has been broadened by increments of material thrust up the continental slope by pressures between the continental blocks and the deep ocean floor[16].

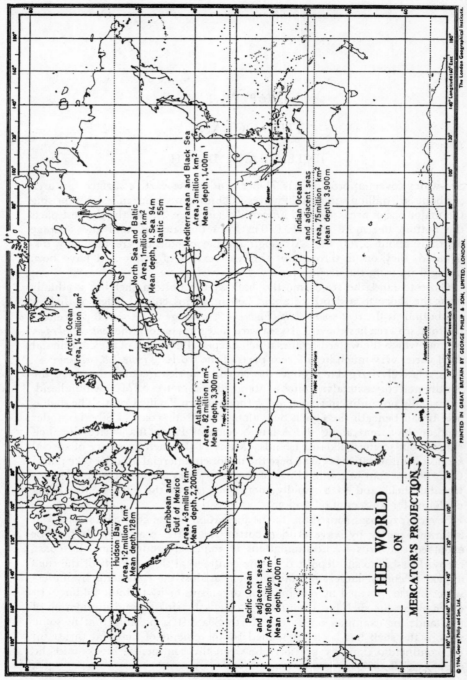

The World on Mercator's Projection

Pacific Ocean and adjacent seas
Area, 180 million km²
Mean depth, 4,000 m

Hudson Bay
Area, 1·2 million km²
Mean depth, 128 m

Caribbean and Gulf of Mexico
Area, 4·3 million km²
Mean depth, 2,200 m

Arctic Ocean
Area, 14 million km²

North Sea and Baltic
Area, 1 million km²
Mean depth, N. Sea 94 m, Baltic 55 m

Atlantic
Area, 82 million km²
Mean depth, 3,300 m

Mediterranean and Black Sea
Area, 3 million km²
Mean depth, 1,400 m

Indian Ocean and adjacent seas
Area, 75 million km²
Mean depth, 3,900 m

180° Longitude 160° West   140°   160°   The London Geographical Institute.   140° Longitude 160° East   180°

20° Meridian of 0° Greenwich. 20°

© 1966, George Philip and Son, Ltd.

PRINTED IN GREAT BRITAIN BY GEORGE PHILIP & SON, LIMITED, LONDON.

Figure 1.1 Areas and mean depths of major oceans and seas
(Based on a Map of the World by courtesy of G. Philip and Son Ltd.)

EXTENT AND DEPTH

Beyond the continental edge the gradient of the sea bottom becomes steeper (the *continental slope*) and descends to the floor of the ocean basins, often reaching a depth of 3,000–6,000 m and even deeper in some places. The gradient of the continental slope varies with locality, averaging about 1 in 15, but may be as steep as 45 degrees. The slope is seldom an even descent, and is much fissured by irregular gullies and steep-sided submarine canyons.

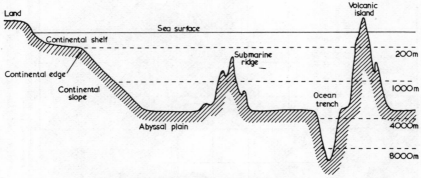

Figure 1.2. Terms applied to parts of the sea bottom

Beyond the continental slope, the deep ocean floor may be virtually flat over great areas, forming an *abyssal plain* extending for hundreds of miles with only slight changes of level. But in places the ocean floor rises to form ranges of submarine mountains with many summits ascending to within 2,000–4,000 m of the surface and the highest peaks breaking the surface as oceanic islands. Submarine ridges and plateaux cover a great area of the earth's surface (*Figure 1.3*), approximately equal to that of the continents.

One of these submarine ridges lies between the North Atlantic and Arctic basins, extending from the north of Scotland and the Orkneys and Shetlands to Rockall and the Faeroes (the Wyville–Thompson ridge), and then extends to Iceland (the Iceland–Faeroes rise), and across to Greenland and Labrador (the Greenland–Iceland rise). Much of the crest of this ridge lies within 500 m of the surface and forms a barrier which separates the deep levels of the Arctic basin from those of the Atlantic.

The bottom of the Atlantic is divided into two basins by the mid-Atlantic ridge which extends from the Arctic through Iceland, and then follows a roughly S-shaped course from north to south, touching the surface at the islands of the Azores, St. Paul, Ascension, Tristan da Cunha and Bouvet. A branch of the mid-Atlantic ridge, the Walvis ridge, extends from Tristan da Cunha to Walvis Bay on the west coast of Africa. South of South Africa the mid-Atlantic ridge links with a north-south submarine ridge, the mid-Indian ridge, bisecting the Indian Ocean between Antarctica and the Indian and Arabian Peninsulas, and extending into the Arabian Sea (the

3

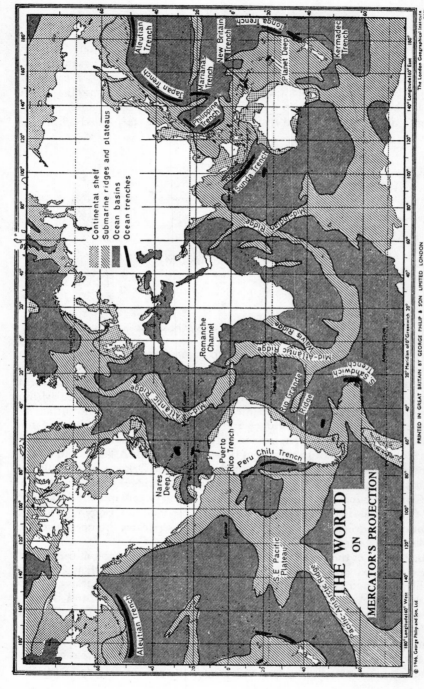

Figure 1.3. *Main areas of continental shelf, submarine ridges and plateaux, and ocean trenches*
(Based on a Map of the World by courtesy of G. Philip and Son Ltd.

PRINTED IN GREAT BRITAIN BY GEORGE PHILIP & SON LIMITED LONDON

4

Carlsberg ridge). In the Pacific a very broad submarine plateau extends in a north-easterly direction from Antarctica to the west coast of North and Central America, its peaks forming some of the East Pacific Islands. The more numerous islands of the Central and West Pacific appear mainly to be of separate volcanic origin. A peculiar feature of the Pacific Ocean is the large number of flat-topped, underwater hills known as guyots. Although the summits of some of them now lie beneath as much as 800 m of water, they have the appearance of having been worn flat by wave erosion. It seems likely that at some earlier time these volcanic mounds reached above the surface, but have subsequently subsided.

In other parts the ocean bottom is furrowed by deeper troughs, the ocean trenches (*Figure 1.3*), where the bottom reaches depths of over 7,000 m. These trenches occur mainly beneath the western Pacific Ocean close to oceanic islands; for example, east of the Mariana Islands is the Mariana Trench, where the deepest known soundings have been made at over 11,000 m. This is part of a great line of trenches extending north from the Philippines, along the east of Japan and on to the Aleutians. The bottom is also very deep in the New Britain Trench near the Solomon Islands, and in the Tonga Trench and Kermadec Deep to the north-east of New Zealand. In the Indian Ocean the deepest water has been found in the Sunda Trench south of Java and also in an area south-east of the Cocos Islands. In the Atlantic, water of comparable depth occurs in a pit north-east of Puerto Rico, and a trench near the South Sandwich Islands.

Submarine ridges and trenches are of special geological interest as areas of volcanic activity, probably associated with movements of the earth's crust. The ridges are essentially different from mountain ranges on land because they are formed entirely of extrusions of igneous rock into the sea floor, whereas mountains on land consist mainly of folded upthrusts of sedimentary rock. According to the current theory of seafloor spreading, the outer crust of the earth (the *lithosphere*) is made up of a number of separate plates which cover the molten *mantle* rather like a cracked shell. The continents lie upon the plates of the lithosphere. These plates are not fixed in position, but are moved over the mantle by forces not yet understood but probably due to inequalities of temperature in different parts. The submarine ridges are believed to mark the lines where lithosphere plates are moving apart. To fill the gaps between the separating plates there is an upward flow of mantle basalts to the surface, forming a submarine ridge which gradually subsides laterally to become a new ocean floor. Along the centre line of each ridge there is a depression which marks the actual line of division from which lateral spreading is taking place (*Figure 1.4*). Where the edges of moving plates collide they may fold or buckle to cause the upthrust of a mountain range along the line of collision, or one plate may be forced below the other to form a deep oceanic trench with adjacent volcanic islands. On this theory, the Atlantic and part of the Indian Oceans

5

are thought to be younger than the Pacific, and to have originated in Triassic times when a splitting of the lithosphere was followed by a break-up and separation of continental blocks. The Atlantic and Indian Oceans may still be enlarging at the expense of the Pacific by a westward drift of North and South America and by a combination of northward and eastward drifts of the crustal plates bearing Africa, Eurasia and Australia.

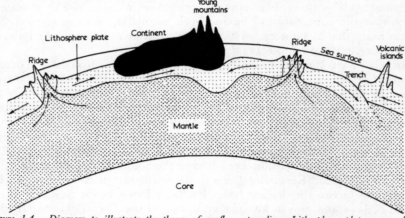

Figure 1.4. Diagram to illustrate the theory of seafloor spreading. Lithosphere plates, on which continents lie, are believed to move over the mantle. Mountains or ocean trenches form at lines of collision. Submarine ridges mark lines of separation where mantle material moves to surface

OCEAN CURRENTS

The major currents of the oceans are caused by the combined effects of wind action on the surface, and density differences between different parts of the sea. The density differences exist mainly because of inequalities of heat exchange between atmosphere and water at various parts of the sea surface, and also because of differences of evaporation and dilution. The course taken by currents is influenced by the rotation of the earth and by the shape of the continents and ocean floor. Their flow is meandering rather than steady, complicated by innumerable ever-changing eddies, comparable in some respects with the movements of the atmosphere but proceeding very slowly.

Wind action on the surface does not simply blow the water in the same direction as the wind, except in very shallow depths. The earth's rotation causes a deflecting effect so that surface water is moved at an angle to the wind. The deflection acts to the right of wind direction in the northern hemisphere and to the left in the southern hemisphere, and is theoretically a deflection of 45 degrees at the surface in deep water of uniform density. The deflection increases with depth, but the speed of the wind-generated current decreases logarithmically with depth and becomes almost zero at the depth at which its direction is opposite to the surface movement. The

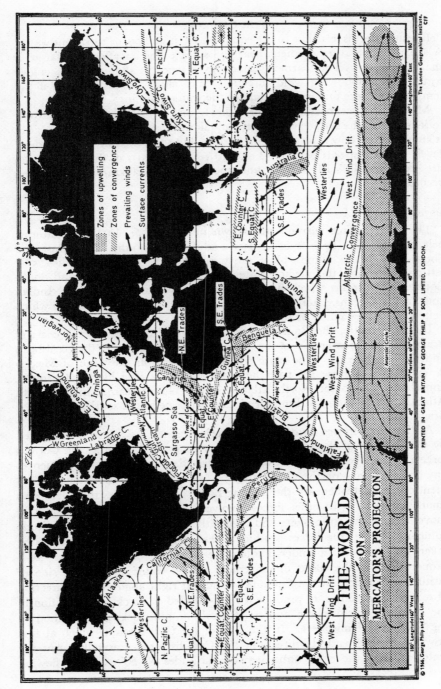

Figure 1.5 Prevailing winds and major surface currents of the oceans, with zones of upwelling and convergence
(Based on a Map of the World by courtesy of G. Philip and Son Ltd.)

7

deflecting effect is less close to the equator and in shallow or turbulent water. A sharp temperature gradient near the surface has an effect similar to a shallow bottom. The warmer surface layer tends to slide over the colder water below, following wind direction more closely than it would if the temperature were uniform throughout the water column.

Mostly, the ocean currents move slowly and irregularly. In the Equatorial currents the surface water generally flows at some 8–14 km/day. Little is known about flow rates below the surface, but some measurements at deep levels indicate speeds between 2 and 10 km/day. Parts of the Gulf Stream are exceptionally fast moving and speeds up to 180 km/day have been recorded.

Surface Currents

The chief surface currents are shown in *Figure 1.5*. In the Equatorial belt between the Tropics of Cancer and Capricorn, the North-East and South-East Trade Winds blow fairly consistently throughout the year, setting in motion the surface water to form the great North and South Equatorial Currents which flow from east to west in the Atlantic, Indian and Pacific Oceans. Across the path of these currents lie continents which deflect the water north or south.

In the Atlantic, the Equatorial currents are obstructed by the coast of Brazil, and the greater part of the water flows northwards into the Caribbean and Gulf of Mexico. The main surface outflow from the Gulf of Mexico flows strongly northwards past the coast of Florida (the Florida Current), and then out into the North Atlantic as the Gulf Stream. In the South Atlantic, water from the Equatorial Current is deflected southwards as the Brazil Current.

As the water moves away from equatorial regions, its course is influenced by the rotation of the earth. This effect, generally known as the Coriolis force after the French physicist who first derived an equation for it, influences any object moving on the earth's surface, and is due to the rotational movement of the earth beneath the moving body. In most cases the effect is so small compared with other forces involved that it can be ignored, but in movements of the atmosphere and oceans the Coriolis force has a magnitude comparable with the forces producing the motion, and must certainly be taken into account in understanding the course of ocean currents. It is proportional to the speed of movement of the water and to the sine of latitude, being zero at the Equator. It acts at right angles to the direction of movement, tending to produce a right-handed or clockwise deflection in the Northern Hemisphere and a leftwards or anticlockwise deflection in the Southern Hemisphere. In the case of the Gulf Stream, which at first flows in a north-easterly direction, the Coriolis force gradually turns the water towards the right until, between latitudes 40–50° N., it is flowing

8

eastwards across the Atlantic and becomes known as the North Atlantic Current. By the time it reaches the eastern part of the Atlantic it has been further deflected until it is flowing towards the south as the Canaries and Madeira Currents, eventually merging once again with the North Atlantic Equatorial Current. This vast circulation of surface water in a clockwise gyre surrounds an area of relatively little surface movement, the Sargasso Sea.

Where the North Atlantic Current moves eastwards across the Atlantic it is influenced by winds, the Westerlies, blowing from the south-west. These winds deflect some of the surface water towards the north-east to form the North Atlantic Drift, which flows into the Bay of Biscay and along the west and north of the British Isles, some eventually entering the northern part of the North Sea. North Atlantic Drift water also flows far up into the Arctic along the west and north coasts of Norway (the Norwegian Coastal Current), while some turns westwards south of Iceland (the Irminga Current). Part of this water flows clockwise around Iceland, and part flows on to the west and eventually reaches the west coast of Greenland (the West Greenland Current).

The inflow of water into the Arctic must be balanced by an equivalent outflow. Some of the surface water cools and sinks, and leaves the Arctic as a bottom current (page 11). There is also an outflow of cold surface water from the Arctic which enters the Atlantic as the East Greenland and Labrador Currents, and flows along the coast of Labrador and down the eastern seaboard of the United States, eventually sinking below the warm waters of the Gulf Stream flowing in the opposite direction.

In the South Atlantic, water from the Brazil Current under the influence of the Coriolis force makes a counter-clockwise rotation, flowing eastwards across the Atlantic between latitudes 30–40° S., and turning in a northerly direction along the west coast of Africa. Here it is known as the Benguela Current which eventually merges with the Equatorial Current. In the southern hemisphere the Westerlies blow from the north-west, and deflect surface water into the Southern Ocean where there are no intervening land masses to interrupt the flow. Here the sea is driven continually in an easterly direction by the prevailing winds and becomes a great mass of moving water, the Antarctic Circumpolar Current, which encircles the Southern Continent. The surface current is termed the West Wind Drift.

The surface movements of the Pacific Ocean have a broadly similar pattern to those of the Atlantic. The Kuro Siwo Current, flowing in a north-easterly direction past the south island of Japan, is the counterpart of the Gulf Stream in the Atlantic. This water later moves eastwards across the North Pacific towards the coast of British Columbia (The North Pacific Current), and then mostly turns south as the California Current. A cold current, the Oyo Siwo, flows down the western side of the Pacific towards the north Japanese Island.

The surface circulation of the northern part of the Indian Ocean is

9

complicated by seasonal changes in the direction of the monsoons, and will not be considered here. In the South Indian and South Pacific Oceans there is a counter-clockwise surface gyre and a deflection of water into the Southern Ocean, similar to the South Atlantic.

In the equatorial belt between the latitudes of the N.E. and S.E. Trade Winds there is a calm zone, the Doldrums, where the effects of wind are minimal. In the Pacific and Indian Oceans, a certain amount of backflow of surface water towards the east occurs in this region, forming the Equatorial Countercurrents. There is relatively little backflow of surface water in the Atlantic, but a short distance below the surface there is an appreciable movement of water from west to east. In the Pacific, there is an even more extensive subsurface current, the Cromwell Current, which transports a large volume of water in an easterly direction between latitudes 2° N.–2° S. at depths of 20–200 m.

The distribution of warm or cold surface water has a great influence on climate, and accounts for the climatic differences at equal latitudes on the east and west sides of the oceans. For example, the British Isles lie to the north of Newfoundland but here we have a temperate climate whereas Newfoundland is subarctic. This is because warm water from low latitudes moves across to high latitudes towards the eastern side of the ocean, and cold water moves towards low latitudes along the western side. Consequently, mild conditions extend further north in the east.

Water Movements Below the Surface

Oceanic circulation should be visualized in three dimensions. We have already mentioned that wind action on the surface sets different layers of water in movement in different directions (page 6). Where the wind causes a surface current, the moving water must be replaced by a corresponding inflow from elsewhere. This may be surface water from other regions or deep water rising to the surface, often both. Also, when surface water flows from low to high latitudes, cooling leads eventually to sinking and this causes movement of water at deep levels.

The replacement of the water of the North and South Equatorial Currents is derived partly from surface water from higher latitudes and partly from upwelling deep water. The temperature of the Canaries and Benguela Currents is low compared with other surface water at these latitudes because of mixing with enormous volumes of cold water from below. Similarly, upwelling into the California, Peru and West Australia Currents cools the surface water (*Figures 1.5* and *5.3*).

Water movements also arise from density changes due to differences of temperature or salinity. In low latitudes the surface water is warm, and has a low density. We have already seen how this water is carried by surface currents into high latitudes, and there it loses heat and increases in density until it eventually becomes heavier than the underlying water. It then

10

sinks and tends to return equatorwards at deep levels. There are complications due to salinity changes which occur as the water moves from place to place. Although warming in low latitudes reduces the water density, this is offset to some extent by evaporation which raises salinity and increases density. But there are tropical areas where heavy rainfall dilutes the water and decreases the density. In high latitudes, water density is increased by cooling and also by the greater salinity which occurs when ice crystals separate in the formation of sea ice. On the other hand, the density is lowered by dilution of the water by snow, rainfall, land drainage and the melting of ice. However, in general terms, high density water is formed at the surface in high latitudes and flows to low latitudes below the surface.

Several factors influence the course of subsurface currents. They are subject to the Coriolis force. They may be deflected or obstructed by submarine ridges. Their direction may be modified by the presence and movements of other masses of water, and the interactions are complex and not well understood. In many areas it is possible to distinguish three main systems of subsurface water movements, the Bottom current, the Deep current, and the Intermediate current.

In the Atlantic, Indian and Pacific Oceans the Bottom currents result mainly from the sinking of cold water around the Antarctic continent. The spread of cold bottom water from the Arctic Ocean is obstructed by the series of submarine ridges between Scotland and Labrador (page 3) and by the shallow Bering Straits. Cold water sinking in the Arctic is therefore trapped in the Arctic Basin. Beneath the Southern Ocean the cold water can escape, and creeps slowly northwards along the bottom, initially at a temperature of about 0°C but gradually becoming warmer as heat is gained by admixture with other warmer water, and perhaps a little by conduction through the sea-bed.

In the Atlantic, this Bottom current flows mainly up the western basin to the west of the mid-Atlantic ridge, being held back from the eastern basin by the Walvis ridge. A corresponding ridge between Tristan da Cunha and the Brazil coast, the Rio Grande ridge, is incomplete and permits the passage of the bottom water. Just south of the Equator the mid-Atlantic ridge is cut by the Romanche Channel, through which some bottom water from the Antarctic eventually enters the eastern basin (*Figure 1.3*).

The bottom current in the Atlantic, deriving from the Antarctic and usually termed the Antarctic Bottom current, flows northwards across the Equator and has been traced to about latitude 40° N. Here it gradually loses its identity as it merges with water flowing in the opposite direction, the North Atlantic Deep current (*Figure 1.6*). This water comes mainly from the cooling and sinking of surface water carried into the Arctic by the North Atlantic Drift. We have seen that the coldest water is held back within the Arctic basin by submarine ridges, but a large volume of water spills over the crest and this becomes the North Atlantic Deep current. It has at first a

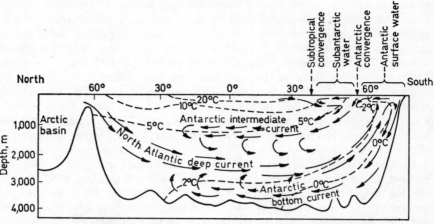

Figure 1.6 Some of the main movements of water below the surface of the Atlantic

temperature of 7–8°C and is characterized by relatively high salinity and oxygen content. It flows down the southern face of the ridge and along the bottom until it meets the colder Bottom current moving northwards from Antarctica. It continues its progress southwards getting colder as it goes, flowing above the Antarctic Bottom current at levels between about 1,500 and 3,000 m, and is joined by part of the deep water outflow from the Mediterranean.

Near latitude 60° S., the Deep current rises to the surface, upwelling to replace surface water, some of which is spreading northwards under wind influence as a surface drift while some is sinking due to cooling to become the Antarctic Bottom current. The cold surface water spreading to the north has a temperature of 0–4°C and the salinity is reduced to about 34 per mille by melting ice. At approximately latitude 50° S. it reaches an area of warmer and lighter surface water and sinks below it, continuing to flow northwards as the Antarctic Intermediate current at depths between about 800 and 1,200 m. This water can be traced to about 20° N.

Regions where surface currents meet, and surface water consequently sinks, are termed Convergences. The Antarctic Intermediate water sinks at the Antarctic Convergence, and this occurs all round the Southern Ocean, mainly between latitudes 50° and 60° S. Further north at about 40° S. is the Subtropical Convergence, another zone where surface water sinks and mixes with the Intermediate water. In the North Atlantic, a southward-flowing intermediate current is formed where the Labrador Current dips beneath the Gulf Stream (*Figure 1.5*).

In the Indian and Pacific Oceans the Intermediate and Bottom currents both flow in a northerly direction, as in the Atlantic, but are not traceable

12

so far north. The Deep currents seem to be derived largely from the back-flow of the Intermediate and Bottom currents, and flow southwards until they rise to the surface in the Southern Ocean. All around Antarctica the deep water upwells and the surface water sinks, and at the same time most of the water of the Southern Ocean at all depths is also flowing east-wards as the Antarctic Circumpolar Current. This brings about a continual transport of water from the Atlantic to the Indian Ocean, from the Indian to the Pacific and the Pacific to the Atlantic, whereby the waters of the major ocean basins are intermixed.

ECOLOGICAL ASPECTS OF THE MARINE ENVIRONMENT

Ecology is the study of relationships between organisms and their surround-ings. This study is fundamental to an understanding of biology because organisms cannot live as isolated units. The activities which comprise their lives are dependent upon, and closely controlled by, their external circum-stances, by the physical and chemical conditions in which they live and the populations of other organisms with which they interact. In addition, the activities of organisms have effects on their surroundings, altering them in various ways. Organisms therefore exist only as parts of a complex entity made up of interacting inorganic and biotic elements, to which we apply the term *ecosystem*.

All living processes involve energy changes. Energy for life is drawn primarily from solar radiation (*see* Chapter 9), transformed into the chemical energy of organic compounds by the photosynthetic processes of plants and transferred through the ecosystem by movements of materials within and between organisms. An ecosystem is therefore essentially a working, changing and evolving sequence of operations, powered by solar energy. In the long term, the intake of energy to the system is balanced by energy loss as heat.

An ecosystem may be considered on a large or a small scale. All life on earth constitutes a single ecosystem divisible into innumerable parts. In this book we are concerned with one of the major subdivisions, the sea and its populations comprising the marine ecosystem. This can be further sub-divided into many smaller sections. The aim of marine ecological studies is to understand the marine ecosystem as a working process, but at present this can only be done in a very uncertain way because of insufficient knowledge. This book provides general information about how marine organisms are influenced by, and have effects on, their environment, and describes some of the methods of investigation which may eventually provide the neces-sary information for a better understanding of marine ecosystems. In the concluding chapter (page 284) a tentative, elementary analysis of the working of part of the system is attempted.

We shall start by summarizing some general biological features of the marine environment. Sea-water is obviously an excellent medium for an abundance and variety of life. We know from geological evidence that the

13

seas have been well populated since the earliest time for which we have fossil records. It is widely thought that life originated in the sea, most likely in pools on the sea-shore where many different solutions of varying composition and concentration could accumulate in various conditions of temperature and illumination. The seas have now been populated for so long that it has been possible for marine life to evolve in great diversity.

Probably all natural elements are present in solution in the sea, and all the constituents needed for the formation of protoplasm are present in forms and concentrations suitable for direct utilisation by plants (page 80). The transparency of the water and its high content of bicarbonates and other forms of carbon dioxide (page 78) provide an environment in the upper layers of the sea in which plants can form organic materials by photosynthesis, and in this way great quantities of food become available for the animal population. Because light penetrates only a short distance into the water, marine plants must be able to float close to the surface or, if attached, are limited to shallow water.

In an aquatic environment very simple and fragile forms of life can exist because the water affords them support, flotation, transport and protection, thereby permitting very simple reproductive processes, and minimizing the need for structural complications, such as locomotor organs, skeletons or protective coverings. In aquatic organisms there are several advantages in small size. For example, a large surface-to-volume ratio retards sinking, facilitates absorption of solutes at great dilution and favours light absorption. Also, small organisms can usually reproduce rapidly to take advantage of favourable conditions.

We shall discuss later (in Chapter 4, page 62) how organisms in the sea may be influenced by some of the parameters of the environment, notably the temperature, composition, specific gravity, pressure, illumination and movements of the water. However, we have already said enough about the circulation of the oceans to indicate that the water is kept well mixed, and this ensures a generally homogeneous environment. The composition of sea-water (pages 72 ff.) remains almost uniform throughout its extent despite considerable differences in the rates of evaporation and addition of fresh water in different localities. The composition of present-day sea-water may differ in some respects from that of the remote past; but if so, marine organisms have been able to evolve and adjust to changing conditions. The body fluids of all the major groups of marine invertebrates are virtually isotonic with sea-water, and of a generally similar composition (page 75).

The high specific heat of water and the great volume of the oceans provides a huge thermal capacity, and the thorough mixing of the water ensures a fairly even distribution of heat. Consequently the temperature range of the oceans is relatively restricted and temperature changes occur slowly (page 63). The sinking of surface water due to cooling at high latitudes carries well-oxygenated water to the bottom, and thereby makes animal life possible at all depths. Despite biological activity, the buffer properties of the water (page 78) are sufficient to keep the pH stable. It is therefore a notable feature of the marine environment that conditions are remarkably constant over great areas, and many marine plants and animals have correspondingly wide distributions. Such changes as do occur take place slowly, giving time for some organisms to acclimatize. However, stable conditions permit the evolution of a diversity of forms whose environmental requirements are very precise and whose range is limited by quite slight changes in their surroundings.

Although certain parameters do not vary throughout enormous volumes of water, there are strong contrasts between some parts of the sea. The cold, dark, slowly-moving bottom layer of the deep ocean is obviously a very different environment from the well-illuminated, wave-tossed waters of the sea surface, or the strong currents and fluctuations of temperature and salinity that often occur near the coast. We therefore need a classification of subdivisions of the marine environment which takes account of different conditions of life in different parts of the oceans.

ELEMENTARY CLASSIFICATION OF THE MARINE ENVIRONMENT (*Figure 1.7*)

There are broadly two ways in which organisms live in the sea; they float or swim in the water, or they dwell on or within the sea bottom. We can correspondingly make two major divisions of the environment, the *Pelagic* and the *Benthic*, the Pelagic Division comprising the whole body of water forming the seas and oceans, and the Benthic Division the entire sea bottom.

In shallow water there is usually more movement and greater variations of composition and temperature than occur where the water is deep. We can therefore subdivide the Pelagic Division into (*a*), the Neritic Province, the shallow water over the continental shelf, and (*b*), the Oceanic Province, the deep water beyond the continental edge.

In deep water, conditions change with level and it is sometimes useful to distinguish three zones as follows. (*a*) The Epipelagic Zone from the surface to 200 m depth, in which there are sharp gradients of illumination, and often temperature, between the surface and the deeper levels; and also diurnal and seasonal changes of light intensity and temperature. In many areas the temperature gradient is irregular, involving discontinuities or thermoclines (page 64). Water movements may be relatively rapid. (*b*) The Mesopelagic Zone from 200–1,000 m depth, where very little light penetrates, and the temperature gradient is more even and gradual without much seasonal variation. An oxygen-minimum layer (page 77) and the maximum concentrations of nitrate and phosphate (pages 79–82) often occur within this zone. (*c*) The Bathypelagic Zone below 1,000 m, where darkness is virtually complete except for bioluminescence, temperature is low and constant and water pressure very great.

15

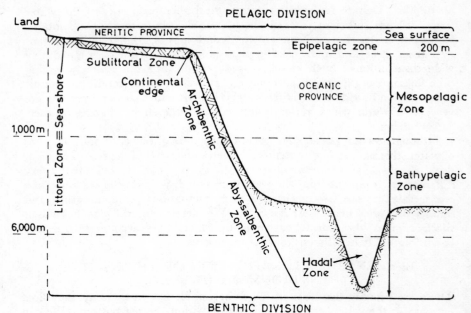

Figure 1.7 Main divisions of the marine environment

The sea bottom and the sea-shore together make up the Benthic Division which comprises three major zones, the Littoral, the Sublittoral and the Deep Sea Zones. The Littoral Zone includes the greater part of the sea-shore together with the wave-splashed region above high tide level (pages 185–186). The Sublittoral Zone is the shallow sea bottom extending from the lower part of the shore to the continental edge. The Deep Sea Zone lies below the continental shelf, and can be subdivided into an Archibenthic Zone between the continental edge and a depth of about 1,000 m, and an Abyssalbenthic Zone from 1,000 m downwards. The deepest parts of the ocean floor within the trenches below some 6,000–7,000 m are termed Hadal Zones.

Subdivisions of the marine environment with respect to temperature and light are mentioned later (pages 96–97).

Organisms of the Pelagic Division comprise two broad categories, *plankton* and *nekton*, according to their powers of locomotion. The plankton consist of floating plants and animals which drift with the water, and whose swimming powers, if any, serve mainly to keep them afloat rather than to carry them from place to place. A brief account of the constituents of marine plankton appears in the next chapter. The nekton comprises the more powerful swimming animals, vertebrates and cephalopods, which are capable of travelling from one place to another independently of the flow of the water.

The populations of the Benthic Division, the sessile and attached plants and animals and all the creeping and burrowing forms, are known collectively as *benthos*.

FURTHER READING

Books

[1] Defant, A. (1961). *Physical Oceanography*. Vols. 1 & 2. Oxford; Pergamon

[2] Fairbridge, R. W. (1966). *Encyclopaedia of Oceanography*. New York; Reinhold

[3] Sverdrup, H. U., Johnson, M. W. and Fleming, R. H. (1946). *The Oceans. Their Physics, Chemistry and General Biology*. 'Chap. 2. The Earth and the Ocean Basins. Chap. 8. The Sea as a Biological Environment. Chap. 11. General Character of Ocean Currents. Chap. 15. The Water Masses and Currents of the Oceans.' New York; Prentice-Hall

[4] Weyl, P. K. (1970). *Oceanography. An Introduction to the Marine Environment*. London and Chichester; Wiley

Papers

[5] Dietz, R. S. and Holden, J. C. (1970). 'The Breakup of Pangea' *Scient. Am.* October

[6] Fisher, R. L. and Revelle, R. (1955). 'The Trenches of the Pacific' *Scient. Am.* November

[7] Gordienko, P. A. (1961). 'The Arctic Ocean' *Scient. Am.* May

[8] Heezen, B. C. (1956). 'The Origin of Submarine Canyons' *Scient. Am.* August

[9] Heezen, B. C. (1960). 'The Rift in the Ocean Floor' *Scient. Am.* October

[10] Knauss, J. A. (1961). 'The Cromwell Current' *Scient. Am.* April

[11] Kort, V. G. (1962). 'The Antarctic Ocean' *Scient. Am.* September

[12] McDonald, J. E. (1952). 'The Coriolis Effect' *Scient. Am.* May

[13] Menard, H. W. (1961). 'The East Pacific Rise' *Scient. Am.* December

[14] Munk, W. (1955). 'The Circulation of the Oceans' *Scient. Am.* September

[15] Stommel, H. (1955). 'The Anatomy of the Atlantic' *Scient. Am.* January

[16] Emery, K. O. (1969). 'The Continental Shelves' *Scient. Am.* September

CHAPTER 2

MARINE PLANKTON

DEFINITIONS

The word plankton is taken from a Greek verb meaning to wander and is used to refer to those pelagic forms which are carried about by the movements of the water rather than by their own ability to swim. These organisms are called *planktonts*. The plants of the plankton are the *phytoplankton*, the animals the *zooplankton*.

Some planktonts can only float passively, unable to swim at all. Others are quite active swimmers but are so small that swimming does not move them far compared to the distance they are carried by the water, but serves chiefly to keep them afloat, alter their level, obtain food, avoid capture, find a mate or set up water-currents for respiration. Although the majority of planktonts are small, mainly of microscopic size, a few are quite large; for example, the tentacles of *Physalia* sometimes extend 15 m through the water, and there are Scyphomedusae which grow to over 2 m in diameter.

Terms in wide use for referring to different components of the plankton include the following:

Macroplankton—Large planktonts visible to the unaided eye, retained by a coarse net (gauge 00), with a mesh-aperture of approximately 1 mm. Exceptionally large planktonts are sometimes termed Megaloplankton.

Microplankton—Planktonic organisms less than 1 mm in maximum dimension, but retained by a fine-mesh plankton net (gauge 21), mesh-aperture approximately 0·06 mm.

Nanoplankton—Organisms too small to be retained by fine-mesh bolting silk (less than 60 μ) but larger than 5 μ maximum dimension.

Ultraplankton—Less than 5 μ maximum dimension.

Epiplankton—Plankton of the epipelagic zone, i.e. within the uppermost 200 m. Organisms which inhabit the extreme surface layer and surface film are termed *neuston*.

Bathyplankton—Plankton of deep levels.

Hypoplankton—Plankton living near the bottom.

Protoplankton—Pelagic bacteria and unicellular plants and animals.

Holoplankton, or permanent plankton—Organisms whose entire life span is planktonic.

Meroplankton or temporary plankton—Planktonic organisms passing through a pelagic phase which is only part of the total life-span; for example, planktonic spores, eggs or larvae of nektonic or benthic organisms.

Euphausids, mysids and other strongly swimming animals of intermediate size, though generally regarded as part of the macroplankton, are sometimes termed *micronekton*. The description *tychopelagic* refers to organisms of normally benthic habit which occasionally become stirred up from the bottom and carried into the water.

MARINE PHYTOPLANKTON

Marine phytoplankton is made up of small plants, mostly microscopic in size and unicellular. Two orders of algae commonly predominate in the phytoplankton, Diatoms (Bacillariophyceae) and Dinoflagellates (Dinophyceae). The phytoplankton often also includes a numerous and diverse collection of extremely small, motile plants collectively termed microflagellates.

Floating masses of large algae are found living and growing in some areas, notably the Sargassum weed of the Sargasso area of the North Atlantic, but these are not generally regarded as phytoplankton because they derive from the fragmentation of benthic plants growing on the sea bottom in shallow water.

Many of the phytoplankton species of the North Atlantic are described in detail in volumes by M. V. Lebour[5,6] and N. I. Hendey[4]. Wood and Lutes[14], provide a phytoplankton key to the northwest Atlantic coast.

Diatoms

The majority of diatoms are unicellular, uninucleate plants with a size range of about 15 μ to 400 μ in maximum dimension, although some smaller and a few considerably larger forms exist. The diatom cell, known as a frustule, has a cell wall of unusual composition and structure. It is impregnated with siliceous material giving a glassy quality and consists of two parts, the valves. At its simplest, for example in *Coscinodiscus*, the cell wall is

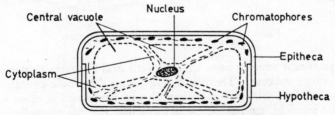

Figure 2.1. Diagrammatic section of a pillbox diatom

like a transparent pillbox (*Figure 2.1*), the larger valve or epitheca overlapping the smaller hypotheca much as the lid of a pillbox overlaps the base.

The valves are often very elaborately ornamented with an intricate sculpturing of minute depressions, perforations or tiny raised points which are sometimes arranged in beautiful symmetrical patterns of great variety. In some, the cell wall has larger projections forming spines, bristles and knobs. Ornamentation increases the surface area and also strengthens the cell wall, which in the majority of planktonic diatoms is very thin. In some species, growth occurs by elongation of the valves at their margins forming a number of intercalary bands, for example *Guinardia* (*Figure 2.2a*). Internal thickenings of these bands may form septa which partially divide the interior of the frustule.

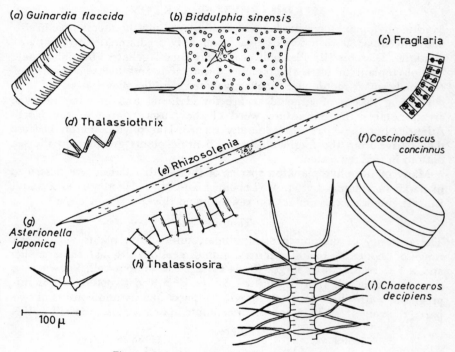

(a) *Guinardia flaccida*

(b) *Biddulphia sinensis*

(c) Fragilaria

(d) Thalassiothrix

(e) Rhizosolenia

(f) *Coscinodiscus concinnus*

(g) *Asterionella japonica*

(h) Thalassiosira

(i) *Chaetoceros decipiens*

100 μ

Figure 2.2 Some common diatoms from shallow seas

The cytoplasm usually lines the cell wall and contains numerous small, brown chromatophores. There is a large central vacuole containing a cell sap. The nucleus with an enclosing film of cytoplasm is often suspended within the vacuole, supported by cytoplasmic threads extending from the peripheral layer. In planktonic diatoms the cell sap is probably lighter than sea-water and may confer some buoyancy to support the heavier protoplasm and cell wall. In many diatoms the cytoplasm is not confined to the interior of the frustule, but exudes through small perforations to cover the surface or form long thin threads, and these may join the cells together in chains.

Planktonic diatoms present a considerable variety of shape, each in its way well adapted to provide a large surface/volume ratio (*see* page 95). They may be grouped into four broad categories as follows:

Pillbox shapes—usually circular and radially symmetrical when seen in top or bottom view, for example *Coscinodiscus* (*Figure 2.2f*), *Hyalodiscus*. Sometimes they are connected by protoplasmic strands to form chains, for example *Thalassiosira* (*Figure 2.2h*) *Coscinosira*.

Rod or needle shapes—The division between the valves may be at right angles to the long axis of the cell, for example *Rhizosolenia* (*Figure 2.2e*), and these are often joined end to end to form straight chains. In others the division runs lengthways, for example *Thalassiothrix*, *Asterionella* (*Figure 2.2d and g*), and these may be joined to form starlike clusters or irregular zig-zag strands.

Filamentous shapes—Cells joined end to end by the valve surfaces to form stiff, cylindrical chains (*Guinardia*, *Figure 2.2a*) or flexible ribbons (*Fragilaria*, *Figure 2.2c*).

Branched shapes—Cells bearing various large spines or other projections, and sometimes united into chains by contact between spines (*Chaetoceros*, *Figure 2.2i*), or by sticky secretions (*Biddulphia*, *Figure 2.2b*).

In addition to the planktonic forms there are numerous benthic species of diatom occurring on the shore or in shallow water. These may grow on the surface of sediments or form a slimy covering on rocks and stones. Some project above the surface of the substrate on short stalks. Diatoms are also commonly found attached to the surface of other plants or animals. Benthic diatoms usually have appreciably thicker and heavier cell walls than the planktonic species.

The usual method of reproduction is by simple asexual division. Under favourable conditions this may occur three or four times a day, so that rapid increase in numbers is possible. The protoplast enlarges and the nucleus and cytoplasm divide. The two valves become gradually separated, the daughter cells each retaining one valve of the parent cell. The retained valve becomes the epitheca of each daughter cell, and a new hypotheca is secreted, the margin of which fits inside the old valve. The new cell formed within the parent epitheca is therefore the same size as the parent cell, but the cell formed inside the original hypotheca is smaller. Because of this, it is a peculiarity of diatoms that the average size of the individuals in a population tends to decrease as division continues (*Figure 2.3*). This process of size-reduction does not go on indefinitely. Eventually the valves of the smaller individuals separate, the protoplasm flows out and the valves are shed. The naked protoplasm, known as an auxospore, enlarges and grows new, larger valves.

Some diatoms can form resistant spores to carry them over unfavourable periods, for example, during the winter months in neritic water when the

21

temperature falls and salinity may fluctuate appreciably. The cell vacuole disappears and the protoplasm becomes rounded, secreting a thick wall around itself. Probably many resistant spores sink to the bottom and are lost, but in shallow water some may be brought to the surface again later

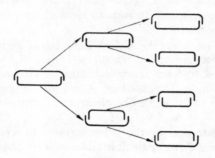

Figure 2.3. Reduction of mean cell-size in diatoms following cell division

by wave action, currents and turbulence, and then germinate. In high latitudes, diatom spores become enclosed in sea ice during the winter months and germinate the following year when the ice melts.

Sexual reproduction has been observed in certain diatoms. In some species it precedes auxospore formation, the protoplasts of two diatoms fusing to form a single auxospore. In other cases, fusion of protoplasts appears to give rise to two or more auxospores. The formation of microspores has also been observed, the protoplast dividing numerous times to form minute biflagellate structures which are thought to act as gametes.

Dinoflagellates

These are unicellular, biflagellate organisms with a range of size similar to diatoms but with a larger proportion of very small forms that escape through the mesh of fine plankton nets. The arrangement of the flagella is characteristic of the group. Typically (*Figure 2.4*) the cell is divided into anterior and posterior parts by a superficial encircling groove termed the girdle, in which lies a transverse flagellum wrapped around the cell and often attached to it by a thin membranelle. Immediately behind the origin of the transverse flagellum, a whip-like longitudinal flagellum arises in a groove known as the sulcus, and projects behind the cell. The longitudinal flagellum performs vigorous flicking movements and the transverse flagellum vibrates gently, the combined effects driving the organism forwards along a spiral path. There are various departures from this characteristic form. For example, *Amphisolenia* has a thin, rod-like shape. *Polykrikos (Figure 2.5f)*, has several nuclei and a series of girdles and sulci, usually eight, each provided with transverse and longitudinal flagella.

Many dinoflagellates have no cell wall. In these non-thecate forms the cytoplasm is covered only by a fine pellicle. Others are thecate, covered with a strong wall of interlocking cellulose plates. In certain species the cell wall is elaborated into spines, wings, or parachute-like extensions, and these are specially complex in some of the warm water forms, perhaps assisting flotation.

Dinoflagellates reproduce by asexual fission, but not as rapidly as diatoms. In thecate dinoflagellates the process is somewhat similar to fission in diatoms, each daughter cell retaining part of the old cell wall and secreting the other part, but the old and new cell plates do not overlap, and there is consequently no size reduction as occurs in diatoms. The daughter cells do not always separate completely and repeated divisions then form a chain. Resistant spores may be produced during adverse periods.

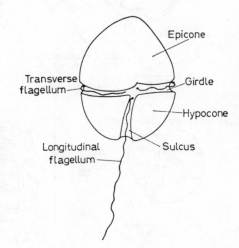

Figure 2.4. A simple non-thecate dinoflagellate

Many dinoflagellates contain small chromatophores and perform photosynthesis. The group is certainly important as primary producers of food materials. Some of them are highly pigmented, and are sometimes so numerous that the water appears distinctly coloured, different species producing green, red or yellow tints. There are also many colourless dinoflagellates. Thecate forms without chromatophores are presumably saprophytes, but some of the non-thecate, non-coloured forms are holozoic, feeding on various other small organisms including dinoflagellates, diatoms, microflagellates and bacteria. *Noctiluca* (*Figure 2.5e*) devours copepod larvae and other small metazoa. Some holozoic dinoflagellates possess tentacles, amoeboid processes or stinging threads for capturing their food.

Dinoflagellates are mainly a marine planktonic group occurring in both oceanic and neritic water. They are most numerous in the warmer parts of the sea, where they sometimes outnumber diatoms, but are also found in cold areas. Around the Atlantic coasts they are scarce in the winter months, and reach their greatest abundance in the midsummer period when the low concentration of nutrients seems to have less effect in limiting the growth

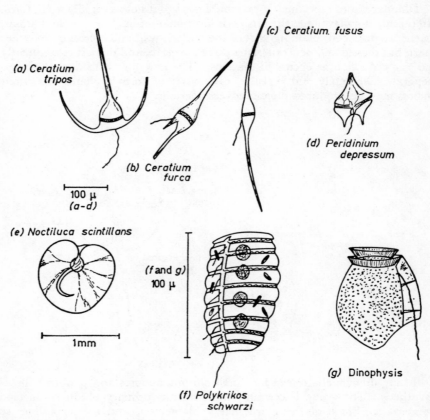

Figure 2.5 *Some dinoflagellates of the north Atlantic*

of dinoflagellates than of diatoms. Dinoflagellates also occur in fresh and brackish water, and are sometimes abundant in estuaries. Some are found in sand in the interstitial water between the particles. There are also many parasitic dinoflagellates infecting a variety of planktonic organisms including radiolaria, copepods, pteropods, larvaceans and fish eggs.

Certain species of dinoflagellates secrete highly toxic substances into the water which cause the death of other marine creatures; for example, the 'red tide' effect (page 84). Some species are luminous, and can be the cause of remarkable displays of phosphorescence in sea-water. *Noctiluca* is an example which sometimes occurs in swarms around the Atlantic coast, visible at night as myriads of tiny pin-points of light.

Some other Planktonic Plants

Halosphaera (*Figure 2.6*) is a large, spherical plant cell, often nearly 1 mm in diameter, with a tough elastic wall. There is a large central vacuole in which a single nucleus is usually suspended, but fully grown cells may contain as many as eight nuclei. In the peripheral cytoplasm are numerous small, yellowish-green chromatophores giving the cell a vivid colour.

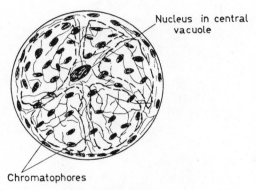

Nucleus in central vacuole

Chromatophores

Figure 2.6. Halosphaera

Asexual reproduction involves a motile phase, repeated divisions of the protoplasm leading to the liberation of numerous 4-flagellate spores. Sexual reproduction has not been recorded. Halosphaera is commonest in warm oceans but is sometimes carried into colder waters in great numbers by the North Atlantic Drift.

Phaeocystis (*Figure 2.7*) is a minute brownish-green biflagellate cell which develops into a colonial structure. The cells repeatedly divide, and extensive mucilaginous capsules form around the cells, binding them together in large gelatinous clumps. These are sometimes so numerous that they give the water a slightly slimy consistency, which is apparently distasteful to some animals; for example, herring shoals seem to avoid this water and the catches are poor when Phaeocystis occurs in quantity on the fishing grounds.

Silico-flagellates are a marine planktonic group of small unicellular plants, often about 100 μ in diameter, in which the protoplasm is supported by a

25

skeleton of interconnecting siliceous rods forming a capsular structure with outwardly-radiating spines. There is usually a single flagellum. They are sometimes present in considerable numbers, mainly in the colder parts of the seas.

The phytoplankton also sometimes includes filaments of blue-green algae, the Cyanophyceae (Myxophyceae). These filaments are short chains of minute spherical or oval cells with no definite chloroplasts, the pigments

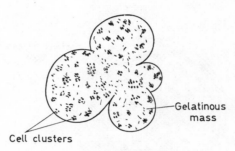

Figure 2.7. Phaeocystis

being diffused in the cytoplasm. In addition to chlorophyll, phycocyanin and phyco-erythrin are present, giving the cells their bluish-green colour. These plants have occasionally been found in large numbers at deep levels (page 166) and may be able in some circumstances to feed saprophytically.

Nanoplankton

The smallest planktonic organisms, less than 60 μ in diameter, are termed nanoplankton. These escape through fine-mesh plankton nets but may be collected by centrifuging, membrane filtration, or sedimentation (page 49). This diverse group of minute organisms includes the smallest diatoms and dinoflagellates, coccosphaeres, various other small flagellates known collectively as microflagellates, and also fungi and bacteria.

It is only in fairly recent years that the nanoplankton has been much studied, but it now appears that the quantity of living material in the water in this form sometimes exceeds that present as diatoms and dinoflagellates. The nanoplankton is thought to make a considerable, sometimes major, contribution to the primary production of organic food in the sea, and is specially important as the chief food for many larvae.

26

Coccolithophoridae

The coccolithophoridae (*Figure 2.8*) are minute. unicellular plants, mostly some 5–20 μ in diameter. Some are non-motile, but others are uni- or bi-flagellate. Not much is known of their life history, but some have been shown to have both motile and non-motile pelagic phases, and some have

Figure 2.8. A coccolithophore

a benthic filamentous phase. They contain a few brown chromatophores, and in the pelagic phases the surface is characteristically covered by a number of calcareous plates, usually extremely finely and elaborately sculptured. At their simplest, these plates are oval discs, but in some coccoliths the plates form long projections from the surface of the cell, often of bizarre design.

Coccoliths are widely distributed and are sometimes so numerous near the surface that they impart a slight coloration to the water. The appearance which herring fishermen call 'white water', regarded as indicating good fishing, is sometimes due to swarms of coccoliths. The calcareous plates of disintegrated coccoliths are a conspicuous component of the deep sea sediment in some areas.

Coccoliths have occasionally been found in surprisingly large quantities far below the photosynthetic zone, sometimes very numerous between 200 and 400 m, and even in considerable abundance at depths of 1,000–4,000 m. This deep-water distribution suggests that some coccoliths can feed to some extent by methods other than photosynthesis, perhaps by absorption of organic solutes or even by ingestion of organic particles. In some areas they may be an important source of food for some of the animals at deep levels (*see* page 166).

Microflagellates

The seas contain many minute unicellular organisms which swim by means of one or more flagella, and are loosely termed microflagellates. The majority are within the size range 1–20 μ in diameter, but there are a few larger species up to about 100 μ. Most of them contain chlorophyll, and there is no doubt that they are important as primary food producers. There

27

are also many colourless saprophytic forms. The life histories are not well known. Some have been observed to reproduce by fission, while others are spores of larger algae. Non-motile cells, similar in general appearance to microflagellates but lacking flagella, are also known; and in some cases the life history includes both motile and non-motile pelagic stages.

MARINE ZOOPLANKTON

The zooplankton includes a very wide variety of organisms. Every animal phylum contributes at least to the meroplankton, and it is beyond the scope of this book to attempt a comprehensive survey of the range of zooplankton organisms. Some excellent general accounts are now available, and several are listed at the end of this chapter. The purpose of this section is simply to mention the major groups of holoplanktonic animals with a few examples of species common in the north Atlantic.

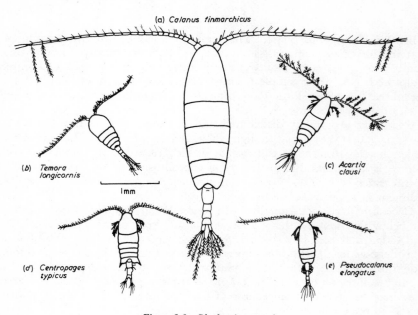

Figure 2.9. Planktonic copepods

The most conspicuous element in the permanent zooplankton, commonly amounting to at least 70 per cent of the total, is the Crustacea, and the predominant class is the Copepoda. These often outnumber all the other animal groups, and are represented by many species. In the plankton around the United States, species of *Calanus*, *Acartia*, *Centropages*, *Temora*, *Oithona*, *Pseudocalanus* and *Paracalanus* are common (*Figure 2.9*).

Another abundant group of planktonic Crustacea is the Euphausiacea. A large species, *Euphausia superba* (*Figure 2.10a*), occurs in enormous numbers in the Southern Ocean south of the Antarctic Convergence and constitutes

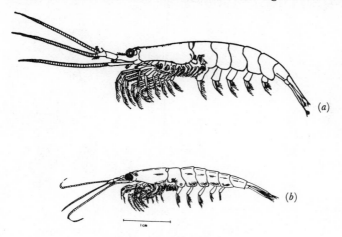

Figure 2.10. (a) Euphausia superba, *the 'krill' of Antarctic seas.* (b) Meganyctiphanes norvegica, *a common euphausid of the North Atlantic*

the 'krill' upon which the giant whales of the Antartic feed. In the north Atlantic, species of *Nyctiphanes*, *Meganyctiphanes* (*Figure 2.10b*) and *Thysanoessa* are common.

Other crustacean groups that are sometimes numerous in marine plankton are the Cladocera, for example *Podon* and *Evadne* (*Figure 2.11*), the Ostracoda, for example *Conchoecia*, *Philomedes*, and the Amphipoda, for example *Parathemisto* (*Figure 2.12*). Mysids mostly live close to the bottom but are sometimes found in coastal plankton, especially in estuarine regions. In addition to these holoplanktonic forms, great numbers of larvae are contributed to the plankton by benthic crustacea.

The phylum Chaetognatha is widespread and well represented in most plankton samples. Around the United States there are several species of *Sagitta* (*Figure 2.13d*), and *Eukrohnia hamata* occasionally enters the region from the Arctic. Chaetognaths are of special interest as 'indicator species' (page 110). They are an important group of planktonic predators.

The zooplankton includes a variety of Protozoa. We have mentioned earlier that some flagellates are holozoic. Foraminifera and Radiolaria are sufficiently numerous for their skeletons to form a conspicuous part of some deep-water sediments (page 153). Ciliates are found mainly in coastal plankton.

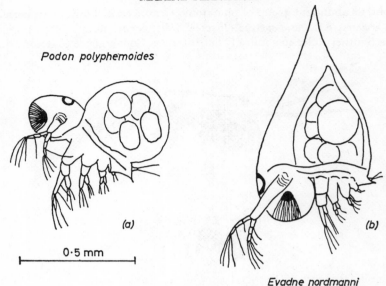

Podon polyphemoides

(a)

0·5 mm

(b)

Evadne nordmanni

Figure 2.11. Two species of Cladocera common in spring and summer

There are some holoplanktonic Cnidaria, for example, the Trachylina and Siphonophora (*Figure 2.13a* and *b*). The pelagic Scyphozoa mostly have a benthic stage, the scyphistoma, for example *Aurelia, Cyanea*, but this stage is omitted in some, for example *Pelagia*. The meroplankton includes great numbers of medusae set free by benthic hydroids.

The Ctenophora are holoplanktonic predators of widespread distribution. *Pleurobrachia* (*Figure 2.13c*) and *Mnemiopsis* are common around the United States.

Several families of polychaetes include holoplanktonic species, the most conspicuous being the family Tomopteridae. The genus *Tomopteris* (*Figure 2.13e*) occurs throughout the oceans. The majority of benthic annelids start life as planktonic larvae, and some also have adult pelagic phases, usually associated with spawning.

1 cm

Figure 2.12. Parathemisto gaudichaudi, *a hyperiid amphipod which is epipelagic in colder parts of North and South Atlantic*

30

A numerous and widespread group of planktonic molluscs is the Pteropoda, small opisthobranch gastropods. These are of two types, the tiny thecosomatous forms which have a very lightly-built shell, for example *Limacina* (*Figure 2.13h*) and the larger gymnosomatous forms which have no shell, for example *Clione*. Pelagic prosobranchs, the Heteropoda, for example *Carinaria*, *Pterotrachea*, occur in warm oceans. Benthic molluscs produce innumerable planktonic larvae. (*Limacina≡Spiratella*, pages 113–115.)

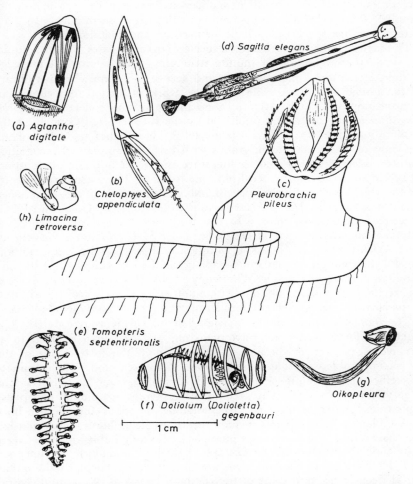

Figure 2.13. Various zooplanktons from around the United States. The following groups are represented in the figure; (a) Trachymedusa (b) Siphonophora (c) Ctenophora (d) Chaetognatha (e) Polychaeta (f) Thaliacea (g) Larvacea (h) Pteropoda thecosomata

Urochordates are sometimes abundant. There are three planktonic orders; the Larvacea, for example *Fritillaria, Oikopleura (Figure 2.13g) Appendicularia;* the Thaliacea, for example *Doliolum (Figure 2.13f), Salpa;* and the Lucida, for example *Pyrosoma.* Species of the Larvacea are common all round the world. The Thaliacea and Lucida in this area are mainly associated with the warmer water from the south and west.

Several other phyla, while not featuring prominently in the plankton, are sometimes represented. Small nematodes are occasionally seen in tow-net samples. Marine rotifers are sometimes collected close inshore. In deep water there are species of pelagic nemertines and holothurians.

The identification of zooplankton species is in many cases difficult, requiring detailed examination of minute structures, a considerable knowledge of the systematics of many groups, and access to a comprehensive library of taxonomic literature. The reference work of major importance for identification of North Atlantic zooplankton is the collection of *Fiches d'Identification du Zooplankton* published by the International Council for the Exploration of the Sea (ICES), Copenhagen[13]. These 'fiches' are leaflets of line drawings of many planktonic groups with brief descriptions of recognition features, references to taxonomic literature and information on distribution. They are added to and revised from time to time. There are two students' books which are useful for provisional identification of the commoner planktonts, namely, *Marine Plankton. A Practical Guide,* by G. E. and R. C. Newell[8], which relates mainly to the north-east Atlantic and includes a helpful bibliography, and *An Introduction to the Study of Tropical Plankton* by J. H. Wickstead[10]. Both books, in addition to many line drawings, contain general information on plankton sampling, analysis of samples and the distribution of some species. Dr. Wimpenny's book, *The Plankton of the Sea*[11], is an excellent general account of the plankton around the British Isles, and its many clear drawings are useful for identification, particularly for students.

The zooplankton includes both vegetarian and carnivorous feeders[17]. The vegetarian forms feed upon phytoplankton, and are often referred to as 'herbivores' or 'grazers' because their position in the food chains of the sea is comparable with that of herbivorous animals on land. These animals have efficient filtration mechanisms for sieving microscopic food dispersed in large volumes of water. The planktonic herbivores are mainly copepods, euphausids, cladocera, mysids, thecosomatous pteropods and the urochordates. Planktonic carnivores include medusae, ctenophores, chaetognaths, polychaetes and gymnosomatous pteropods. Feeding habits differ between quite closely related forms; for example, although the majority of copepods feed chiefly on phytoplankton, some are carnivorous, particularly those that live at deep levels, and some of the common copepods of Atlantic coastal waters appear to be omnivorous. The euphausids are also predominantly vegetarian but there are carnivorous and omnivorous species. In many cases, food requirements change with age. The majority of invertebrate larvae at first rely mainly upon phytoplankton for their food but become omnivorous or carnivorous later. Many fish larvae are at first omnivorous, but later take only animal food.

REFERENCES AND FURTHER READING

Books
1 Fraser, J. H. (1962). *Nature Adrift. The Story of Plankton.* London; Foulis
2 Hardy, A. (1956). *The Open Sea, Part 1. The World of Plankton.* London; Collin.
3 Hardy, A. C. (1967). *Great Waters.* A voyage of natural history to study whales, plankton and the waters of the Southern Ocean in the old Royal Research Ship, Discovery, with the results brought up to date by the findings of R.R.S. Discovery II. London; Collins
4 Hendey, N. I. (1964). An Introductory Account of the Smaller Algae of British Coastal Waters. Part 5. Bacillariophyceae. *Fish. Investig. Lond. Series.* 4, Pt. 5
5 Lebour, M. V. (1925). *The Dinoflagellates of Northern Seas.* Marine Biological Assoc. U.K.
6 Lebour, M. V. (1930). *The Planktonic Diatoms of Northern Seas.* London; Ray Society
7 Marshall, S. M. and Orr, A. P. (1955). *The Biology of a Marine Copepod, Calanus finmarchicus.* Edinburgh and London; Oliver and Boyd
8 Newell, G. E. and Newell, R. C. (1956). *Marine Plankton. A Practical Guide.* Revised edition. London; Hutchinson
9 Sverdrup, H. U., Johnson, M. W. and Fleming, R. H. (1946). *The Oceans.* 'Chap. 9. Populations of the Sea'. New York; Prentice-Hall
10 Wickstead, J. H. (1965). *An Introduction to the Study of Tropical Plankton.* London; Hutchinson Tropical Monographs
11 Wimpenny, R. S. (1966). *The Plankton of the Sea.* London; Faber
12 Wood, E. J. F. (1965). *Marine Microbial Ecology.* London; Chapman and Hall
13 *Fiches d'Identification du Zooplankton.* Ed. by Jesperson, P. and Russell, F. S. Cons. perm. int. Explor. Mer. Copenhagen. For identification of many North Atlantic zooplankton species.
14 Wood, R. D. and Lutes, J. (1967). *Guide to the phytoplankton of Narragansett Bay, Rhode Island.* West Kingston; Kingston Press

Papers
15 Digby, P. S. B. (1950). 'The Biology of small Planktonic Copepods off Plymouth.' *J. mar. biol. Ass. U.K.* **29,** 393
16 Gauld, D. T. (1966). 'The Swimming and Feeding of Planktonic Copepods.' *Some Contemporary Studies in Marine Science,* p. 313. Ed. by H. Barnes. London; Allen and Unwin
17 Wickstead, J. H. (1962). 'Food and Feeding in Pelagic Copepods.' *Proc. Zool. Soc. Lond.* **139,** 545–55

CHAPTER 3

MEASURING AND SAMPLING

In this chapter we will outline briefly some of the measuring and sampling techniques used at sea to obtain information of interest to ecologists. To evaluate the interactions between organisms and their environment, both oceanographic and biological data are required. Oceanographic data relate to the inorganic parameters of the environment, including measurements of water movement, temperature, composition, illumination and depth, and the nature of the sea bottom. Biological data relate to the distribution, numbers, activities and relationships of organisms in different parts of the sea. In order to collect this range of information, a great variety of apparatus has been devised. Only a few widely used methods are described here to give a general indication of available techniques.

I. OCEANOGRAPHIC DATA

Currents

Drift Bottles

The major surface currents of the oceans became known during the days of sailing ships when this knowledge was needed for successful ocean voyages. Information was accumulated by noting the course of drifting objects such as becalmed ships, drift-wood or pieces of wreckage. Over the past century, scientists have extended and refined this type of observation by making use of drift bottles.

A drift bottle has a long narrow neck and is usually ballasted to float with only the tip of the neck projecting above the surface, so that its course follows the movement of the water and is not much influenced by direct wind action on the bottle. The effect of the wind may be further reduced by attaching a small sea-anchor to the bottle. Inside the bottle is a postcard bearing an identification number and a request printed in several languages for the return of the card with details of time and place of finding, for which a small fee is paid.

Drift bottles can be used in several ways[11]. They may simply be thrown overboard from an anchored vessel and their direction and speed of movement directly noted. This provides immediate information about the surface current in that locality at the time of observation. Usually they are set adrift in large numbers in the hope that some of them may eventually be found stranded, and their location reported. In this way, information regarding the general course of currents over wide areas and considerable periods may be obtained.

Drift bottles may also be used to investigate bottom currents by weighting the bottle so that it just sinks but is held clear of obstructions on the bottom by a long trailing wire. These bottles are sometimes recovered in the nets of trawlers. Recently the movements of the bottom water of the North Sea and Irish Sea, the north-west Atlantic and the Pacific coast have been investigated using mushroom-shaped plastic floats which drift along the sea-bed.

Much information about the water movements around the United States has been obtained by the use of drift bottles. Interpreting drift-bottle strandings requires the collation of many records, together with consideration of various other sources of data on water movements. The course of a bottle between launching and stranding is unlikely to be direct, and allowance must be made for its wanderings.

Recently Dr. J. N. Carruthers of the National Institute of Oceanography in England, working in north-east Atlantic waters, has invented a new device which is cheap and simple and gives information on both bottom and surface currents. Nicknamed an 'oyster'[12], it is thrown over a ship's side and sinks to the bottom. Here it records the speed and direction of bottom current, determined from the angle of deflection of a pendulum inside the instrument. Later it floats to the surface and travels thereafter like a drift bottle.

As alternatives to surface drift bottles, plastic envelopes or patches of dye on the water have been used, the latter having the advantage of easy visibility from the air. As these float entirely on the surface, however, they are more likely to be influenced by wind direction than the more deeply submerged drift bottles.

Current Meters

The rate and direction of flow of water can be measured by various ingenious meters. An example is the Ekman Current Meter (*Figure 3.1*). This apparatus can be suspended beneath a moored vessel, and the flow of water past the large vane (V) turns the instrument so that the propeller (P) faces the oncoming current. The current rotates the propeller, and the number of revolutions is recorded on a meter (M), from which the speed of flow of the water can be determined by relating the number of revolutions to the duration of the recording. The upper part of the instrument also contains a number of small balls; each time the propeller completes 33 revolutions, one ball is automatically discharged into a circular compass-box in the base of the apparatus. The ball falls into a groove on the upper surface of the compass needle and, depending on the direction of the needle, rolls down into one or other of the thirty-six 10 degree compartments into which the compass box is divided. After use, the distribution of balls in the compartments of the compass box indicates the direction in which the instrument was pointing while in operation.

During lowering and hauling, the propeller of the instrument is locked. When the required depth is reached, a slotted weight known as a *messenger*

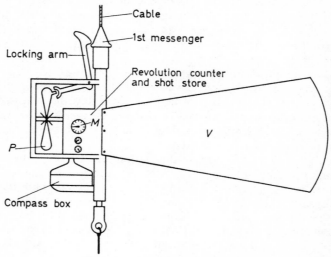

Figure 3.1. The Ekman Current Meter

is sent sliding down the suspending wire. On reaching the instrument, the messenger strikes a release arm which frees the propeller allowing it to rotate, and the instrument starts to record. After a measured period, a second messenger is sent down to relock the propeller prior to hauling.

The Ekman Meter is intended for use during periods of an hour or so, but there are other meters working on a similar principle which can be used over longer periods, for example the Carruthers Meter. Some meters are connected to the vessel by an electric cable, and the revolutions of the propeller and the direction of the current are electrically recorded.

Apart from propeller instruments, there are several other ways in which measurements of flow-rate can be made. For example, some instruments work on the principle of measuring the angle through which a vane suspended in the water is deflected by the force of the current. Other techniques estimate flow-rate from measurements of the slight electric currents induced in the water by its movements relative to the earth's magnetic field. The von Arx geomagnetic electrokinetogram uses electrodes towed from a ship[27]. The flow of water through the Strait of Dover has been estimated by measuring the potential difference between the two ends of the outer conductor of a cross-Channel, submarine, coaxial telephone cable[9].

I. OCEANOGRAPHIC DATA

The direct measurement of water currents by meters suspended in the water involves difficulties connected with obtaining a fixed reference point, particularly in deep water. An anchored vessel is by no means stationary, and movements of the ship may falsify the measurements. These difficulties can be avoided to some extent by using a buoy attached to a taut anchor wire as the reference point, movements of the ship relative to the buoy being allowed for in the interpretation of the meter readings.

Submerged Floats

It is possible to detect and measure water movements at middle depths by using neutrally-buoyant floats, the weight of which can be accurately adjusted to match the density of the water so that they sink to a predetermined depth and then drift with the current. Drift bottles have been used in this way, but tracking and recovery present obvious difficulties. Recently, sound-emitting floats have been developed for this purpose, their course being tracked by means of hydrophones aboard ship. An example is the 'Pinger' invented by Dr. J. C. Swallow of the British National Institute of Oceanography. This is an aluminium tube containing a battery and acoustic transmitter. It can be ballasted to sink to a required depth, and emits an intermittent 'ping' sound as it drifts.

Water Samples and Temperature Measurements

Because of the difficulties of measuring ocean currents directly, especially the slow movements of water at deep levels, much of our knowledge of the deep water circulation has been obtained indirectly. The waters of different parts of the oceans are to some extent distinguishable by virtue of their physical and chemical characteristics, in particular their temperature, salinity, and content of oxygen, nitrate or phosphate. By studying the distribution of these quantities throughout the oceans, the movements of the water can be inferred. Temperature measurements and the analysis of water samples therefore provide much of the basic data of oceanography. The foundations of this science were laid during the voyage of H.M.S. Challenger, 1872–76, when a large amount of this information was first collected from a series of depths at each of some 360 stations spread over the major oceans[7].

Surface Measurements

Temperature readings and water samples are usually taken together. Oceanographic work requires thermometers accurate to not less than $\pm$ 0·01°C because small variations in temperature produce considerable changes in water density. Measurements of surface temperature can be made from samples collected in a bucket lowered from a ship's side, but there are obvious inaccuracies where air temperature differs appreciably from water temperature. For more accurate measurements, heat-insulated surface samplers and thermometers are used. An example is The Lumby Surface Sampler which contains a thermometer within a water-bottle designed for towing. Water flushes through the bottle as it moves, ensuring that the sampler reaches exact water temperature, and it can then be hauled aboard without appreciable change of temperature of the contents. Surface temperatures are also measured by continuous-reading thermographs attached to the water intake of ships, or on the side of the hull. Recently, radiation thermometers mounted on aircraft have come into use for making wide-scale aerial surveys of sea surface temperature.

Subsurface Measurements

Although there is some variety of design, sampling bottles for collecting water from below the surface mostly take the form of an open-ended cylinder with spring-loaded valves for closing the ends. The sampler is lowered with the ends open so that water flows freely through it. When the required depth is reached, a release mechanism is operated by a messenger (page 36), causing the valves to snap shut firmly closing the ends of the cylinder. The bottle with its enclosed sample is then hauled up.

For samples and temperature measurements within the upper few hundred metres, a heat-insulated water-bottle is used. The Nansen-Pettersson bottle (*Figure 3.2*) contains a thermometer from which the temperature is read

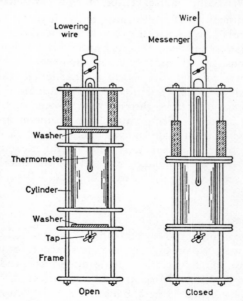

Figure 3.2.   The Nansen-Pettersson water-bottle

after hauling to the surface. The bottle is subdivided into a number of concentric cylinders around a central compartment containing the thermometer, in this way providing heat insulation from the exterior by several separate jackets of water.

For samples and temperature measurements at deeper levels, reversing water-bottles have been widely used in conjunction with reversing thermometers. The reversing water-bottle is a cylindrical container attached to the lowering wire by a hinged frame, the cylinder lying above the hinge when the bottle is in the open position. The bottle is closed when a messenger strikes a release-mechanism which causes the cylinder to swing downwards through 180 degrees until it lies below the hinge. As this occurs, valves automatically close the ends of the cylinder (*Figure 3.3*).

The purpose of this reversing mechanism is to allow a reversing thermometer to be used with the sampler. The reversing thermometer is usually

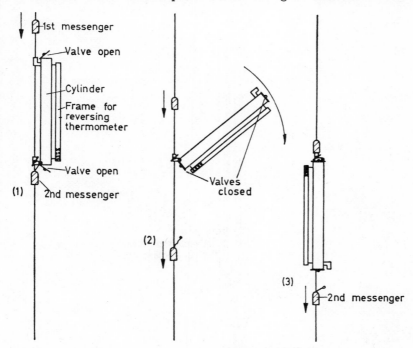

Figure 3.3. Diagram of the action of a reversing water-bottle

mounted on the side of the reversing water-bottle, although the two instruments can be used independently. This type of thermometer (*Figure 3.4*) has a mercury reservoir at both ends of the capillary tube, a large reservoir (*LR*) at one end and a small one (*SR*) at the other. The capillary tube incorporates a loop with a constriction and branch (*C*) just above the large reservoir. Alongside this thermometer, an auxiliary thermometer (*Aux*) of ordinary type is mounted with its reservoir (*AR*) adjacent to the small reservoir of the reversing thermometer.

Temperature measurements at deep levels are made by letting down the thermometer to the required depth in the 'set' position, that is, with the large reservoir downwards. In this position, mercury fills the large reservoir and extends up the capillary above the constriction (*C*) a distance depending upon the temperature. When the water-bottle on which it is mounted reverses, the thermometer becomes inverted and the small reservoir now lies below the capillary. The effect of reversal is to break the mercury thread in the constriction and branch of the capillary (*C*). The mercury which was initially above the constriction runs down to fill the small reservoir and

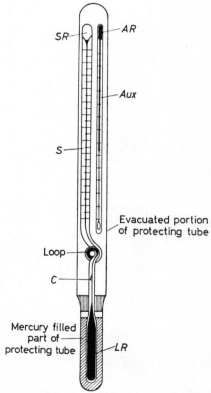

Figure 3.4. A protected reversing thermometer

extend up the capillary along the scale (S), the reading on this scale indicating the temperature. Correction must be made for the difference between the temperature at the instant of reversal and the temperature when the reading is taken after hauling. This correction is made from tables which relate the readings of the reversing and the auxiliary thermometers.

Reversing thermometers are of two types, protected and unprotected, having different uses. The protected reversing thermometer is used simply for temperature measurements. It is enclosed within a strong glass tube to protect the thermometer from water pressure which would falsify the temperature reading. The protecting tube is partially evacuated except for the section enclosing the large reservoir, which is filled with mercury to provide heat conduction between water and reservoir.

The unprotected type is used for measuring depth, and is attached to the reversing water-bottle in addition to the protected thermometer. It is of similar construction to the protected thermometer except that the protective cylinder is open to the water. Temperatures recorded by this thermometer are influenced by the water pressure at the moment of reversal, and therefore differ from the readings simultaneously obtained on a protected thermometer. From the difference in readings of the two thermometers it is possible to calculate the water pressure, and hence the depth, at which the thermometers were reversed. This method of depth measurement is usually more reliable than measuring the length of wire paid out because of the difficulty of knowing the angle of the wire in the water (*see* page 43).

When temperatures and water samples are required from several depths at one station, the procedure is speeded by attaching a series of reversing water-bottles and thermometers at intervals along the wire. As the uppermost bottle reverses, it automatically releases a second messenger which slides down the wire to operate the next bottle, and so on until every bottle has reversed. Even with this method, the process of lowering, reversing and hauling in deep water is a lengthy one, often taking several hours.

Continuous records of temperature while lowering or towing can be obtained with instruments known as thermosounds and bathythermographs. Temperature is recorded as a trace scratched on a smoked-glass plate by a pen controlled by a temperature-sensitive transducer, the plate being moved either by hydrostatic pressure or by means of a propeller which rotates as the instrument moves through the water. These instruments do not equal the accuracy of reversing thermometers, but provide a general picture of temperature distribution simply and quickly. They are sometimes used in conjunction with multiple water-samplers containing several small water-bottles which close automatically at predetermined depths by means of pressure-operated valves. Electrical thermistor thermometers are also coming into use for oceanographic work.

Illumination

The rate of decrease of illumination with depth is expressed as the Extinction Coefficient, k_λ.

$$k_\lambda = \frac{2 \cdot 30 \ (\log I_{\lambda d_1} - \log I_{\lambda d_2})}{d_2 - d_1}$$

where $I_{\lambda d_1}$ and $I_{\lambda d_2}$ are the intensities of light of wavelength λ at depths d_1 and d_2 metres. For accurate measurements of illumination beneath the sea surface, photomultipliers and photoelectric photometers are used[21]. But for rough measurements, the Secchi Disc (*Figure 3.5*) provides a simple method which has often been used by biologists. This is a white disc, 30 cm in diameter, which is lowered into the water to the depth at which it just disappears from sight. The extinction coefficient can then be roughly determined from the empirical relationship $k = 1 \cdot 7/d$ where d is the maximum depth in metres at which the disc is visible.

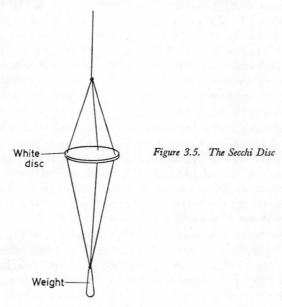

White disc

Weight

Figure 3.5. The Secchi Disc

Depth Measurements

Sounding Lines

The classical method of measuring the depth of the sea is by means of sounding weights and lines. The weight is lowered from the vessel until it strikes bottom and the length of line measured, usually by means of a meter attached to a sheave through which the line passes. In very deep water a tapered line is used, which gradually increases in diameter along its length to bear the weight of line already lowered. To indicate when the bottom is reached, the line is suspended from a dynamometer to show the reduction of load when the weight strikes the sea-bed. The undersurface of the sounding weight often contains a small depression filled with grease, and this provides a small sample of the bottom sediment which adheres to the grease.

Measurements by sounding line involve inaccuracies due to the difficulty of knowing the angle of the wire in the water. This can be overcome to some extent by attaching small tubes of hot gelatin solution at intervals along the wire as it is paid out. When the gelatin cools and sets, the angle of the wire is recorded by the slope of the gelatin surface. Correction can then be made to the depth reading registered on the cable meter. In deep water the use of sounding lines is very time-consuming and the method is now largely superseded by sonic sounding.

Sonic Sounding (Echo Sounding)

The depth of water may be measured by timing the interval between the emission of a sound impulse at the surface and its return to the surface as an echo reflected from the sea-bed, the speed of propagation of sound through water being known. The speed of sound through sea-water varies with temperature, salinity and pressure, usually lying between the limits of about 1,450 to 1,550 m/sec. For accurate sonic sounding it is therefore necessary to have knowledge of the hydrography of the underlying water column.

In early models of the sonic sounder, sound-waves were produced by a gong attached to the ship's hull. An operator listened on earphones connected to a hydrophone and amplifier, and measured with a stop-watch the interval between the striking of the gong and the sound of the echo. Modern sonic sounders combine the sound-emitter and hydrophone in a single transducer fitted in the underpart of the ship's hull. These transducers make use of the piezo-electric properties of quartz or the magnetostriction properties of nickel. In either case, rapid dimensional changes are produced by electrical excitation, causing brief pulses of vibration to be emitted into the water. The returning echo vibrates the transducer and sets up electrical signals which can be amplified and recorded, the transducer thus functioning as both transmitter and receiver.

The frequencies emitted by modern sonic sounders lie above the audible range, usually between 15 and 50 kc/sec. The use of ultrasonic frequencies has several advantages. They can be focused into fairly narrow directional beams, giving a more precise echo than is obtainable from the audible part of the sound spectrum, and enabling a more detailed picture of the bottom profile to be drawn. There is also less interference from natural sounds.

These instruments automatically record the echo as a trace on a cathode-ray tube or as a line drawn on a paper chart. In the paper-recording instrument, a strip of sensitized paper bearing a printed scale is drawn slowly through the recorder. A moving stylus in contact with the paper scans to and fro across the scale at a speed which can be adjusted in relation to the depth of water. A short pulse of sound is emitted from the ship's hull in a downward-directed beam as the stylus passes the zero mark of the scale. The stylus continues to move across the scale, and the echo signals are amplified and applied as an electric current to the stylus, marking the sensitized paper electrochemically. The position of this mark on the scale indicates the depth from which the echo is received. As the paper moves through the instrument, the repeated scanning of the stylus produces a series of marks which build up a line on the paper corresponding with the profile of the sea-bed (*Figure 3.6*).

Sonic techniques have further applications in marine biology. Echoes may be returned by objects floating or swimming in the water in the path of the sound beam. Fish shoals, or even individual fish, may be detected in this way, and different species may to some extent be identified by their characteristic echoes. The sonic scattering layers (page 99) were discovered in the course of investigations with sonic equipment.

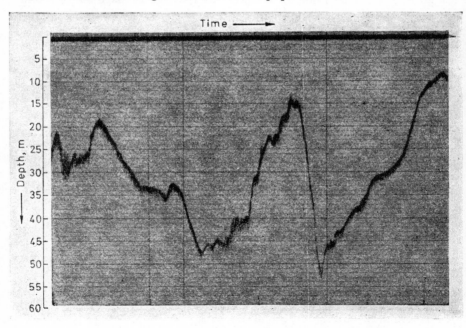

Figure 3.6. A typical sonic sounding trace over a widely varying sea-bed

The technique of echo-ranging makes use of a horizontally-directed beam of sound which scans to and fro ahead of a vessel, and is now increasingly used for fish location. Echo-ranging also finds applications in surveying the sea-bed, different types of substrate reflecting different patterns of echo trace[26].

Sediment Samples (*see* page 52)

II. BIOLOGICAL SAMPLING
Plankton

Plankton Nets

Samples of plankton are usually collected by plankton nets. These are of many designs, but all consist essentially of a long cone of fine-mesh net. The mouth of the net is usually some 50–100 cm in diameter, and is held open by a strong hoop to which the tow-rope is attached by three bridles (*Figure 3.7*). The narrow end of the net is firmly tied to a small metal or plastic vessel in which much of the filtered material collects. After hauling, the net is washed into a suitable receiver to collect any material left on the mesh.

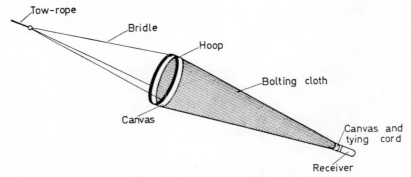

Figure 3.7. A simple plankton net

To filter efficiently, plankton nets must be towed quite slowly, not faster than about 1–1½ knots. Fine mesh presents high resistance to the flow of water through it, and if towed too fast, the net sets up so much turbulence in the water that floating objects are deflected away from the mouth. In many designs of plankton net the aperture is reduced by a tapering canvas sleeve (*Figure 3.8*) which cuts down the volume of water entering the net to give more effective filtering.

The net is usually made of bolting cloth, a silk or nylon material used in industry for sieving flour and other fine powders, and woven in such a way that the mesh remains virtually constant under strain. Bolting cloth is made in many grades of mesh size, and for plankton nets three grades commonly used are gauge 3, a coarse net of mesh aperture 0·324 mm;

gauge 15, a medium net of aperture 0·0925 mm; and gauge 21, a fine
net of aperture 0·063 mm.

Coarse mesh is more effective than fine mesh for catching the larger
planktonts because it offers less resistance and allows a faster flow of water
through the net. For collecting organisms over a wide range of sizes the
three grades of net are often used together. When only one type of organism
is sought, it is best to select the largest mesh size that will retain it. For
collecting macroplankton, larger nets known as young fish trawls (YFT's)[25]
are sometimes used, having a mesh of 1 mm approx. and an aperture of
1–2 m diameter.

Plankton nets can be towed behind a slowly moving vessel, or lowered
from a stationary vessel and hauled vertically. It is sometimes of special
interest to have samples of plankton from particular levels and, to avoid
contamination of the samples by organisms entering the net while it is being
lowered or raised, there must be some method of opening and closing the
net at the required depth. The simplest method of closure is to encircle
the mouth of the net with a noose which can be drawn tight, as in the Nansen

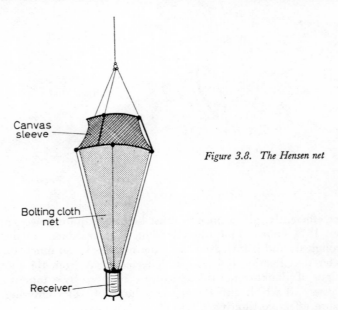

Figure 3.8. The Hensen net

Canvas sleeve

Bolting cloth net

Receiver

net (*Figure 3.9*). In more elaborate nets, for example the Clarke-Bumpus
net, the mouth is provided with valves which can be opened and closed by
a messenger.

Many planktonic organisms are very sensitive to temperature, and to
keep a sample alive for any length of time it must be kept at an even

temperature as close as possible to that of the water from which it was filtered. Vacuum flasks provide a means of doing this. For most purposes, however, preserved samples are needed. The addition to the sample of

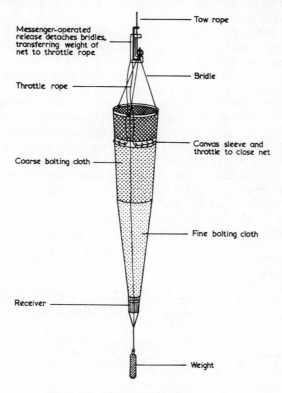

Figure 3.9. The Nansen closing net

sufficient neutral formalin to produce a 4–5 per cent concentration will preserve the majority of planktonts satisfactorily.

Quantitative Plankton Studies

The aim of quantitative plankton studies is usually to estimate numbers or weights of organisms beneath unit area of sea surface or in unit volume of water. There are many difficulties in these studies. For instance, plankton is often very patchy in distribution, and it is difficult to obtain any clear picture of the amount and variety of plankton unless samples are taken at numerous stations spread over the area of investigation. Modern navigational aids enhance the precision of the sampling grid. There is also the difficulty of knowing how much of the plankton is actually retained in a plankton net, for some may be displaced from the path of the net by turbulence; small organisms may escape through the meshes and the larger active forms may avoid capture by swimming. A further difficulty is to know the volume of water filtered. If a net could filter all the water in its path, the volume passing through it would be $\pi r^2 d$, r being the radius of net aperture and d the distance of the tow, but in practice this formula can only give an approximate measurement. A net does not filter all the water in its path and the filtering rate reduces as material collects on the mesh and the resistance of the net increases. There is a further difficulty in knowing precisely the distance a net has moved during towing.

To measure the filtered volume more accurately, a flow-meter can be added to a plankton net. A flow-meter has a multi-bladed propeller which is rotated by the flow of water, and a simple counter records the number of revolutions. This can be placed in the aperture of the net to measure the volume of water entering, for example Currie–Foxton[14] and Clarke-Bumpus nets; or the net may be surrounded by an open-ended cylinder and the flow-meter placed behind the net to measure the volume filtering through, for example the Gulf III Sampler (*Figure 3.10*).

Another method of collecting plankton is the plankton-pump. This draws water up a hose and pumps it through nets or filters to trap the

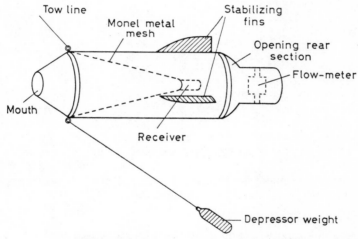

Figure 3.10. A high-speed plankton sampler of the Gulf III type

plankton. With this method it is possible to measure quite accurately the volume of water filtered, to retain the smallest organisms that escape through nets, and to sample the actual water from which the plankton is filtered. Despite these advantages the method has some drawbacks. Large creatures are prone to damage as they pass through the pump, the stronger-swimming planktonts may escape being sucked into the hose and there are difficulties in the use of pumps to obtain samples from deep levels.

Nanoplankton and Ultraplankton

The smallest planktonts such as bacteria and micro-flagellates escape through the meshes of ordinary bolting-silk nets. Materials of finer mesh are now becoming available which retain much smaller organisms than hitherto, but the sampling of nanoplankton and ultraplankton is usually done by collecting samples of sea-water in sterile bottles and then concentrating the organisms by allowing them to settle, by centrifuging or by fine filtration.

Plankton Counting

Having filtered a sample of plankton from a known volume of water, the quantity of plankton still remains to be measured. The method of measurement depends upon the type of study. The most detailed investigations are by direct counting. Large organisms are usually few in number and can be individually picked out and counted. Smaller organisms may be so numerous that the sample must be sub-sampled to reduce them to a number that it is practicable to count. The sub-sample can be spread out in a flat glass dish and examined, with a microscope if necessary, against a squared background. The count is then made, a square at a time. For very small organisms present in large numbers, the haemocytometer used by physiologists for counting blood-cells can be used for plankton counting.

Sometimes an estimate is wanted of the gross quantity of plankton of all types. A rough volume estimate can be made very simply by allowing the sample to settle in a measuring cylinder and reading the volume directly from the scale. Measurements of displacement volume are probably rather more accurate, and estimates can also be made by weighing, either as a rough wet weight, or, better, by drying to constant weight. Other methods applicable to phytoplankton are mentioned on pages 124–125.

High-Speed Plankton Samplers

We have previously pointed out that ordinary plankton nets must be towed slowly to be effective. They can therefore be used only from vessels operating for scientific purposes. Plankton samplers that can be used at higher speeds have some advantages; for example, they interfere little with the normal cruising of the ship and can therefore be towed behind commercial vessels proceeding on their normal routes, extending the scope of plankton studies. Also, they are probably more effective than slow-moving nets in capturing the more actively-swimming creatures.

For high-speed sampling the net area must be very large in relation to the aperture to reduce the high back-pressure developed when a fine-mesh net is towed rapidly through water. A simple sampler of this type is the Hardy Plankton Indicator consisting of a torpedo-shaped cylinder with stabilizing fins. Water enters a small opening at the front, filters through a disc or cone of bolting cloth and leaves through a rear aperture. A rather larger apparatus, the Gulf III Sampler[24] (*Figure 3.10*), incorporates a flow-meter and has been used in recent surveys of the plankton around the British Isles.

The Hardy Continuous Plankton Recorder (*Figure 3.11*) is a high-speed sampler which provides a continuous record of the plankton collected over a long-distance haul. During towing, the flow of water rotates a propeller, *P*, and drives an internal mechanism which gradually draws a long strip of bolting cloth, *B*1, across the path of the water flowing through the apparatus. Organisms trapped on the cloth are secured in place by a second strip of cloth, *B*2, the two strips being wound together on a take-up spool enclosed

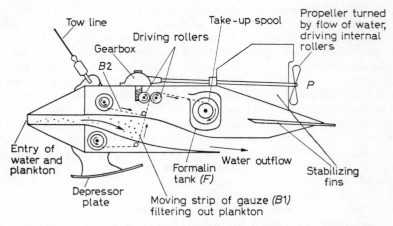

Figure 3.11. Diagrammatic section of a Hardy Continuous Plankton Recorder

in a formalin reservoir, *F*, where the plankton is preserved undisturbed. At the end of a voyage, the instrument is sent to the Marine Laboratory at Edinburgh where the spool of bolting cloth is unwound and the position of plankton organisms noted. By reference to the ship's log, a general picture of the distribution of plankton can be built up. Several of these instruments have been in regular use for some years, towed by commercial vessels on many routes around the Atlantic and Pacific, and so provide a means of mapping the plankton of these areas. The results of these investigations are published at intervals in the *Bulletin of Marine Ecology* and elsewhere.

Nekton

Various types and sizes of midwater trawl have been designed to attempt the capture of nekton at middle depths down to 1,000 m or more[18]. The

II. BIOLOGICAL SAMPLING

Isaacs–Kidd net is an elongate conical bag, usually with a mouth aperture of 8 m, and with an angled depressor plate to keep the net below the surface while towing. It can be fitted with a depth recorder, and its depth during towing can be monitored on hydrophones by attaching to the net a pressure-sensitive sound-emitter with a pulse frequency which varies with the depth. A recent modification is a double-ended net with the opening to the two receivers controlled by a pressure-operated valve. Above a pre-selected depth the captured material goes into one container, while below this level a flap deflects the catch into the other container, thereby providing a deep-level sample separate from material collected during lowering and hauling.

A new midwater trawl[13], recently developed for use from the R.R.S. Discovery, and operated to depths of over 1,500 m, has a mouth which can be opened and closed by remote control (*Figure 3.12*). With the mouth

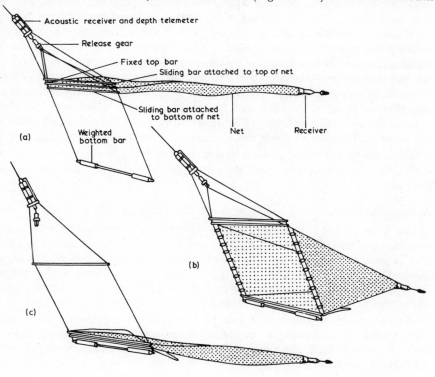

Figure 3.12. Midwater trawl for mesopelagic organisms, used on R.R.S. Discovery. The net can be opened and closed by sonic signals sent from the ship, operating the release mechanism. The depth of the net is indicated by the pulse-frequency of sonic signals generated by the depth telemeter. (a) Net closed for lowering; (b) net open; (c) net closed for hauling

51

closed the net is lowered to the required depth, as indicated by the pulse frequency of a pressure-sensitive sound-emitter on the net. The mouth is then opened by a release mechanism activated by an acoustic signal from the ship. At the end of fishing a second acoustic signal from the ship causes the mouth of the net to close before hauling.

Large active bathypelagic fish and cephalopods have proved extremely difficult to catch, and little is known about their distribution. Probably very large pelagic trawls are needed for their capture. Some abyssal fish have been taken by line, either laid on the bottom or simply suspended from the surface, and some success has also been achieved with deep water fish traps.

The Benthos

There are several branches of science which seek information about the sea bottom; for example, oceanography, geology and palaeontology as well as marine biology. Each makes use of apparatus designed primarily to collect information needed in that particular field of study, but there is so much overlap between these various aspects of marine science that data relevant to one may well be of interest to another, and there is consequently a variety of instruments for studying the sea bottom which produce information relevant to marine biology. We shall refer to only a few. Those selected for mention here are of two general types; instruments which are intended mainly for collecting samples of sediment, and those for collecting benthic organisms. The distinction is not a firm one because sediment samplers are likely to include small organisms in the material brought up, and apparatus designed to catch bottom-dwelling creatures may also retain some of the deposit.

Sediment Samplers

We have already mentioned (page 42) the early method of collecting specimens of surface sediment on a dab of grease under a sounding weight. For larger samples, various small spring-loaded, snapper grabs have been devised which take a shallow bite out of the sea-floor (*Figure 3.13*). Some investigations seek information about the deeper layers of deposit, and instruments known as corers are used for this purpose.

A corer is a long tube which can be driven down into the sea-floor, and then withdrawn enclosing a core of sediment. The coring tube contains a separate liner to facilitate removal of the core. Considerable force is required to drive a corer far into the deposit, and several methods are used for this purpose.

The corer may be heavily weighted and allowed to descend at speed, penetrating the sediment under its own momentum. An explosive charge can be detonated to drive the tube downwards. The Kullenberg piston corer makes use of hydrostatic pressure to assist deep penetration. This corer consists of a weighted coring-tube with brass liners, inside which fits a sliding piston attached to the lowering cable. The corer is lowered with the piston at the lower end of the tube, and the apparatus is slung from a release mechanism held in the closed position by counter-weights suspended below the nose of the coring tube. When these counter-weights touch bottom, the release mechanism opens to let the coring-tube fall under its own weight. At this moment, the reduced strain on the lowering cable is indicated on the vessel by a dynamometer, and the cable-winch is stopped immediately so that the piston attached to the cable is held stationary as

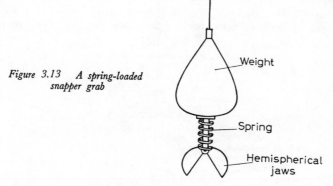

Figure 3.13   A spring-loaded snapper grab

Weight

Spring

Hemispherical jaws

the coring-tube plunges downwards. This creates a tremendous suction inside the tube which helps to overcome the resistance of the substrate to penetration. Undisturbed cores over 20 m long have been obtained from very deep water with this device. These cores contain the remains of planktonic organisms deposited on the sea-bed over a period of several hundred thousand years, and something may be learnt of oceanic conditions in the past by studying the variations in composition at different levels of the core.

For taking short cores up to about 1 m long from deep water, a recent development is the 'free-fall corer'. This is a weighted coring tube which has no cable for lowering and hauling, but is simply thrown over the side of the vessel to sink freely. When the corer has sunk into the bottom an automatic release frees the tube, and a pair of glass floats carries the tube and enclosed sample to the surface. Retrieval is aided by a flashing light on the instrument.

Collecting Organisms from the Bottom

Methods of sampling the benthic population vary with the types of organisms under study, and the type of bottom. Hypoplankton can be collected in a plankton net attached to a sledge[10] (*Figure 3.14*) and dragged over the sea-floor. Demersal fish and many other creatures that live on, rather than within, the sea bottom can be captured by the trawls and seines used by commercial fisheries, described later (page 209 ff.). For research purposes, the nets are usually of smaller mesh than is permitted for commercial fishing.

A net much used for biological work is the Agassiz trawl (*Figure 3.15*), which has the advantage of very easy handling because it does not matter which side up it reaches the bottom. The mouth of this net is held open by a metal frame, and it can be fitted with fine-mesh net to retain small creatures. It is simply dragged along the bottom.

To capture animals that live beneath the surface, the sampling device must be capable of digging into the deposit. The naturalists' dredge is a simple device which can be operated from a small boat. It consists of a bag of strong sacking or wire mesh held open by a heavy, rectangular

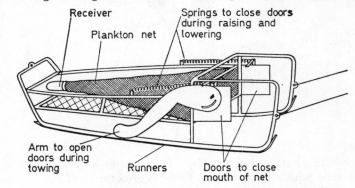

Figure 3.14 The Bossanyi hypoplankton net

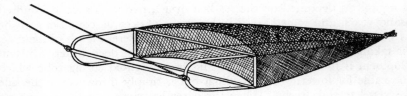

Figure 3.15. The Agassiz trawl

metal frame. This can bite a few inches into a soft sediment as it is hauled along, but tends to fill mainly with material lying on the bottom. The leading edges of the frame can be angled and sharpened to increase the tendency to dig rather than to ride along the surface, but it does not catch the deeper burrowing creatures.

An example of an instrument that takes a considerably deeper bite is the Forster[16] anchor-dredge (*Figure 3.16*). This requires a sizable vessel for its operation. The net is attached to a strong rectangular metal frame with

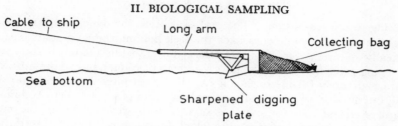

Figure 3.16 *The Forster anchor dredge*
(Based on Forster, G. R.[16])

a long, forward-projecting upper arm and a lower, downward-sloping digging-plate. The dredge is lowered to the bottom and remains stationary as the ship moves slowly astern paying out a long length of cable, at least three times the depth of the water. The winch brakes are then applied so that the strain is taken on the cable, causing the dredge to tilt and bite deeply into the substrate. Instead of sliding along the bottom the dredge should dig in like an anchor, bringing the ship to a stop. Finally, the cable is winched in until the dredge eventually breaks out, the contents being retained within the net.

Quantitative Bottom Sampling[20]

Quantitative studies of benthic populations require samplers which take a standard bite of known area and depth. For small organisms (micro- and meiobenthos), most of which live close to the surface, short coring-tubes can provide satisfactory samples from soft deposits. Capturing larger creatures presents more difficulty because some can escape the sampling gear by crawling away or moving deeper down their burrows.

On soft sediments the Petersen grab (*Figure 3.17a*) has been much used. It consists of a pair of heavy metal jaws which are locked wide apart while

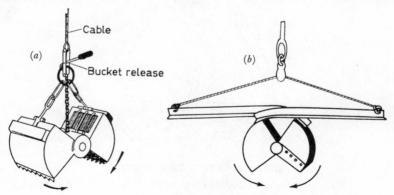

Figure 3.17.   (a) *The Petersen grab.*   (b) *The Van Veen grab*

55

lowering to the sea-bottom. The grab sinks into the deposit under its own weight, and as the cable goes slack the lock holding the jaws apart is automatically released. On hauling, before the grab lifts off the bottom, the tightening cable first draws the jaws together enclosing a bite of the substrate of approximately 0·1 m² surface area. The grab bites fairly well into soft mud, but on sand or gravel it digs only to a depth of some 3–4 cm and many creatures escape. Stones or pieces of shell may wedge between the jaws, preventing complete closure, and much of the catch may then be lost during hauling.

The Petersen grab is less used nowadays than several other samplers which work on similar principles. For example, both the Van Veen grab (*Figure 3.17b*) and the Baird grab have paired jaws for biting into the substrate like a Petersen grab but they also have arms attached to the jaws to give greater leverage for forcing the jaws together.

Another type of bottom sampler for quantitative work is the Knudsen suction sampler. This is a short corer of wide bore with a suction pump in the upper part of the instrument. After reaching the bottom, tension on the hauling cable first turns the pump, thereby generating a suction inside the tube to assist penetration. When the corer breaks out of the substrate it automatically turns upside down to avoid loss of contents.

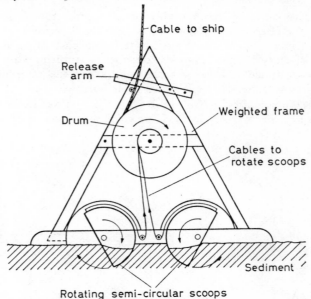

Figure 3.18 Diagrammatic representation of the Holme double scoop sampler

(Based on Holme, N. A.[19])

III. UNDERWATER OBSERVATIONS

An example of a sampler which can be used on rather coarser deposits is the Holme scoop sampler, used in studies of the biomass of the English Channel. This digs by means of semi-circular scoops. Two models have been designed, one having a single scoop and the other a pair of counter-rotating scoops (*Figure 3.18*). The apparatus is lowered with the scoops in a fully open position. On reaching the bottom, a release mechanism operates so that, when hauling commences, the strain on the cable is first applied to the scoops, turning them through 180 degrees so that they dig into the substrate. Each scoop samples a rectangular area of approximately 0·05 m² and in favourable conditions each bite is semi-circular in vertical section with a maximum depth of 15 cm.

Rough quantitative comparisons can also be made using the anchor dredge (page 54), which takes a fairly uniform bite. Accurate position fixing is important in quantitative bottom surveying, and the development of modern radio-navigation equipment has greatly added to the precision of this.

III. UNDERWATER OBSERVATIONS

It would simplify many problems in marine biology if the range of direct observation could be extended. The only marine populations which are easily accessible to close inspection are those of the sea-shore, and then only for a part of each tidal cycle. Our knowledge of the rest of marine life comes almost entirely from the incomplete samples obtained by nets, dredges, grabs and similar devices. Recently, new techniques for visual, underwater exploration have been developed, and have already provided much new information on marine organisms.

Diving

Diving by means of air pumped down a tube from the surface to a man enclosed in a special helmet and diving suit was first introduced in 1819 by Auguste Siebe. Apparatus of this type is still used by commercial divers working in connection with underwater constructions or salvage operations, but has found little application in biological work. Apart from its expense, it has the drawback that it does not permit free movement of the diver over a wide area because he is limited by the length of his breathing tube and the need to keep it free from snags.

The recent development of aqualung equipment provides a diver with a means of moving freely underwater, unencumbered by an airpipe, his air-supply being carried on his back in compressed-air cylinders. This apparatus is simple and inexpensive, and no lengthy training in its use is required. It greatly extends the scope of underwater observation at shallow depths, and simple underwater photographic apparatus is also available to record what is seen.

The special problems of breathing underwater are due to the pressure of water surrounding the body, which increases approximately 1 atm for every 10 m of depth. For a diver to be able to expand his chest against the water pressure, he must be supplied with air at a pressure equal to that of the water. Whereas at the surface we breath air at atmospheric pressure, at a depth of 10 m the diver must have air at double this pressure, 2 atm; at 20 m, 3 atm; and so on. The cylinder of the aqualung contains air at very high pressure, and this passes through a pressure-reducing valve which regulates the pressure of air supplied to the diver so as to balance exactly the surrounding water pressure.

Diving is without serious physiological hazards down to about 15 m depth. Below this, precautions must be taken to avoid the dangerous condition of 'decompression sickness' during ascent[6]. This occurs if blood vessels become blocked by bubbles of gas when air dissolved under pressure in the blood comes out of solution too rapidly as the pressure falls, a process similar to the fizzing of soda-water when the bottle is unstoppered. Decompression sickness can produce a variety of severe symptoms including intense joint pains (bends), paralysis or, in extreme cases, death. It can be avoided by making a very gradual ascent involving a series of pauses, or 'decompression stops', to give ample time for the excess air dissolved in the blood and tissues to be eliminated in the expired breath.

If adequate precautions are taken to ensure gradual decompression, descents are possible with safety to about 30 m. Below this, a diver breathing compressed air faces the additional hazard of nitrogen narcosis due to the large quantity of nitrogen dissolved under pressure in the blood. This has an effect on the brain producing a condition of rapturous inebriation in which the diver loses control of his actions, with possibly fatal results.

The deepest descent and successful return yet made by a diver is to about 200 m. This record was first achieved in 1956 by a Royal Naval diver using helmeted diving equipment with an air-line to the surface. Nitrogen narcosis was avoided by breathing a mixture of $7\frac{1}{2}$ per cent oxygen in helium. The dive lasted 6 h 33 min. The descent took 3 min, 4 min were spent on the bottom and the ascent took 6 h 21 min for gradual decompression.

At any particular pressure the body can only absorb a certain amount of gas before becoming saturated. Once the saturation point is reached, no more gas will be absorbed and the time required for decompression will then be the same, however long the duration of the dive. There are now experiments with underwater support systems for divers which enable them to have a long working period at depth and to avoid time spent on frequent re-surfacing. Several types of underwater chamber have been devised in which divers can live at pressure, returning to this submerged 'house' to feed or sleep between working periods without decompression. In this way, French divers have lived for three weeks at a depth of about 100 m. Although the decompression period was finally about $3\frac{1}{2}$ days, this was no longer than would have been needed after 24 hours at that depth.

Attempts to dive much deeper than 200 m encounter additional dangers: the breathing of oxygen and inert gases at high pressure causes various physiological disturbances which are not well understood, and in some cases leads to convulsions and death. Successful experimental sea dives have now been made to about 230 m, whereas in a simulated laboratory dive a depth of about 500 m has been reached. Recent predictions put the lowest limit at which divers exposed to pressure may be able to work at between 500 and 1,000 m.

Underwater Observation Cabins

Below the depth that can be safely reached by divers, exploration is possible in observation cabins where air can be breathed at normal pressure. For deep descents the cabin must be of great strength to withstand the enormous water pressure and must be provided with lighting equipment to illuminate the surroundings.

Between 1930–1934, the Barton-Beebe Bathysphere broke all previous records for descent into the deep sea. This was a spherical observation cabin lowered and raised on a cable from a winch on a surface vessel. This device was the first to reach the deep sea bottom at a depth of nearly 1,000 m, the limit of length of cable then available. In 1950, in a later version known as the Benthoscope, Barton made a deeper descent to 1,300 m.

The Bathyscaphe, invented by Professor Auguste Piccard and first used by him in 1953, is a combined observation cabin and underwater float, which can be likened to an underwater balloon. The gondola is a spherical, pressure-resisting cabin with portholes, and is suspended beneath a large, lightly-built float filled with petrol. Being much lighter than water and virtually incompressible, the petrol provides adequate buoyancy to support the heavy cabin, and the float does not need to be constructed to withstand great pressure. The Bathyscaphe carries ballast and sinks freely under its own weight. To ascend, sufficient ballast is shed for the vehicle to float up again to the surface. In some models, electrically driven propellers provide a limited amount of horizontal movement when submerged.

Several bathyscaphes have been built and many dives made. In 1960, the bathyscaphe 'Trieste', originally built by Professor Piccard with money largely provided by the Italian Government but later bought by the United States, made a successful return voyage from nearly 11,000 m in the Challenger Deep of the Mariana Trench, virtually the deepest known part of the ocean floor.

Deep Water Photography

Automatically-operating underwater cameras provide a means of making visual observations in deep water without the complications of personal descent. The apparatus requires two main parts; a camera unit with a mechanism for winding the film, and a lighting unit to provide illumination. Some instruments include an acoustic signaller which emits an intermittent 'ping' enabling the camera to be located by hydrophones. The apparatus must be sealed in strong containers to withstand the pressures of deep water. Some deep water cameras are fitted with small scoops which automatically take a sample of the sediment as the camera operates.

An instrument developed by Laughton of the British National Institute of Oceanography for deep water photography has a camera holding sufficient 35 mm film to take a sequence of one hundred photographs, a 100 joule electronic flash-light and an acoustic unit. The camera needs no shutter because, being used in darkness, an exposure is made only when the flashlight operates. The instrument is lowered from the ship until a trip-weight hanging beneath the apparatus touches bottom. This discharges the flash-bulb so that a photograph is taken, operates a small motor to wind on the film to the next frame and causes the acoustic unit to emit a rapid series of 'pings'. As soon as these are heard on the ship's hydrophone the apparatus is raised slightly above the sea-bed and then lowered again. As the weight touches bottom a further picture is taken, and the process is repeated for each photograph. Many other cameras have also been developed.

The appearance of the sea-bed shown in photographs provides information about the nature of the sediment and the speed of movement of the bottom water. Organisms can be seen undisturbed in their natural environment, or their presence known from their tracks or burrows. Where creatures can be easily recognized, quantitative information from photography is probably more reliable than that obtained by grab samples.

Underwater Television

Apparatus for closed-circuit underwater television has been developed, and first achieved a notable success in locating a sunken submarine on the bottom of the English Channel in 1951. Underwater television has found some applications in biological work, having advantages over photography in allowing immediate, continuous observation. The apparatus is more complex and costly than photographic cameras, and there are greater difficulties in its use at very deep levels.

REFERENCES AND FURTHER READING

Books

[1] Sverdrup, H. U., Johnson, M. W. and Fleming, R. H. (1946). *The Oceans*. 'Chap. 10. Observations and Collections at Sea'. New York; Prentice-Hall
[2] Murray, J. and Hjort, J. (1912). *The Depths of the Oceans*. London; Macmillan
[3] Bruun, A. F. *et al.*, (1956). *The Galathea Deep Sea Expedition*. London; Allen and Unwin

REFERENCES AND FURTHER READING

[4] Barnes, H. (1959). *Oceanography and Marine Biology, A Book of Techniques.* London; Allen and Unwin

[5] Hill, M. N. (Editor) (1963). *The Sea. Ideas and Observations in the Study of the Seas.* Vol. 2. 'The Composition of Sea Water. Comparative and Descriptive Oceanography. Chap. 5. Chemical Instrumentation. Chap. 6. Water Sampling and Thermometers. Chap. 13. Deep-Current Measurements using Neutrally Buoyant Floats. Chap. 14. Drogues and Neutral-Buoyant Floats. Chap. 23. Bathyscaphes and other Deep Submersibles for Oceanographic Research.' New York and London; Interscience.

[6] Bennett, P. B. and Elliott, D. H. (Editors) (1969). *The Physiology and Medicine of Diving and Compressed Air Work.* London; Bailliere, Tindall and Cassell

Papers

[7] Bailey, H. S. (1953). 'The Voyage of the 'Challenger'.' *Scient. Am.* May

[8] Barnes, H. (1963). 'Underwater Television.' *Oceanogr. Mar. Biol. Ann. Rev.* **1**, 115

[9] Bowden, K. F. (1956). 'Flow of water through the Strait of Dover.' *Phil. Trans. Roy. Soc. Lond.* A. **248**, 517–551

[10] Bossanyi, J. (1951). 'An Apparatus for the Collection of Plankton in the Immediate Vicinity of the Sea Bottom.' *J. mar. biol. Ass. U.K.* **30**, 265

[11] Carruthers, J. N. (1930). 'Further investigations on the water movements in the English Channel.' *J. mar. biol. Ass. U.K.* **17**, 241–75

[12] Carruthers, J. N. (1969). 'The plastic seabed "oyster" for measuring bottom currents.' *Fiskdir. Skr. Ser. Havunders.* **15**, 163–171

[13] Clarke, M. R. (1969). 'A new midwater trawl.' *J. mar. biol. Ass. U.K.* **49**, 945–960

[14] Currie, R. I. and Foxton, P. (1957). 'A New Quantitative Plankton Net.' *J. mar. biol. Ass. U.K.* **36**, 17

[15] David, P. M. (1965). 'The Neuston Net.' *J. mar. biol. Ass. U.K.* **45**, 313

[16] Forster, G. R. (1953). 'A New Dredge for Collecting Burrowing Animals.' *J. mar. biol. Ass. U.K.* **32**, 193

[17] Fraser, J. H. and Corlett, J. (Editors) (1961). Symposium on Zooplankton Production; Copenhagen, 1961. *Rapp. P.-v. Réun. Cons. perm. int. Explor. Mer.* Vol. 153

[18] Harrisson, C. M. H. (1967). 'On methods for sampling mesopelagic fishes.' *Symp. Zool. Soc. Lond.* **19**, 71–126

[19] Holme, N. A. (1953). 'The Biomass of the Bottom Fauna of the English Channel.' Part II. *J. mar. biol. Ass. U.K.* **32**, 1

[20] Holme, N. A. (1964). 'Methods of sampling the benthos.' *Adv. Mar. Biol.* **2**, 171

[21] Kampa, E. M. (1970). 'Underwater daylight and moonlight measurements in the eastern North Atlantic.' *J. mar. biol. Ass. U.K.* **50**, 397–420

[22] McIntyre, A. D. (1956). 'The use of Trawl, Grab and Camera in Estimating the Benthos.' *J. mar. biol. Ass. U.K.* **35**, 419

[23] Smith, W. and McIntyre, A. D. (1954). 'A Spring-loaded Bottom Sampler.' *J. mar. biol. Ass. U.K.* **33**, 257

[24] Southwood, A. J. (1962). 'The Distribution of some Plankton Animals in the English Channel and Western Approaches. II Surveys with the Gulf. III High-Speed Sampler 1958–60.' *J. mar. biol. Ass. U.K.* **42**, 275

[25] Southward, A. J. (1970). 'Improved methods of sampling post-larval young fish and macroplankton.' *J. mar. biol. Ass. U.K.* **50**, 689–712

[26] Stride, A. H. (1963). 'Current-swept sea floors near the southern half of Britain.' *Quart. Journ. Geol. Soc. Lond.* **119**, 175–199

[27] Vaux, D. (1965). 'Current measuring by towed electrodes.' *Fishery Investig. Lond.* Ser. 2, **23**, No. 8

CHAPTER 4

SOME PARAMETERS OF THE ENVIRONMENT

Although many features of the marine environment are virtually uniform over wide areas, different parts of the sea are populated by different communities of organisms. The aim of marine ecological studies is to discover what these differences are and why they exist and to evaluate the factors responsible for them. These investigations encounter many difficulties.

There are the obvious problems of working in an environment to which we have no easy access. Observations and measurements have mostly to be made with remotely-controlled instruments. Some of the physical and chemical parameters can be measured with precision; but biological measurements involve many uncertainties because sampling apparatus such as nets, dredges and grabs are not instruments of high accuracy. Measurements of the activities of marine organisms in their natural surroundings are virtually impossible. They can be brought into the laboratory and kept alive for a time, but here their behaviour may not be the same as in natural surroundings because it is obviously impossible to simulate closely in a tank all the conditions of the open sea.

Because several properties of the marine environment usually vary together, the effects of variation in single factors are seldom evident in natural conditions. There are two major zonations of distribution in the sea—between the tropics and the poles, and between the surface and the depths. Both are associated with differences of penetration and absorption of solar radiation, and therefore with gradients of temperature, illumination, and to a lesser extent salinity. Vertical distribution is also influenced by pressure. The distribution of a species is consequently associated with a complex of variables and it is not easy to assess the role of each parameter independently.

The effects of variation in single factors can be studied to some extent in controlled conditions in the laboratory but in this unnatural environment the responses may be abnormal. There is also the complication that several factors often interact in their effects; for example, in some species the tolerance to salinity change is modified by temperature, and temperature tolerance may itself vary with salinity. Furthermore, observations on specimens from one locality may not hold for an entire population of wide distribution because each species exhibits a range of variation for each character, and these may be related to the geographical situation due to selection or acclimatization.

TEMPERATURE

Apart from the effects of the inorganic environment, there are also many ways in which organisms influence each other. Even where physical and chemical conditions seem suitable, a species may not flourish if the presence or absence of other species has an unfavourable effect. Predation may be too severe. Other competing forms may be more successful in the particular circumstances. The environment may be lacking in some essential resources contributed by other species, such as food, protection, an attachment surface or some other requirement. These biological factors are obviously of great importance, but their evaluation is extremely difficult.

Generally, the distribution of a species is an equilibrium involving many complex interactions between population and environment which are at present very incompletely understood. None the less, a start can be made in tracing the complicated web of influences which control the lives of marine organisms by first studying the individual parameters of the environment, noting the extent to which each can be correlated with the distribution and activity of different species, and observing the effects of change both in natural conditions and in the laboratory. The parameters of obvious biological importance which we shall refer to in this chapter are temperature, the composition of the water, pressure, illumination and water movements.

TEMPERATURE

The continual circulation of the oceans and their enormous heat capacity ensure that the extent of temperature variation in the sea is small despite great geographical and seasonal differences in absorption and radiation of heat. Except in the shallowest water, the temperature range in the sea is less than that which occurs in most freshwater and terrestrial habitats, and the relative stability of sea temperature has a profoundly moderating effect on atmospheric temperature change.

The highest sea temperatures are found at the surface in low latitudes where much of the oceanic surface water is between 26–30°C. In shallow or partly enclosed areas like the Persian Gulf, the surface temperature may rise as high as 35°C during the summer, and conditions are extreme on the shore where inter-tidal pools sometimes exceed 50°C*. At the other extreme, the freezing-point of sea-water varies with the salinity, and is depressed below 0°C by the dissolved salts. At a salinity of 35 per mille (page 73) sea-water freezes at approximately − 1·91°C.

Excluding the shore and shallow water, the extreme temperature range between the hottest and coldest parts of the marine environment is therefore in the order of 30–35°C, but in any one place the range of temperature variation is always much less than this. In high and low latitudes, sea temperature remains fairly constant throughout the year. In middle

* Within pits in the floor of the Red Sea remarkably high temperatures (up to 56°C) have been recorded in water of abnormally high salinity (up to nearly 300‰) and unusual composition, rich in trace metals.

latitudes, surface temperature varies with season in association with climatic changes, and the range of seasonal temperature change depends upon locality, commonly about 10 deg. C. Off the south-west coast of the British Isles, the temperature usually varies between about 7°C in winter and 16°C in summer, while off the Maine coast and in the North-West the range is 4°C in winter to about 13°C in summer. The greatest seasonal variations of sea temperature are about 18–20°C, this range being recorded in the China Sea and the Black Sea.

Whereas surface water varies in temperature from place to place and time to time, the deep layers throughout the major ocean basins remain fairly constantly cold. The coldest water is at deep levels of the Arctic where the temperature is between 0°C and − 1·9°C. In the Atlantic, Pacific, Indian and Southern Oceans, the temperature of the bottom water lies between 0°C near Antarctica and 2–3°C at lower latitudes.

In high latitudes, heat passes from the sea to the atmosphere. Surface cooling of the water produces convectional mixing, and there is, therefore, little difference in temperature between the surface and the deep layers. Through the whole depth of water the temperature range is usually within the limits of −1·8°C to 1·8°C. There is often an irregular temperature gradient within the top 1,000 m because the surface is diluted by fresh water from precipitation or melting ice. This forms a low-density layer of colder water above slightly warmer, but denser, water of higher salinity entering from middle latitudes(*Figure 4.1 (1)*). Below 1,000 m the temperature is almost uniform to the bottom, decreasing only slightly with depth.

At low latitudes, heat absorption at the sea surface produces a warm, light surface layer overlying the cold, denser, deep layers. Here the temperature gradient does not descend steadily but shows a distinct step, or *thermocline*, usually between about 100–500 m (*Figure 4.1 (2)*), where temperature falls quite sharply with depth. This zone is termed a *discontinuity layer*. Above it, surface mixing maintains a fairly even warm temperature, a stratum referred to as the *thermosphere*. Below the thermocline is the *psychrosphere* where the water is cold, and there is only a slight further decrease of temperature towards the bottom. To a considerable extent the thermocline acts as a boundary between a warm water population above and a cold water population below.

In middle latitudes, the surface water becomes warm during the summer months and this leads to the formation of temporary, seasonal thermoclines near the surface, commonly around 15–40 m depth (*Figure 4.1 (3)*). In winter, when the surface water cools, these temporary thermoclines disappear and convectional mixing may then extend to a depth of several hundred metres. Below the level to which convectional movements mix the water, there is usually a permanent but relatively slight thermocline between about 500 and 1,500 m.

Water temperature exerts a major control over the distribution and

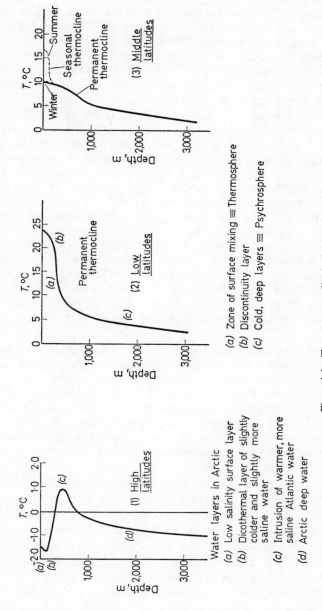

Figure 4.1. *Temperature profiles in the deep oceans*

Water layers in Arctic

(a) Low salinity surface layer

(b) Dicothermal layer of slightly colder and slightly more saline water

(c) Intrusion of warmer, more saline Atlantic water

(d) Arctic deep water

(a) Zone of surface mixing ≡ Thermosphere

(b) Discontinuity layer

(c) Cold, deep layers ≡ Psychrosphere

65

activities of marine organisms. Temperature tolerances differ widely between species, but each is restricted in distribution within its particular temperature range. Some species can only withstand a very small variation of temperature, and are described as *stenothermic*. *Eurythermic* species are those of wide temperature tolerance. Strict stenotherms are chiefly oceanic forms, and their distribution may alter seasonally with changes of water temperature. Eurytherms are typical of the more fluctuating conditions of shallow water. Sessile organisms have generally a rather wider temperature tolerance than free-living creatures of the same region.

Because water temperature has so great an effect on distribution, the extent of marine biogeographical regions can be related more closely to the course of the isotherms than to any other factor. The definition of bio-geographic subdivisions of the sea is inevitably somewhat vague because the marine environment contains few firm ecological boundaries. Land barriers account for some differences between oceanic populations, and wide expanses of deep water prevent the spread of some littoral and neritic species; but for the most part the transition between one fauna and another is gradual, with a broad overlap of populations. However, in a general way the populations of the surface waters fall into three main groups associated with differences of water temperature; namely, the warm-water populations, the cold-water populations, and populations which inhabit waters of inter-mediate temperature where the temperature of the surface layers fluctuates seasonally, i.e. temperate waters. These major divisions of the marine population may be almost endlessly subdivided to take account of local conditions.

Warm-water populations are mainly to be found in the surface layers of the tropical belt where the surface temperature is above about 18–20°C (*Figure 4.2*). This warm-water zone corresponds roughly with, but is rather more extensive than, the zone of corals which have their main abundance in clear shallow water where the winter temperature does not fall below 20°C. Within the warm-water regions of the oceans there is little seasonal variation of temperature. At the equator, the temperature of the surface water in most areas is between 26 and 27°C, and does not change appreciably throughout the year.

Cold-water populations are found in the Arctic and Southern oceans where the surface temperature lies between about 5°C and a little below 0°C. In the Southern Ocean the cold water has a well-defined northern boundary at the Antarctic Convergence (page 12) where it sinks below the warmer sub-Antarctic water. The sharp temperature gradient at this convergence effectively separates many species of plant and animal, and forms a distinct northern limit to the Antarctic faunal and floral zones. The southern boundary of the Arctic zone is less distinct except at the con-vergences of the Labrador Current and Gulf Stream in the Atlantic, and of

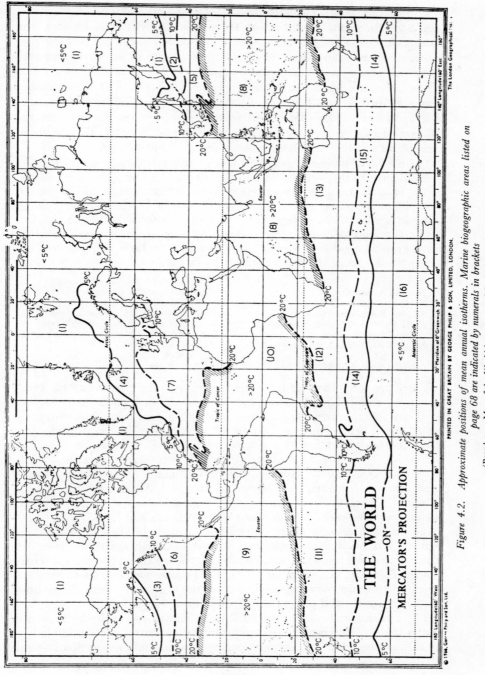

Figure 4.2. Approximate positions of mean annual isotherms. Marine biogeographic areas listed on
page 68 are indicated by numerals in brackets
(Based on a Map of the World by courtesy of G. Philip & Son Ltd.)

67

the Oyo-Shiwo and Kuro-Shiwo currents in the Pacific. Broadly, the Arctic zone comprises the Arctic Ocean and those parts of the Atlantic and Pacific Oceans into which Arctic surface water spreads, the limiting temperature being a summer maximum of about 5°C.

The temperate sea-areas lie between the 5°C and 18°C mean annual surface isotherms, and here the surface water undergoes seasonal changes of temperature. The colder part of the temperate regions between the 5°C and 10°C isotherms are termed the Boreal zone in the Northern hemisphere and the Antiboreal zone in the Southern hemisphere.

The course of the surface isotherms is determined largely by the surface circulation. On the western sides of the oceans the warmest water reaches higher latitudes, and the coldest water lower latitudes, than on the eastern sides. The temperate zones are therefore narrow in the west and much wider in the east, where they extend further to both north and south. On the basis mainly of water temperature we can designate some of the chief biogeographic subdivisions of the littoral and epipelagic zones as follows, their positions being indicated numerically in *Figure 4.2*.

1. Arctic region.
2. East Asian Boreal region.
3. Northwest American Boreal region.
4. Atlantic Boreal region.
5. North Pacific warm temperate region.
   East Asian province.
6. North Pacific warm temperate region.
   West American province.
7. Atlantic warm temperate region.
8. Tropical Indo-West-Pacific region.
9. Tropical East Pacific region.
10. Tropical Atlantic region.
11. South Pacific warm temperate region.
12. South Atlantic warm temperate region.
13. Indo-Australian warm temperate region.
14. Antiboreal region.
15. Kerguelan region.
16. Antarctic region.

There are some cases of the same species, or very closely related forms, occupying zones of similar temperature in middle or high latitudes in both northern and southern hemispheres, although absent from the intervening warm-water belt. Such a pattern of distribution is termed 'bipolar'. The bipolar distribution of a pelagic amphipod *Parathemisto gaudichaudi* is shown in *Figure 4.3*, approximating to the distribution of surface water between 5°C and 10°C. Among numerous examples of bipolarity, Ekman[3] mentions the following inhabitants of the North-East Atlantic, *Balanus balanoides* (North Atlantic, Tierra del Fuego and New Zealand), *Botryllus schlosseri* and *Didemnum albidum* (both North Atlantic and New Zealand), the genus *Engraulis* and the entire order Lucernariida. In some cases apparent bipolarity is really a continuous distribution through the colder layers of water underlying the warm surface layers of the tropics, i.e. tropical submergence. *Eukrohnia hamata* (*Figure 4.4*), *Parathemisto abyssorum* and *Dimophyes arctica* are examples of species found at the surface in both Arctic and Antarctic waters, and present at deeper levels at low latitudes.

The British Isles lie across the 10°C mean annual surface isotherm, and in winter the 5°C isotherm moves south along these coasts. It is possible here to distinguish certain species as belonging to a northern group of arctic and boreal forms, and others as a southern group of Mediterranean and temperate water species. There are seasonal changes in distribution and a broad overlap of populations, but the 10°C isotherm lies approximately between the two groups. Among the fishes of the area, the northern group includes cod (*Gadus morhua = callarias*), haddock (*Melanogrammus aeglefinus*), plaice (*Pleuronectes platessa*), halibut (*Hippoglossus hippoglossus*), and herring (*Clupea harengus*). Examples of southern forms are pollack (*Pollachius pollachius*), European hake (*Merluccius merluccius*), Dover sole (*Solea solea*), turbot (*Scophthalmus maximus*), pilchard (*Sardina pilchardus*), anchovy (*Engraulis encrasicolus*), mackerel (*Scomber scombrus*) and tunny (*Thunnus thynnus* and *T. alalunga*). On the sea-shore almost all boreal species can occur all round the British coast but a few (e.g. *Balanus balanoides*, *Acmaea tessulata*, *Zoarces viviparus* and *Neptunea antiqua*), which are common in the north and east, become scarce or absent towards the south-west. There are a larger number of species which are abundant in the south-west but absent in the north and east, the British Isles being the northernmost limit of their range. These southern forms include *Chthamalus stellatus*, *Balanus perforatus*, *Monodonta lineata*, *Gibbula umbilicalis*, *Patella depressa*, *Anemonia sulcata*, *Leander serratus* and *Asterina gibbosa*. In recent years (since 1961) mean annual sea temperatures around the British Isles have fallen slightly compared with those of the previous 25 years. At the time of writing it is rather early for the effects of this slight temperature change upon the distribution of marine organisms to have become obvious, but already it appears that in the western part of the English Channel the numbers of cod and herring have increased, and pilchards have become fewer over the same period. On shores of the south-west peninsulas of England and Wales (page 194) the range of *Balanus balanoides* has extended westwards, and it now outnumbers *Chthamalus stellatus* in some places where the latter was previously the dominant intertidal barnacle. The topshell, *Monodonta lineata*, no longer extends eastwards as far along the Channel coast as it did before the unusually cold winter of 1962–1963.

Knowledge of the distribution of species dwelling at deep levels is very incomplete, but it appears that the system of submarine ridges between the Arctic basin and the North Atlantic forms a boundary which many species do not cross. The deep levels of the Atlantic, Indian and Pacific Oceans are interconnected via the deep water of the Southern Ocean. Throughout this vast extent of abyss the water temperature is fairly uniform between about 0°C and 4°C, and some deep water species are correspondingly widely distributed. North of the North Atlantic ridge, much of the bottom water of the Arctic basin is below 0°C, and relatively few species are common to deep levels of both the Arctic and the other deep oceans.

With growing knowledge of the abyssal fauna it is becoming apparent that there is much greater diversity of abyssal species than was originally thought, and that many species are relatively restricted in distribution. Studies of

69

abyssal zoogeography now suggest that several distinct faunistic regions exist at deep levels.

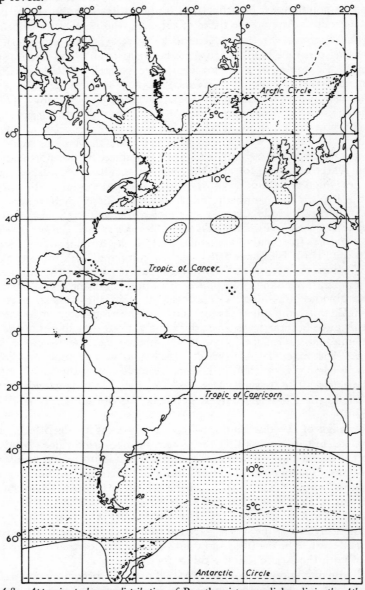

Figure 4.3. Approximate known distribution of Parathemisto gaudichaudi *in the Atlantic, and mean annual isotherms for 5°C and 10°C*

(Based on a Map of the World by courtesy of G. Philip & Son Ltd.)

Except for sea birds and mammals, marine organisms are poikilothermic, i.e. their body temperature is always close to that of the surrounding water and varies accordingly, although some large and active forms often have a

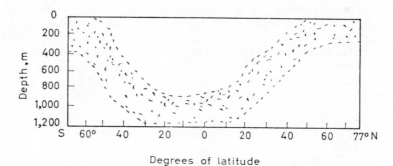

Figure 4.4   Distribution in depth of Eukrohnia hamata in the Pacific, from the Bering Sea to MacMurdo Sound

(From Alvarino, A.[18] by courtesy of Allen & Unwin)

body temperature slightly higher than water temperature due to the release of heat by metabolism. The physiological effects of temperature change are complex but, in simple terms, rates of metabolic processes increase with rising temperature, usually about 10 per cent per 1°C rise over a range of temperatures up to a maximum (*Figure 4.5*) beyond which they fall off rapidly. Death occurs above and below certain limiting temperatures, probably because of disturbances of enzyme activity, water balance and other aspects of cellular chemistry. Marine creatures usually succumb more rapidly to overheating than to overcooling. The limits of distribution of a species in the sea do not coincide closely with the normal occurrences of rapidly lethal temperatures but are much more restricted. In freak climatic conditions, extremes of heat or cold may have devastating effects on marine populations, especially those of the shore, but in normal circumstances temperature probably controls distribution mainly through its influence on several major processes including feeding, respiration, osmoregulation, growth and reproduction, especially the latter.

Temperature regulates reproduction in several ways. It controls the maturation of gonads and the release of sperms and ova, and in many cases the temperature tolerance of embryonic and larval stages is less than that of the adults. Temperature has therefore a major influence on the breeding range and period, and on mortality rates during early stages of development and larval life. Along the fringes of distribution there are usually non-breeding zones where the adults can survive but cannot reproduce, the population being maintained by spread from the main area of distribution within which breeding is possible.

In temperate seas many species virtually cease feeding during the winter. In some cases reduced feeding is simply the result of shortage of food, but many creatures definitely stop eating below a certain temperature. Food requirements are reduced during cold periods because the respiration rate is low and growth ceases. However, despite the depressing effects of cold on growth, it is nevertheless generally observed that where the distribution of a marine species covers a wide range of temperature the individuals living in colder areas attain considerably larger adult sizes than those in the warmer parts of the distribution. This trend is associated with a longer growing period, later sexual maturation and a longer life in cold water. There are exceptions to this trend, and some species reach larger sizes in warmer water.

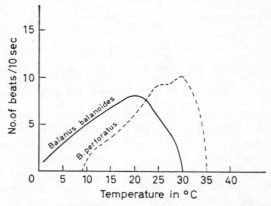

Figure 4.5   Relation of rate of ciliary beat to temperature in two species of barnacle

Apart from direct physiological effects, changes of temperature have certain indirect effects by altering some of the physical properties of the water, notably density, viscosity and the solubility of gases, which in turn influence buoyancy, locomotion and respiration. There are instances, e.g. the summer and winter forms of the diatom *Rhizosolenia hebetata* (*Figure 4.9*), where the morphology of a species appears to vary with changes of temperature, possibly because of alterations in viscosity and buoyancy. The viscosity of water falls considerably with increasing temperature, which may partly account for the increased setation of the appendages of many warm-water planktonts as compared with cold-water forms.

THE COMPOSITION OF SEA-WATER

Sea-water is an extremely complex solution, its composition being determined by an equilibrium between rates of addition and loss of solutes, evaporation and the addition of fresh water. Solutes are continually added to the water mainly by solution of the earth's rocks by processes of weathering and erosion, and may eventually be lost from the water by precipitation on the sea bottom[65]. Short-term, minor fluctuations of composition occur through biological processes involving absorption and release of solutes by organisms. There are also interchanges of gases between sea and atmosphere.

It is uncertain to what extent the composition of sea-water may have changed during geological time, but it is not thought to have varied very widely over the period that life has existed. At present, the principal cations are sodium, magnesium, calcium, potassium and strontium, and the chief anions are chloride, sulphate, bromide and bicarbonate. These make up over 99·9 per cent of the dissolved material, forming approximately a 3·5 per cent solution. The amount of inorganic material dissolved in sea-water expressed as weight in grammes per kilogramme of sea-water is termed the salinity (S), and usually amounts to about 35 g/kg, i.e. $S = 35$ per mille (generally written 35‰). The quantities of the major constituents of a typical sample of ocean water are shown are Table 4.1.

TABLE 4.1.

Major Constituents of an Ocean Water. $S°/_{oo} = 35·00$.
(From H. W. Harvey[6]—by courtesy of Cambridge University Press)

Constituent	g/kilo
Sodium	10·77
Magnesium	1·30
Calcium	0·409
Potassium	0·388
Strontium	0·010
Chloride	19·37
Sulphate as SO_4	2·71
Bromide	0·065
Carbon, present as bicarbonate, carbonate and molecular carbon dioxide	0·023 at pH 8·4 to 0·027 at pH 7·8

The relative proportions of the major ionic constituents in ocean water remain virtually constant despite some variation in total salinity. Estimation of the concentration of any of these ions therefore enables the total salinity to be calculated. Most salinity determinations have been made by titrating sea-water with silver nitrate solution. This precipitates the halides, mainly chloride with a trace of bromide, and their total weight in grammes per kilo of sea-water is termed the chlorinity, Cl. The salinity is then determined from the empirical relationship known as Knudsen's Formula:

$$S°/_{oo} = 0·030 + (1·805 \times Cl°/_{oo})$$

A convenient method is to titrate 10 ml of sea-water with silver nitrate solution containing 27·25 g/l, using a chromate or fluorescein indicator.

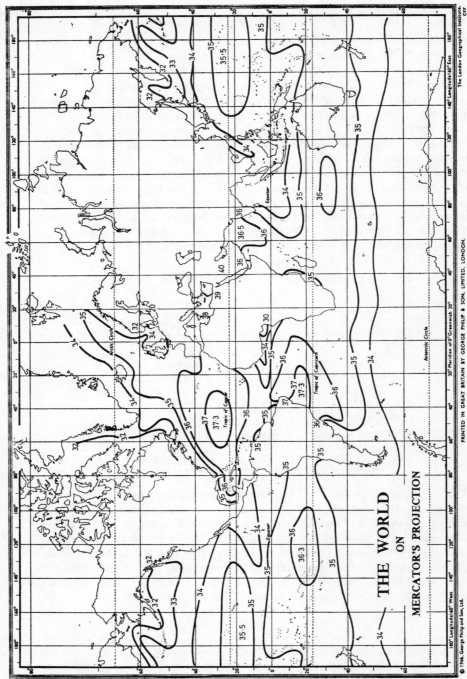

THE WORLD
ON
MERCATOR'S PROJECTION

PRINTED IN GREAT BRITAIN BY GEORGE PHILIP & SON, LIMITED, LONDON.

© 1966, George Philip and Son, Ltd.

The London Geographical Institute.

Figure 4.6 *Approximate positions of mean annual isohalines*
(Based on a Map of the World by courtesy of G. Philip and Son Ltd.)

The added volume of silver nitrate in millilitres is approximately equal to the salinity in grammes per kilo, and a small correction is made from tables to allow for the slight differences in weight of unit volume of sea-water at different salinities.

Greater accuracy is obtained by using 'Standard Sea-water' for comparison. This is water of very accurately known chlorinity available from the Hydrographic Laboratory at Copenhagen. By comparing the titrations of silver nitrate against both sample and Standard Sea-water the calculation of chlorinity becomes independent of the concentration of silver nitrate solution, and all measurements are made to the same standard.

There are various objections to using chlorinity measurements as a basis for all determinations of salinity. It assumes a constant ratio between chlorinity and total amount of dissolved material, which obviously cannot be true for all dilutions of sea-water with other waters of differing compositions. Also, silver nitrate is an expensive reagent, and titration is a relatively time-consuming technique. Consequently, attention has turned to other methods of salinity measurement. Several physical properties of sea-water vary with the amount of dissolved salts and can be used for salinity determination[53,54]; for example, electrical conductivity, density, vapour pressure, freezing point, refractive index and sound conductivity. Electrical conductivity measurements are becoming increasingly used for work of high accuracy.

The salinity of most ocean water is within the range 34–36‰. There are slight seasonal variations of salinity, and average positions for the surface isohalines during the northern summer are shown in *Figure 4.6*. High salinities are associated with low rainfall and rapid evaporation, especially where the circulation of the water is relatively poor. Such conditions are found in the Sargasso area of the North Atlantic and in the South Atlantic off the east coast of Brazil, where the surface salinities are about 37. In high latitudes, the melting of ice, heavy precipitation and land drainage together with low evaporation reduce the salinity of the surface water. In the Arctic, the surface salinity fluctuates between 28 and 33·5‰ with alternate melting and freezing of ice.

In land-locked areas there are appreciable departures from the normal oceanic range of salinities. For instance, in the Baltic, dilution by fresh water reduces the salinity from 29 in the Kattegat region to below 5 in the Gulf of Bothnia. In the Black Sea, rainfall and the outflow of the Danube, Dnieper and Dniester lower the surface salinity to 18 or below. This low-salinity water forms a low-density layer overlying the more saline, deep layers with little mixing between them, and cuts off the depths of the Black Sea from the air, producing the peculiar hydrographic conditions mentioned later (page 77). In hot regions, high surface salinities are found in enclosed seas due to rapid evaporation. In the eastern part of the Mediterranean, salinity may reach 39, and in the Red Sea may exceed 40. On the

shore the salinity of evaporating pools is sometimes greater than 100‰. (*See* footnote to page 63.)

The salinity of neritic water is subject to fluctuation due to changes in the rate of dilution by fresh water from the land. River water often contains ions in very different proportions to those of normal sea-water, and this may produce appreciable changes in the composition of sea-water near a river mouth.

Except for the teleosts and higher vertebrates the majority of marine creatures are in osmotic equilibrium with the surrounding water. The ionic composition of their internal fluids has in most cases a close similarity to that of sea-water, containing relatively high concentrations of sodium and chloride and considerably lower concentrations of potassium, magnesium and sulphate. There is commonly, though not invariably, a rather higher proportion of potassium to sodium in body fluids than that which occurs in sea-water, and somewhat less magnesium and sulphate (*see* Table 4.2). External salinity changes usually produce corresponding changes in the concentration of internal fluids by passage of water into, or out of, the body (osmotic adjustment) to preserve the osmotic equilibrium, and these changes are often accompanied by alterations in the proportions of the constituent ions of the internal fluids. Beyond limits, which differ for different species, departures from the normal concentration and composition of the internal medium cause metabolic disturbances and eventual death.

TABLE 4.2

Concentrations of Ions in Body Fluids of some Marine
Invertebrates (g/kilo)

	Na	K	Ca	Mg	Cl	SO$_4$
Sea-water $S°/_{oo} = 34.3$	10·6	0·38	0·40	1·27	19·0	2·65
Aurelia aureta	10·2	0·41	0·39	1·23	19·6	1·46
Arenicola marina	10·6	0·39	0·40	1·27	18·9	2·44
Carcinus maenas	11·8	0·47	0·52	0·45	19·0	1·52
Mytilus edulis	11·5	0·49	0·50	1·35	20·8	2·94
Phallusia mammillata	10·7	0·40	0·38	1·28	20·2	1·42

The majority of organisms of the open sea have very limited tolerance of salinity change, i.e. they are *stenohaline*. *Euryhaline* forms which can withstand wider fluctuations of salinity are typical of the less stable conditions of coastal water[57,58,62]. Extreme euryhalinity characterizes estuarine species.

Organisms which remain in osmotic balance with their surroundings when the salinity varies are termed *poikilosmotic*, and these include some widely euryhaline creatures. *Arenicola marina* is a familiar example from the Atlantic coastline, where it is widely distributed in marine, brackish and estuarine muddy sands and able to survive salinities down to about 18 per mille. In other parts of its range, for example the Baltic, it is found at even lower salinities. Other examples from the American fauna which are poikilosmotic and moderately euryhaline are *Mytilus edulis, Cardium magnum, Mya arenaria, Balanus balanoides, B. improvisus, Nereis pelagica* and many other common shore forms.

Some animals are able to control within limits the concentration of their internal fluids independently of salinity changes in the water. This process is known as osmoregulation, and organisms which maintain this stability of internal environment are described as *homoiosmotic*. The shore crab, *Carcinus maenas*, is a very euryhaline osmoregulator which extends up estuaries to levels where it encounters immersion in fresh water. Some powers of osmoregulation are also present in *Nereis diversicolor, Palaemonetes, Gammarus locusta, G. duebeni* and *Marinogammarus marinus*. The ability to osmoregulate is influenced by temperature and fails above and below certain limiting temperatures.

Change of salinity alters the specific gravity of the water, and this influences organisms indirectly through its effects on buoyancy.

Dissolved Gases

All atmospheric gases, including the inert gases, are present in solution in sea-water.

Oxygen

The oxygen content of sea-water varies between 0 and 8·5 ml/l, mainly within the range 1–6. High values occur at the surface, where dissolved oxygen tends to equilibrate with atmospheric oxygen. Rapid photosynthesis may sometimes produce supersaturation. Because oxygen is more soluble in cold water than in warm, the oxygen content of surface water is usually greater at high latitudes than nearer the equator, and the sinking of cold surface water in polar seas carries oxygen-rich water to the bottom of the deep ocean basins.

Although the deep layers of water are mostly well oxygenated, oxygen is by no means uniformly distributed with depth, and in some areas there is an oxygen-minimum layer at a depth somewhere between 400 and 1,000 m. This is most evident in low latitudes, where the water at 400–500 m has sometimes been found to be almost completely lacking in oxygen. The reasons for this are uncertain; but the oxygen-minimum zone often appears to be well populated, and one cause of the deficiency of oxygen may be depletion by a large amount of animal and bacterial respiration in water where relatively little circulation is taking place.

The exceptional conditions in the Black Sea were mentioned earlier. Cut off from the Mediterranean by the shallow water of the Bosphorus, there is little mixing between the low-density surface water (page 75) and the denser, more saline deep water. The deep levels are virtually stagnant and have become completely depleted of oxygen. Animal life is impossible below some 150–200 m; but anaerobic bacteria flourish in the deep layers, mainly sulphur bacteria which metabolize sulphate to sulphide and produce large quantities of H_2S which give the deep water a very objectionable smell. Comparable conditions sometimes arise in other land-locked areas of deficient circulation, such as deep lochs or fiords.

Carbon dioxide

The preponderance in sea-water of the strongly basic ions, sodium, potassium and calcium, imparts a slight alkalinity and enables a considerable amount of carbon dioxide to be contained in solution. This is of great biological importance because carbon dioxide is a raw material of photosynthesis. Under natural conditions, plant growth in the sea is probably never limited by shortage of carbon dioxide.

Carbon dioxide is present in sea-water mainly as bicarbonate ions, but there are also some dissolved CO_2, undissociated H_2CO_3 and carbonate ions. At the surface, dissolved CO_2 tends towards equilibrium with atmospheric CO_2, the oceans acting as a regulator of the amount of CO_2 in the atmosphere. The overall equilibrium can be represented as follows:

Atmospheric CO_2

$$\Updownarrow$$

Dissolved CO_2 $\rightleftharpoons$ H_2CO_3 $\rightleftharpoons$ $[H]^+ + [HCO_3]^-$ $\rightleftharpoons$ $[H]^+ + [CO_3]^{2-}$

The pH of sea-water normally lies within the range 7·5–8·4, the higher values occurring in the surface layer where CO_2 is withdrawn by photosynthesis. The presence of strong bases together with the weak acids H_2CO_3 and H_3BO_3 confers an appreciable buffer capacity. Addition of acid to sea-water depresses the dissociation of H_2CO_3 and H_3BO_3, and there is not much change of pH while reserves of $[CO_3]^{2-}$ $[HCO_3]^-$ and $[H_2BO_3]^-$ ions remain. Addition of alkali increases the dissociation of H_2CO_3 and H_3BO_3, and the pH remains fairly stable so long as undissociated acid is still present.

The dissociation constants of the equilibrium are influenced by temperature, pressure and salinity. Increase of temperature or pressure causes a slight decrease of pH. At great depths the lowering of pH due to pressure may be sufficient to cause solution of some forms of calcium carbonate, which is not a conspicuous component of sediments below about 6,000 m[74].

Nitrogen

Amounts of uncombined nitrogen in sea-water vary between 8·4 and 14·5 ml/l. Nitrogen-fixing bacteria are known to occur in the sea, but the quantity of nitrates formed by their activity is probably very small. There is also some return of nitrogen from the oceans to the atmosphere by the nitrogen-freeing activity of denitrifying bacteria and blue-green algae. With increasing quantities of atmospheric nitrogen being fixed by industrial processes for fertilisers, the biological freeing of nitrogen from nitrate becomes of increasing importance in maintaining the equilibrium of the nitrogen cycle.

Minor Constituents

In addition to the major constituents listed on page 73, there are many other elements present in sea-water in very small amounts (Table 4.3). The most abundant of the ionized minor constituents are silicate ions at concentrations up to 6 mg/kg, and fluoride ions up to 1·4 mg/kg. The combined weights of all the other minor constituents, numbering nearly fifty, total less than 2 mg/kg, and at this dilution the estimation of many of them is very difficult. Several are known to be present only because they are concentrated in the bodies of marine organisms.

Few marine organisms survive for long in an artificial sea-water which contains only the major constituents in correct proportion. The minor constituents are evidently of biological importance although in many cases their role is uncertain. Some are known to be essential for the normal growth of plants; for example nitrate, phosphate, iron, maganese, zinc, copper and cobalt. Silicon is an ingredient of diatoms, and some marine algae require molybdenum and vanadium. Many of the minor constituents are also necessary for animal life. Silicon is included in the spicules of most radiolaria and some sponges. Iron is required by all animals. Copper is present in the prosthetic group of the blood pigment haemocyanin which occurs in some molluscs and crustacea. Vanadium and niobium occur in the blood pigment of ascidians. The vertebrate hormone thyroxin is an iodine compound.

Certain organisms concentrate the minor constituents to a remarkable extent. Vanadium in ascidians is an outstanding example, occurring at a concentration approximately 50,000 times greater than in sea-water. Iodine, nickel, molybdenum, arsenic, zinc, vanadium, titanium, chromium and strontium are concentrated in the tissues of various marine algae, and some fish concentrate silver, chromium, nickel, tin or zinc.

Whereas the major constituents of sea-water, and some of the minor constituents, remain virtually constant in proportion (conservative constituents), certain minor constituents fluctuate in amount due to selective absorption by organisms (non-conservative constituents). The latter include nitrate, phosphate, silicate, iron and manganese, and the list will probably increase as our knowledge of the requirements of marine organisms grows.

Nitrate and phosphate

Nitrogen in combined form is present in sea-water as nitrate, nitrite, ammonium ions and traces of nitrogen-containing organic compounds.

TABLE 4.3
Geochemical Parameters of Sea-water
(From Goldberg, E. D.[7] by courtesy of Interscience)

Element	Abundance, mg/l	Principal species	Residence time, years
H	108,000	H_2O	
He	0·000005	He (g)	
Li	0·17	Li^+	$2·0 \times 10^7$
Be	0·0000006		$1·5 \times 10^2$
B	4·6	$B(OH)_3$; $B(OH)_2O^-$	
C	28	HCO_3^-; H_2CO_3; CO_3^{2-}; organic compounds	
N	0·5	NO_3^-; NO_2^-; NH_4^+; N_2 (g); organic compounds	
O	857,000	H_2O; O_2 (g); SO_4^{2-} and other anions	
F	1·3	F^-	
Ne	0·0001	Ne (g)	
Na	10,500	Na^+	$2·6 \times 10^8$
Mg	1,350	Mg^{2+}; $MgSO_4$	$4·5 \times 10^7$
Al	0·01		$1·0 \times 10^2$
Si	3	$Si(OH)_4$; $Si(OH)_3O^-$	$8·0 \times 10^3$
P	0·07	HPO_4^{2-}; $H_2PO_4^-$; PO_4^{3-}; H_3PO_4	
S	885	SO_4^{2-}	
Cl	19,000	Cl^-	
A	0·6	A (g)	
K	380	K^+	$1·1 \times 10^7$
Ca	400	Ca^{2+}; $CaSO_4$	$8·0 \times 10^6$
Sc	0·00004		$5·6 \times 10^3$
Ti	0·001		$1·6 \times 10^2$
V	0·002	$VO_2(OH)_3^{2-}$	$1·0 \times 10^4$
Cr	0·00005		$3·5 \times 10^2$
Mn	0·002	Mn^{2+}; $MnSO_4$	$1·4 \times 10^3$
Fe	0·01	$Fe(OH)_3$ (s)	$1·4 \times 10^2$
Co	0·0005	Co^{2+}; $CoSO_4$	$1·8 \times 10^4$
Ni	0·002	Ni^{2+}; $NiSO_4$	$1·8 \times 10^4$
Cu	0·003	Cu^{2+}; $CuSO_4$	$5·0 \times 10^4$
Zn	0·01	Zn^{2+}; $ZnSO_4$	$1·8 \times 10^5$
Ga	0·00003		$1·4 \times 10^3$
Ge	0·00007	$Ge(OH)_4$; $Ge(OH)_3O^-$	$7·0 \times 10^3$
As	0·003	$HAsO_4^{2-}$; $H_2AsO_4^-$; H_3AsO_4; H_3AsO_3	
Se	0·004	SeO_4^{2-}	
Br	65	Br^-	
Kr	0·0003	Kr (g)	
Rb	0·12	Rb^+	$2·7 \times 10^5$
Sr	8	Sr^{2+}; $SrSO_4$	$1·9 \times 10^7$
Y	0·0003		$7·5 \times 10^3$
Nb	0·00001		$3·0 \times 10^2$
Mo	0·01	MoO_4^{2-}	$5·0 \times 10^5$
Ag	0·0003	$AgCl_2^-$; $AgCl_3^{2-}$	$2·1 \times 10^6$
Cd	0·00011	Cd^{2+}; $CdSO_4$	$5·0 \times 10^5$
In	<0·02		
Sn	0·003		$5·0 \times 10^5$
Sb	0·0005		$3·5 \times 10^5$
I	0·06	IO_3^-; I^-	
Xe	0·0001	Xe (g)	
Cs	0·0005	Cs^+	$4·0 \times 10^4$
Ba	0·03	Ba^{2+}; $BaSO_4$	$8·4 \times 10^4$

TABLE 4.3 (continued)

Element	Abundance, mg/l	Principal species	Residence time, years
La	0·0003		$1·1 \times 10^4$
Ce	0·0004		$6·1 \times 10^3$
W	0·0001	$WO_4{}^{2-}$	$1·0 \times 10^3$
Au	0·000004	$AuCl_4{}^-$	$5·6 \times 10^5$
Hg	0·00003	$HgCl_3{}^-$; $HgCl_4{}^{2-}$	$4·2 \times 10^4$
Tl	<0·00001	Tl^+	
Pb	0·00003	Pb^{2+}; $PbSO_4$	$2·0 \times 10^3$
Bi	0·00002		$4·5 \times 10^5$
Rn	$0·6 \times 10^{-15}$	Rn (g)	
Ra	$1·0 \times 10^{-10}$	Ra^{2+}; $RaSO_4$	
Th	0·00005		$3·5 \times 10^2$
Pa	$2·0 \times 10^{-9}$		
U	0·003	$UO_2(CO_3)_3{}^{4-}$	$5·0 \times 10^5$

Nitrate ions predominate, but in the uppermost 100 m and also close to the bottom there are sometimes appreciable amounts of ammonium and nitrite formed by biological activity.

Phosphorus is present almost entirely as orthophosphate ions $[H_2PO_4]^-$ and $[HPO_4]^{2-}$ with traces of organic phosphorus. The concentrations of these combined forms of nitrogen and phosphorus generally fall within the following wide ranges:

NO_3-N = 1–600 µg/l (0·1–43 µg atoms N/l)
NO_2-N = 0–15 µg/l
NH_3-N = 0·4–50 µg/l
Organic-N = 30–200 µg/l
Phosphate-P = < 1–100 µg/l (0·01–3·5 µg atoms P/l)
Organic-P = < 1–30 µg/l

Nitrogen and phosphorus are essential requirements for plants. Together with other essential trace elements they are commonly referred to as 'nutrients', and the amount of plant growth in the sea is largely controlled by the availability of nutrients in the surface layers.

Quantities of nitrate and phosphate vary greatly with depth. They are generally low and variable at the surface, reflecting the uptake of these ions by plants. Surface values usually fall within the range 1–120 µg/l NO_3-N and 0–20 µg/l phosphate-P, with highest values in winter and lowest in summer. At deeper levels the concentration of nutrients is considerably greater, and where there are high values in the surface layers they are generally due to admixture with water from below by convectional mixing or upwelling. Near the coast, high values often occur due to the stirring up of bottom sediments or to large amounts of nitrate and phosphate present in some river water.

In deep water below the level of surface mixing there is usually a gradient of increasing concentration with depth, and a zone of maximum concentration between about 500 and 1,500 m, with quantities in the range 200–550 µg/l NO_3-N and 40–80 µg/l phosphate-P. Below this there may be a slight decrease but values remain high and fairly uniform. There may be some increase close to the sea-bed due to release of nutrients by bacterial decomposition of organic matter deposited on the bottom.

Despite fluctuations in total amount, the relative concentrations of nitrate and phosphate remain fairly constant, the nitrate/phosphate ratio usually being about 7/1 by weight and 15/1 by ions. This close relationship indicates that the two ions are probably absorbed by living organisms, and subsequently released, in much the same proportions as they are present in the water.

Silicate

Silicon is present in sea-water chiefly as silicate ions and possibly sometimes minute traces of colloidal silica. It is a constituent of the diatom cell wall, some radiolarian skeletons and some sponge spicules.

The concentration of silicate at the surface is usually low, but increases with depth to between 1 and 5 mg Si/l in deep ocean water, the highest values being in the deep Pacific. In the English Channel, surface concentrations of 200–400 µg Si/l have been recorded during the winter, falling rapidly to as low as 10µg/l during the spring diatom peak. Similar fluctuations are recorded in the coastal waters around the United States.

Although much of the silicate incorporated in the diatom cell wall is probably returned to the water fairly quickly after death, siliceous deposits of planktonic origin cover large areas of the sea-bed (page 153).

Iron and Manganese

Ferric hydroxide is almost insoluble within the pH range of sea-water. The amount of iron in true solution is probably not more than 2 µg Fe/l, but there are appreciable quantities of iron in particulate form as colloidal micelles, mainly ferric hydroxide and traces of ferric phosphate, ferricitrate or haematin. This particulate iron can be removed from sea-water by ultrafiltration.

Estimates of total iron vary considerably with time, place, and different techniques of measurement, usually within the limits of 3–70 µg Fe/l. In the English Channel, fluctuations from about 3 to 21 µg Fe/l have been recorded, high values being obtained at the surface and near the bottom. Surface values usually show marked seasonal reductions following peak periods of diatom growth.

There is probably a continual loss of iron from sea-water, and accumulation at the bottom, due to adsorption on sinking detritus and sedimentation.

Iron is an essential plant nutrient, and also has various roles in animal physiology. It is a component of the cytochrome enzyme system and the blood pigments haemoglobin (vertebrates, some annelids and some molluscs), haemerythrin (some molluscs and crustacea) and chlorocruorin (some annelids). The amount of iron in solution seems inadequate to support rapid plant growth, and it is possible that marine plants can utilize particulate iron in some way, perhaps by gradual solution of particles adsorbed on the cell wall, or even by actual ingestion by certain plants which have exposed protoplasm.

Manganese is a plant nutrient which, like iron, is probably present mainly in particulate form as oxide micelles, in amounts between 0·3 and 10 μg Mn/l.

In deep water the surface layers may become depleted of particulate and adsorbed iron and manganese by losses through sinking, and this may limit the amount of plant growth that can be supported. In experimental cultures, enrichment of samples of ocean water by addition of iron and manganese sometimes results in a considerably increased growth. There is often more iron and manganese in neritic water due to replenishment by land drainage.

Dissolved organic matter

Small and varying quantities of organic compounds are present in solution in sea-water. The estimation of minute quantities of organic solutes is difficult, but it appears that ocean water commonly contains about 2 mg C/l in dissolved organic forms, in some of which nitrogen, phosphorus, sulphur, iron or cobalt are included. Although the concentration of these compounds is small, the total quantity in the ocean is very large, and it has been calculated that there are on average about 15 kg of dissolved organic matter beneath each square metre of the ocean surface. This greatly exceeds the amount of organic matter present in particulate form, either living material or organic debris.

Little is known about the source and nature of these solutes, but they are presumably of biological origin, probably derived largely from excretion or breakdown of dead tissues. Possibly traces of organic matter are leached from living tissues, and some organisms are known to liberate organic secretions into the water as part of their normal metabolism. Such external metabolites are termed exocrines.

There is little precise information regarding the quantities and chemical nature of these compounds, but a growing list of substances identified in sea-water includes various carbohydrate materials, amino acids, lipids, and vitamins such as ascorbic acid and components of the vitamin B complex, for example thiamin and cobalamin.

The way in which organic solutes influence the biological properties of sea-water is a matter for speculation, but it is clear that at least some are absorbed by organisms and may have various effects. Artificial sea-water made from all known inorganic constituents including the trace elements does not have the same biological quality as natural sea-water, and organisms do not thrive in it. If, however, small quantities of organic material are added in such forms as soil extracts, urine, extracts of various tissues or even small volumes of natural sea-water, the quality of the solution as a satisfactory medium for marine life is greatly enhanced. The fact that the concentrations of organic solutes in natural sea-water are very low suggests that they may be continuously absorbed from the water by organisms.

Putter in 1907 was one of the first to direct attention to a possible biological importance of dissolved organic matter in the sea. He suggested that it provided a source of food for marine animals because there was thought to be insufficient phytoplankton and particulate organic matter in sea-water to support the animal population. Since that time, improved techniques for the quantitative estimation of phytoplankton, especially the nanoplankton, indicate that Putter's figures for phytoplankton were considerably too low, and that there is probably, overall, adequate particulate food to accord with the observed numbers of animals. None the less, we now know that sea-water often contains many saprophytic forms, bacteria and microflagellates, that utilize dissolved organic matter, so these solutes must certainly contribute something to the general economy of the sea. At deep levels far below the zone of photosynthesis, where quantities of particulate food are very small, organic solutes probably constitute a significant part of the food supply. There is also some evidence that certain forms of dissolved organic matter in sea-water may become aggregated into particles at air–water interfaces, especially on the surface of bubbles, thereby adding to particulate food supplies[76].

In recent years the work of a number of marine biologists, notably Lucas[63,64], has re-aroused interest in the role of dissolved organic substances in sea-water. It seems likely that various interactions between organisms are brought about by external secretions. The olfactory sense is highly developed in many marine creatures and probably serves several useful functions; for example, the recognition and location of other individuals, the detection of food, and in certain species it is important in connection with navigation and migration. A remarkable effect of the secretions of certain organisms is the phenomenon known as the 'red tide', which occasionally causes the death of enormous numbers of fish. It is observed mainly in warm waters, and is usually associated with the sporadic occurrence of swarms of flagellate organisms, sufficient to produce a red coloration of the water and liberating a lethal toxin[77]. Several different species can produce this effect, principally dinoflagellates, for example *Gymnodinium brevis*, *Goniaulax polyedra*, *Exuviella baltica*. In the fall of 1972 cases of food poisoning in humans in the U.S. were traced to dinoflagellate infection of clams on the New England coast, which also caused the death of seabirds, sand eels and flounders.

Various other micro-organisms, particularly flagellates and bacteria, are also known to produce substances toxic or repellent to other organisms. It has been demonstrated that bacterial respiration is sometimes depressed in the presence of diatoms performing photosynthesis, presumably due to anti-biotic substances set free by the plants, and it has also been suggested that zooplankton avoids water containing large numbers of phytoplankton because of distasteful external metabolites produced by the plants (page 134).

In some cases, organic solutes have a growth-promoting effect, and seem to have a role in nutrition comparable to that of vitamins, auxins or hormones. Experimental cultures of marine algae indicate that some have specific organic requirements for normal growth, for example cobalamin. Chelat-ing agents in sea-water may also be of some importance, favouring plant growth by bringing into solution essential trace metals which occur in par-ticulate form[55].

The studies of Wilson and Armstrong[86-90] on the rearing of invertebrate larvae, for example *Echinus* and *Ophelia*, demonstrated differences in survival rates of larvae reared in samples of water collected from different areas al-though there were no obvious chemical or physical features of the water to which the differences could be clearly attributed. The factors responsible for these variations in biological quality of the water could not be deter-mined, but some growth-promoting or growth-inhibiting substances were evidently present. The result of mixing sea-waters of different quality suggested that the differences were probably due to the presence of bene-ficial substances in some samples, rather than to harmful substances in others. It has been suggested that differences in biological properties between sea-waters of different areas may depend partly upon their biological his-tories through the effect of metabolites produced by preceding generations of organisms.

In recent years, man-made organic compounds have been produced on a large scale for various purposes, for example, detergents and pesticides, and many of these eventually find their way into the sea. Organic insecticides have already become surprisingly widespread throughout the oceans and contaminate the tissues of marine creatures and sea birds. Traces of DDT have been detected even in the seals and penguins of the Antarctic and PCB contamination is becoming extensive. The possible dangers of these pollut-ants are obvious, and careful monitoring and control of the situation is clearly necessary.

The increasing extent of oil pollution is a matter for concern[15,69]. There are also hazards in the continual use of the oceans as dumping grounds for a variety of highly toxic wastes, notably radioactive materials, war gases and poisonous industrial effluents. Already several instances of fish poisoning in neritic waters have been traced to industrial wastes, and the recent detection of unexpectedly high levels of mercury in Pacific tuna suggests that chemical pollution may be spreading to the deep ocean. Even when dangerous products are enclosed in strong containers before dumping, the contents are likely eventually to seep into the water where their effects are not known with certainty and must not be lightly dismissed. Some toxic substances

85

become concentrated to a remarkable extent in the tissues of marine organisms and the concentration becomes greater at each stage of a food chain, with the effect that even substances present in the water at very great dilution may in time have disastrous effects on some populations. A population may be able to tolerate a certain level of pollution as long as other conditions are favourable, but if circumstances become adverse, for example through food shortage or climatic extreme, constitutions may be so weakened that mortality then becomes very high.

SPECIFIC GRAVITY AND PRESSURE

The specific gravity of sea-water varies with temperature and salinity and very slightly with pressure. At 20°C and atmospheric pressure, sea-water of salinity 35‰ has a specific gravity of 1·025.

At salinities above 24·7‰ the temperature of maximum density lies below the freezing-point. Because of this, cooling of the sea's surface to freezing-point does not produce a surface layer of light water, as it does in fresh water where the temperature of maximum density is 4°C. In contrast to fresh water, surface cooling of the sea below 4°C therefore causes the density of the surface water to increase, and convectional mixing to continue, right up to the point at which ice crystals begin to separate. Consequently, there is no formation of a winter thermocline in the sea as occurs in fresh-water ponds and lakes.

Hydrostatic pressure increases with depth at approximately 1 atm/10 m. In the deepest ocean trenches, pressures exceed 1,000 atm. Although water is only very slightly compressible, such enormous pressures are sufficient to produce a slight adiabatic compression of the deep water, resulting in a detectable increase in temperature. In some areas, temperature readings taken from the 4,000 m level down to 10,000 m show a rise of about 1°C (*Figure 4.7*). The water column is not unstable because this rise of temperature at the lower levels is the result of compression and there is no decrease of density. Deep water temperatures are often expressed as Potential Temperature, θ, i.e. the temperature to which the water would come if brought to atmospheric pressure without heat change.

Although marine creatures are found at all depths, each species is restricted in the range of levels it inhabits. It is usually difficult to know to what extent this limitation is due to pressure because the associated gradients of temperature and illumination are probably in many cases the dominant factors regulating distribution in depth. However, the pressure gradient must certainly play some part. Organisms which live in the surface layers of the sea can be killed by subjecting them to the very high pressure of abyssal depths. Alternatively, organisms brought up from the deep sea bottom require a high pressure, comparable to that of their usual environment, for normal activity; for example, to culture deep-water bacteria there must be a high pressure within the culture vessel. It is evident, therefore, that marine creatures are adapted to suit particular pressure ranges.

The physiological effects of pressure are not well understood[59,75], but it is at least known that pressure has various direct effects on protoplasm, influencing metabolic rates and producing various alterations of behaviour. Experiments on small pelagic animals indicate that pressure-increase commonly causes an increase of swimming activity, and movement upwards towards light. In natural conditions the pressure gradient may therefore be one factor which helps to prevent surface forms from descending too deep.

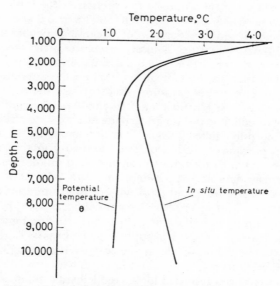

Figure 4.7 The effect of pressure on temperature. In situ *and potential temperatures at deep levels*

Many species are fairly limited in vertical distribution, but some are eurybathic forms found over a great range of depths. Examples from the British fauna which extend from the continental shelf to below 2,000 m, listed by Ekman[3], include:

> *Lumbriconereis impatiens:* at least to 3,000 m
> *Notomastus latericeus:*   „   „   „   „
> *Hydroides norvegica:*   „   „   „   „
> *Pomatoceros triqueter:* 5–3,000 m
> *Amphicteis gunneri:* 20–5,000 m
> *Diastylis laevis:* 9–2,820 m
> *Verruca stroemia:* littoral to 3,000 m

Eudorella truncatula: 9–2,820 m
Henricia sanguinolenta: 0–2,450 m
Ophiopholis aculeata: 0–2,040 m.

Most of the organisms which are found near the surface of the sea seem to have a more restricted depth range than those which inhabit deep levels. The pressure gradient may be partly accountable because the greatest relative changes of pressure with depth occur close to the surface. Between the surface and 10 m the pressure doubles, and on descending a further 10 m the pressure increases by a further 50 per cent. An organism which changes its depth between the surface and 20 m experiences the same relative pressure change as one which moves between 2,000 and 6,000 m.

Earlier we mentioned the dangers of breathing air beneath the sea's surface (page 58). Air-breathing aquatic vertebrates such as seals and whales face some risk of gas embolism when surfacing from deep dives, but the danger is far less than for human divers because the animals are not continually breathing air under pressure while diving. When they dive they take down relatively little air, the lungs being only partially inflated. Water-pressure tends to collapse the lungs, driving the air into the trachea from which little absorption takes place. There is consequently not much gas dissolved under pressure in the blood. These animals contain large quantities of myoglobin from which they draw oxygen during submergence. They can accumulate a considerable oxygen-debt and can tolerate a high level of CO_2 in the blood, the threshold sensitivity of the respiratory centre in the brain being higher than in terrestrial mammals. In seals the heart rate is generally reduced during diving, the peripheral circulation is shut down and the blood-flow mainly restricted to supplying the brain. Little is known of the movements of seals and whales under water. It is thought that seals do not often dive very deeply, probably seldom much over 30 m, although a depth of 600 m has been recorded for a Weddell seal. Some whales go even deeper than this, with sperm whales known to descend to over 1,000 m.

A distinctive feature of certain species living on the deepest parts of the sea floor in the ocean trenches is the exceptionally large size they attain compared with closely related forms in shallow water. This phenomenon of 'gigantism' is specially marked in some groups of small crustacea, notably amphipods and isopods, which seldom exceed lengths of 2–3 cm in shallow and middle depths but grow to 8–10 cm or more in the hadal zones. The explanation of these unusually large sizes is uncertain. The water temperature in ocean trenches is not much lower than on shallower parts of the abyssal floor, so it is possible that gigantism is an effect of pressure on metabolism, perhaps associated with a longer growing period and delayed maturation. Also, because deep-sea sediments are more radioactive than near-shore deposits, the mutation rate may be greater at hadal depths resulting in a higher rate of speciation.

Buoyancy Problems

Most protoplasm, cell walls, skeletons and shells have a density greater than sea-water, and therefore tend to sink. One of the problems facing pelagic organisms is how to keep afloat between whatever levels are suitable for their survival. The phytoplankton must obviously remain floating quite close to the surface because only here is there sufficient illumination for photosynthesis. Plants which sink below the euphotic zone die once their food reserves are exhausted. Although animals are not directly dependent on light, the most numerous pelagic fauna are small herbivorous planktonts feeding on phytoplankton, and these must also remain fairly close to the surface to be within easy reach of their food. However, many of these herbivores do not remain constantly in the illuminated levels but move up and down in the water, ascending during darkness for feeding and retiring to deeper levels during daylight. Many carnivorous animals, too, both plank-tonic and nectonic, make upward and downward movements, often of considerable extent (page 97), and traversing ranges of temperature and pressure. These creatures must therefore be able to control their level in the water, and make the necessary adjustments to suit the conditions at different depths.

There are broadly two ways in which pelagic organisms can keep afloat and regulate their depth; by swimming, or by buoyancy control. In many cases the two methods function together.

A wide variety of marine creatures, both small and large, swim more or less continuously and control their level chiefly by this means. For example, dinoflagellates are said to maintain themselves near the surface by repeated bursts of upward swimming, alternating with short intervals of rest during which they slowly sink. In the laboratory, copepods seem generally to swim almost continuously, mainly with a smooth, steady motion but with occasional darting movements. When there are brief resting periods they usually appear to sink slowly; but Bainbridge[20], using aqualung equipment, observed copepods in the sea and reported *Calanus* apparently hanging motionless in the water as if suspended from its long outstretched antennae, drifting without any obvious tendency to rise or sink. Among the larger crustacean zooplankton are many forms which swim almost ceaselessly, and seem to sink fairly quickly if swimming ceases, for example mysids, euphausids and sergestids. Some of the most abundant pelagic fish also keep afloat by swimming; for example, almost all the cartilaginous fish, and certain teleosts including mackerel and some tunnies.

Apart from flotation, swimming movements also fulfil several other important functions. They are obviously important for pursuit or escape, and in many forms they serve to set up water currents for feeding and respiration.

The alternative to swimming is to float by means of some type of buoyancy device. Some pelagic animals have a gas-filled float, and in a few cases this is so large and light that the animal has positive buoyancy and floats at the surface; for example, the large pneumatophores of some species of siphonophora (*Physalia*, *Velella*), or the inflated pedal disc of the pelagic actinians (*Minyas*). The pelagic gastropod, *Ianthina*, floats at the surface attached to a raft of air bubbles enclosed in a viscid secretion. If detached from its raft, the animal sinks and apparently cannot regain the surface.

Many of the organisms which dwell below the surface have some method of controlling their overall density to match that of the water, in this way achieving neutral buoyancy so as to remain suspended. This is commonly done by means of some form of adjustable gas-cavity. Some siphonophoran colonies (*Stephanomia* and *Forskalia*) float below the surface suspended beneath a gas-filled pneumatophore. *Nautilus*[34] obtains buoyancy from gas secreted into the shell, and in *Sepia* the cuttlebone is partly filled with gas. In numerous families of fish there is a gas-filled swimbladder[56].

The volume of a gas varies inversely with pressure, so the pressure gradient presents special problems to those animals which rely for buoyancy upon gas-filled cavities. When a fish with a gas-bladder moves upwards there is a tendency for the bladder to expand as the hydrostatic pressure falls. If this should occur, the fish would rapidly gain positive buoyancy and bob up to the surface, and the distension of the bladder would be likely to cause damage. Alternatively, during downward movement the bladder must tend to contract under increasing pressure, and this would cause loss of buoyancy and force the fish downward. Normally, of course, such changes of buoyancy are prevented by the ability of the animal to maintain the volume of the float virtually constant despite changes of external pressure. A fish with a closed swimbladder does this by absorbing gas during ascent, or secreting more gas during descent, thereby counteracting the tendency of the float to change volume as the pressure alters. However, these processes of absorption and secretion cannot take place very quickly, and creatures which rely for flotation on this mechanism are therefore unable to make very rapid changes of depth without danger. This effect shows dramatically in fish brought quickly to the surface in trawls, when the swimbladder often expands so greatly that it extrudes through the mouth like a balloon.

Some creatures with gas-filled floats can release gas through an opening; for example, the siphonophoran *Stephanomia* controls the buoyancy of its pneumatophore by secreting gas into it, or by allowing gas to escape through a sphincter-controlled pore. In some fish there is a connection between the swimbladder and the exterior via the gut. This is commoner in freshwater fish than in marine species but is found in the Clupeids (herring, sprat, pilchard, etc.) and in some others. Some of these fish can release pressure during ascent by allowing gas to escape from the bladder through the mouth or anus; at the surface, air can be swallowed and forced into the bladder. However, in most marine teleosts, if a swimbladder is present, it is closed.

In those that live at great depth the gas is at high pressure and the gas-secreting structures are enlarged; for example in Myctophidae (lantern fish), Sternoptychidae (hatchet fish) and some Gonostomatidae. Below 1,000 m few of the pelagic species have a swimbladder, although in some cases it is present in the larval stages, which live nearer the surface. In certain bottom-dwelling fish highly developed swimbladders are found down to about 7,000 m (in some Macrourids, Brotulids and Halosaurs), probably associated with a need for neutral buoyancy to enable the fish to hover just above the sea-floor.

Apart from acting as a buoyancy device, the swimbladder has additional functions. It is a pressure-sensing organ whereby buoyancy is automatically adjusted to changes of level. Also, it probably has a role as a detector of vibrations in the water, and in some species has connections with the ear, indicating a hydrophonic function. In certain fish it is evident that the swimbladder can itself be set in vibration to function as a sound-producing mechanism, perhaps for communication or echo-location.

Some cephalopods obtain flotation from gas but have a different method of regulating buoyancy from that of fish. In the cephalopods the problems of pressure change are avoided by enclosing the gas in a rigid-walled container which cannot appreciably change volume as depth alters. For example, *Sepia* has gas in its cuttlebone[31−33,35], which contains spaces filled partly with gas and partly with liquid. The overall density of the cuttlebone varies between about 0·5 and 0·7, but the mass of gas within it seems to remain constant, and the density of the bone is controlled by regulating the amount of liquid it contains. The siphuncular membrane on the underside of the cuttlebone apparently acts as a salt pump, controlling osmotically the quantity of liquid in the bone. By maintaining the concentration of the cuttlebone liquid below that of the blood, sufficient osmotic pressure is generated to balance the hydrostatic pressure of the surrounding water. There is a similar mechanism in *Nautilus*, which also has gas and liquid in its shell and adjusts its buoyancy by controlling the quantity of liquid[34].

Instead of gas-filled floats with their attendant difficulties of control, many pelagic organisms obtain buoyancy from liquids of lower specific gravity than sea-water in a way similar to the bathyscaphe (page 59). Liquid-filled floats have the advantage of being virtually incompressible; but because of their higher density they must comprise a much greater proportion of the organism's overall volume than is necessary with gas-filled floats if they are to give equivalent lift. We have referred earlier to the probability that the large central vacuole of diatoms contains cell sap of low specific gravity, conferring some buoyancy. In young, fast-growing cultures, diatom cells often remain suspended, or sink only very slowly, although in older cultures they usually sink more rapidly. Studies of the distribution of diatoms in the sea suggest that some species undergo diurnal changes of depth, usually rising nearer the surface during daylight and sinking lower in darkness, possibly due to slight alterations of their specific gravity. In Radiolaria, the vacuolated outer protoplasm is probably of lower specific gravity than the water, giving buoyancy to the whole. Changes in the number of vacuoles cause the cells to rise or sink. Some of these animals possess myonemes with which the outer protoplasm can be expanded or contracted, and this presumably alters the overall density and may provide a means of changing depth.

Many pelagic organisms contain considerable quantities of oil. This may well serve a dual purpose, acting as both a food reserve and a buoyancy device. Oil droplets are common inclusions in the cytoplasm of phytoplankton, and many zooplanktonts also contain oil vacuoles, or oil-filled cavities; for example, radiolaria, some siphonophora, many copepods and many pelagic eggs. The body fluids of marine teleosts are more dilute and less dense than sea-water, and therefore confer some buoyancy, and many species derive additional buoyancy from accumulations of fat. In the abundant oceanic fish, *Cyclothone*, there is fat in the swim-bladder and tissues amounting to about 15 per cent of the total volume of the fish.

The gelatinous tissues which are a feature of many pelagic organisms, for example medusae, siphonophora, ctenophora, salps, doliolids, heteropod and pteropod molluscs, have a slight positive buoyancy which affords flotation to the denser parts. The tissue fluid is isotonic with sea-water but of a lower specific gravity due to the replacement of sulphate ions with lighter chloride ions. The small deep-sea squids of the family Cranchidae, mainly 1–2 cm in length, gain buoyancy from coelomic fluid of slightly lower density than sea-water contained in their capacious coelomic cavities. About two-thirds of their weight is made up of coelomic fluid, isotonic with sea-water but of low density due to its high concentration of ammonium ions. A high concentration of ammonium ions is also reported in the dinoflagellate *Noctiluca*.

In fish which live at deep levels and have no gas-filled bladders there are several ways in which body weight may be reduced. Compared with shallow-water fish, the body fluids are probably more dilute and the tissues contain rather more fat and considerably less protein. Apart from the jaw structures, the greater part of the skeleton is lightly developed and relatively poorly ossified, and the associated muscles correspondingly reduced. Denton and Marshall[36] have drawn up a 'buoyancy balance sheet' comparing a coastal fish, *Ctenolabrus rupestris*, which has a gas-bladder, with a bathypelagic fish, *Gonostoma elongatum*, in which the swim-bladder is degenerate and filled with fat (*Figure 4.8*). According to this reckoning, the deep-water fish is only slightly heavier than the weight of water it displaces, and presumably does not need to make much effort to maintain its level.

In elasmobranchs[23,27], none of which have swim-bladders, the overall density varies considerably in different species in ways related to their different shapes and modes of life. Some species approach or attain neutral buoyancy due to the relative lightness of their skin, skeleton and muscles. Lift is provided by low-density oil, especially in the liver, and also from other non-fatty tissues of low density, notably the subcutaneous layer or masses of gelatinous tissue. Also, when they are swimming, the fish gain dynamic lift

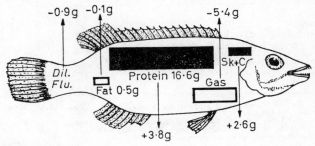

Ctenolabrus rupestris

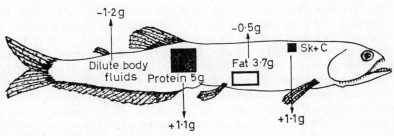

Gonostoma elongàtum

Figure 4.8 Diagram of the 'buoyancy balance sheet' for a bathypelagic fish Gonostoma elongatum *without a swim-bladder (below) and a coastal fish* Ctenolabrus rupestris *with a swim-bladder (above). Positive values are given for those components of the fish which are heavier than the sea-water which they displace and thus tend to 'sink' the fish, whilst negative values are given for those components which displace more sea-water than their own weight and thus tend to 'float' the fish. Weights given per 100 g of fish. Dil. Flu., dilute body fluids;* Sk + C, *skeleton and other components*

(From Denton, E. J. and Marshall, N. B.[36] published by Cambridge University Press)

Balance Sheet for *Ctenolabrus rupestris*

Component	% wet weight	Weight in sea-water/100 g of fish
Fat	0·5	−0·1
Protein	16·6	+3·8
Body fluids	73·3	−0·9
Other components including bone	9·2*	+2·6*

Buoyancy. This fish without its swim-bladder has a weight in sea-water of +5·4 per cent of its weight in air

* These values are given by difference.

SOME PARAMETERS OF THE ENVIRONMENT

Balance Sheet for *Gonostoma elongatum*

Component	% wet weight	Specific gravity	Weight in sea water/100 g of fish
Fat	3·7	0·91*	−0·5
Protein	5·0	1·33†	+1·1
Body fluids (water+dissolved salts)	87·6	1·013	−1·2
Other components including bone	3·2‡		+1·1‡

Buoyancy. These fish had no gas-filled swim-bladder and their average weight in sea-water was approximately +0·5 per cent of their weight in air. (Wet weight)

* *Handbook of Biological Data* (1956). Ed. W. S. Spector. Ohio, U.S.A.: Wright Air Development Centre.
† Höber, (1954). *Physical Chemistry of Cells and Tissues.* London; Churchill. Specific gravity taken as the reciprocal of the partial specific volume.
‡ These values are given by difference.

from the plane surfaces of the pectoral fins, which also function for changing direction or braking.

It appears that many of the bottom-dwelling elasmobranchs of shallow water have a relatively high density, probably associated with a fairly inactive mode of life and a habit of resting on the bottom. But other bottom species are close to neutral buoyancy, especially those of deeper water; these are probably more active fish which cruise above the bottom in search of food, which at deep levels is more scarce.

The pelagic predatory sharks, swimming rapidly in pursuit of prey, have densities which result in nearly neutral buoyancy. This enables them to be fully manoeuverable with fins which are quite small and therefore add little frictional drag to their well-streamlined forms. There are other pelagic species which are slow-moving and feed by filtering plankton, and these are of two shapes with different densities. The shark-shaped form *Cetorhinus* (basking shark) gains little lift from its small fins but can remain afloat while swimming slowly because it has almost neutral buoyancy. This is largely due to the low-density oil *squalene*, found in the liver, and a mass of gelatinous tissue in the nose. On the other hand, the manta rays (*Mobula*) have a relatively high density but gain enormous dynamic lift from their large plane surfaces.

The buoyancy afforded by the water is of course important in reducing the skeletal needs of aquatic organisms, which can be very lightly built as compared with terrestrial forms. This has made possible the evolution of some extremely large marine creatures. The enormous present-day whale-bone whales, for example the Blue Whale, *Balaenoptera musculus* and the Fin Whale, *B. physalus*, may reach weights in the order of 100 tons. In comparison, the largest known terrestrial animals, the huge Mesozoic reptiles, probably did not exceed some 30 tons. The rapid death of large whales on stranding is often caused by the collapse of the rib cage when the massive weight is unsupported by the water.

94

The Sinking Rate

The rate at which a small object sinks in water varies with the amount by which its weight exceeds that of the water it displaces, and inversely with the viscous forces between the surface of the object and the water. The viscous forces opposing the motion are approximately proportional to the surface area, and therefore, other things being equal, the greater the surface area the slower the sinking rate.

There are a number of structural features of planktonic organisms which increase their surface area and must certainly assist in keeping them afloat. The majority of planktonts are of small size, and therefore have a large surface to volume ratio. In many cases, modifications of the body surface increase its area with very little increase in weight. Comparable adaptations are found in wind-dispersed seeds and fruits of land plants, which clearly serve to keep them airborne. These modifications generally take two forms; a flattening of the body, or an expansion of the body surface into spines, bristles, knobs, wings, or fins. Reference has already been made to the range of flattened or elaborately ornamented shapes that occur in diatoms. In dinoflagellates, also, the cell wall is in some cases prolonged into spines (*Ceratium*) or wings (*Dinophysis, see Figure 2.5*). In the zooplankton, flattened shapes are common, for example the pelagic polychaete *Tomopteris* (*see Figure 2.13*), the copepod *Sapphirina* and various larvae (Phyllosoma of *Palinurus*, Cyphonautes of *Membranipora*). Arrays of spines and bristles also occur, for example some species of *Calocalanus* and *Oithona* and many larvae (Mitraria of *Owenia*, zoeas). The pluteus larvae of echinoderms have long ciliated arms. The chaetognaths have flat fins, and in *Tomopteris* the parapodia form a series of flattened, wing-like appendages.

Warm water has a lower density and viscosity than cold, and therefore affords less buoyancy and less resistance to sinking. The plankton of warm water includes several species in which the bristles, wings and other flotation devices are exceptionally elaborate (for example *Calocalanus spp. Ornithocercus*). In temperate waters, there is a tendency in some species for spines to be relatively longer in the warm season than in the colder months (*Figure 4.9*).

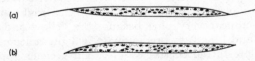

(a)

(b)

Figure 4.9. Rhizosolenia hebetata (*a*) *Summer and* (*b*) *winter forms*

Reduction of the sinking rate is unlikely to be the only, or even the major, advantage gained from modifications which enlarge the surface area. In the phytoplankton, an extensive surface presumably facilitates the absorption of nutrients present only in very low concentration, and also favours light absorption. The shape and distribution of weight determines the orientation of a passively floating body, and it seems probable that diatoms float in a position which presents the maximum surface to light from above. Irregular shapes and long spines may function also as protective devices which small predators find awkward to grapple with. In certain species there are sexual differences in bristles and plumes which indicate some role in sexual display; for example, the long plume of the males of *Calocalanus plumulosus*.

ILLUMINATION

Compared to the depth of the ocean, light does not reach very far into the sea. Illumination of the surface layers varies with place, time and conditions depending upon the intensity of light penetrating the surface and upon the transparency of the water. The strength of the incident light varies diurnally. seasonally and with latitude, and is influenced by cloud conditions and atmospheric absorption. Much of the incident light is reflected from the surface, more light being reflected from a ruffled surface than a calm one, and reflection increases as the sun becomes lower in the sky. Depending on conditions some 3 to 50 per cent of incident light is usually reflected. The light which penetrates the surface is quickly absorbed, partly by the water and dissolved substances but often largely by suspended matter including planktonic organisms, translucent water generally being indicative of a sparse plankton. Extinction coefficients (*see* page 42) vary from about 1·0 to 0·1 between turbid inshore and clear offshore areas, but in exceptionally transparent ocean water may be as low as 0·02. Even in clear water about 80 per cent of the total radiation entering the surface is absorbed within the uppermost 10 m, and in more turbid water the absorption is far more rapid than this. The heating effects of solar radiation are therefore confined to a very thin surface layer.

Different wavelengths of light do not penetrate equally. Infra-red radiation penetrates least, being almost entirely absorbed within the top 2 m, and ultra-violet light is also rapidly absorbed. In clear water the greatest penetration is by the blue-green region of the spectrum, while in more turbid conditions the penetration of blue rays is often reduced to a greater extent than that of the red-yellow wavelengths. This differential absorption of the solar spectrum partly accounts for the colour of the sea's surface by its effect on the spectral composition of reflected light. In bright sunlight, clear ocean water may appear very blue because the yellow and red rays are largely absorbed, and blue rays predominate in light reflected from below the surface. In more turbid coastal waters, their greener appearance may result from the relatively greater absorption of blue light. Sometimes the colour of the water is due to the pigmentation of minute organisms.

Light is of supreme biological importance as the source of energy for photosynthesis. Primary food-production in the marine environment is virtually confined to the illuminated surface layers of the sea where there is sufficient light to support plant life. The depth of this photosynthetic or *euphotic zone* varies with conditions, extending to some 40–50 m from the surface in middle latitudes during the summer months and to 100 m or more in low latitudes if the water is fairly clear. Below the euphotic zone, down to about 200 m, is the dimly-illuminated *dysphotic zone* where light is insufficient for the survival of plants. The water below 200 m is termed the *aphotic zone* because there is little or no light, but in clear tropical waters a small amount of blue radiation penetrates to at least 1,000 m.

Light is certainly one of the major factors controlling the distribution of marine organisms, but in many cases its effects are not easy to understand. Plants are restricted to the euphotic zone by their dependence upon light for energy; and animals are most numerous in or near the surface layers because they derive their food, directly or indirectly, from plants. Below the productive surface zone, animals are almost entirely dependent upon food sinking to them from above, and the deeper they are, the less food is likely to reach them because much of the assimilable material is decomposed on the way down. Broadly, the deeper the level, the less the food supply and the fewer the population. But numbers do not fall off evenly with depth. Although we do not have much knowledge about the distribution of mesopelagic and bathypelagic forms, it appears that populations tend to congregate at certain levels. There is commonly a concentration of animals somewhere between 300 and 1,000 m, sometimes coinciding with the oxygen-minimum zone, and lesser concentrations below this[19,43,46]. Many organisms, however, do not remain consistently at one level but perform vertical movements, often over a considerable distance, which are related to changes of illumination (*Figure 4.10*).

Diurnal changes of level

It has been known from time immemorial that certain fish, for example, herring and mackerel, usually approach the surface in large numbers only during darkness, and fishermen have adapted their techniques to match this feature of fish behaviour. Since the early days of plankton research it has also been observed that tow-net catches of zooplankton from the surface are usually richer and more varied during the night than in day-time. Poor day-time catches of zooplankton at the surface may be partly due to the ability of some of the larger and more active animals to see the slow-moving net and to avoid capture by swimming; but if repeated plankton hauls are made at a series of depths throughout the day and night, it becomes clear that many planktonts alter their depth diurnally, moving nearer the surface during darkness and returning to the deeper water to pass the day. At night, these organisms are not found in samples collected at the deeper levels where they can be captured during the day, but are then found in samples taken nearer the surface from which they are absent in day-time.

Diurnal changes of distribution are shown by a great variety of organisms, both plankton and nekton, including medusae, siphonophores, ctenophores, chaetognaths, pteropods, copepods, cladocerans, amphipods, mysids, euphausids, pelagic decapods, some cephalopods and fish. In shallow water there are many creatures that live on or in the bottom during the day but leave the sea-bed at night and swim in the surface water.

Early studies of these diurnal vertical movements of plankton and other creatures were mainly confined to populations of shallow water, but it is now known that this phenomenon is of very wide occurrence throughout

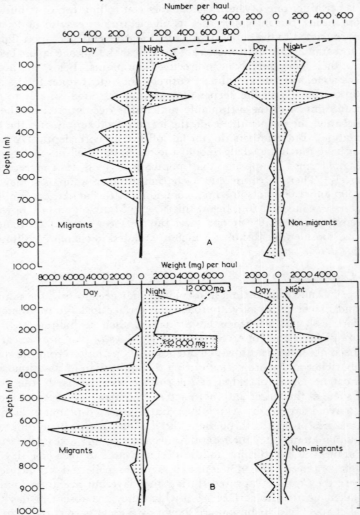

Figure 4.10. Day and night distributions of total migrant and non-migrant mesopelagic euphausids off the Canary Islands taken during a cruise by R.R.S. Discovery, 1965: A Total numbers per haul, B Total weight per haul. Pecked line in upper 100 m of night distribution indicates catches adjusted by addition of species assumed to be missed at night by migrating above 50 m. Pecked line at 250 m in night distribution indicates catch when a large sample of Nematoscelis megalops is omitted

(From Baker, A. de C., J. mar. biol. Ass. U.K., 50, 301–342 (1970) published by Cambridge University Press)

the deep oceans[28]. During World War II, physicists investigating the use of underwater echoes for the location of submarines obtained records during daylight hours of a sound-reflecting layer in the deep water beyond the continental shelf[41]. On echogram tracings this layer gave the appearance to a false bottom at about 300 m where the sea-floor was known to be far below this. Towards the end of the day this sound-reflecting layer was observed to rise until close to the surface. During darkness it was less distinct, but at the dawn twilight the layer formed again near the surface, and descended gradually to its usual daylight level as the sun rose. Further detailed investigations revealed that there are sometimes several of these sound-reflecting layers at depths between 250 and 1,000 m, and these are now usually termed Sonic Scattering Layers (S.S.L's.), or Deep Scattering Layers (D.S.L's.). Their positions depend to some extent upon the frequency of the equipment.

It was first thought that the S.S.L's. were due to discontinuities between layers of water differing in some physical property, for example, temperature, but this would not account for such marked diurnal changes of position. It is now generally accepted that the tracings are caused by echoes returned from marine creatures which change their depth between daylight and darkness. Where the S.S.L's. comprise several sound-reflecting zones, these are probably caused by concentrations of different groups of animals at each depth. Little is known of the creatures which exist in these layers because of the sampling difficulties, but the organisms which are thought to predominate are euphausids, sergestids, and various small oceanic fishes, in particular Myctophids (lantern fish), Sternoptychids (hatchet fish) and Gonostomatids.

None of these creatures are large, the crustacea being only about 2–5 cm in length and the fishes mostly about 4–10 cm. The reflection of so much sound by such small creatures suggests that they are present in great numbers but this is probably not the case. Populations in deep water, even if concentrated at particular levels, are probably small compared with those commonly found near the surface, and it seems likely that the strong echoes are due to structural features of the organisms. The exoskeleton of some crustacea may be a good sound-reflecting material at certain frequencies, and the three families of fish mentioned above possess gas-filled swim-bladders which may give strong echoes through resonance. What is known of the distribution and behaviour of these fish corresponds well with the presence and movements of the S.S.L's. Many planktonts contain vacuoles of various types, and some of these may also reverberate at the frequencies of echo sounders.

The S.S.L's. have been detected in most regions of deep oceans except the Arctic. They vary in distinctness from place to place, being generally indefinite beneath areas of low fertility presumably because of the smaller population of these waters.

It seems clear that, throughout the greater part of the oceans, in both deep and shallow water, a numerous and varied assortment of animals perform vertical migrations with a diurnal rhythm. There are differences in the depths through which different species move, the speeds at which they ascend and descend, and the precise times at which they make their movements, and the same species may behave differently in different areas and at different times, but there is none the less a remarkable consistency in the general behaviour-pattern. During daylight, each species appears to collect near a particular level. Shortly before sunset they commence an ascent which continues throughout the twilight period. During total darkness the population tends to disperse, but shortly before dawn it again congregates near the surface and then makes a fairly rapid descent until, about an hour after sunrise, the day-time level is reached (*Figure 4.11*).

There are many features of the migrations which are not understood[46], but it seems certain that changes in illumination play a dominant role in regulating the activity. No other factors are known to which the movements can be so exactly related. Although the relative durations of night and day vary with latitude and season, the migrations are always closely synchronized with dawn and dusk. In high latitudes they occur during spring and autumn when there are daily alternations of light and dark, but the migrations are not observed when darkness or daylight persists throughout the 24 h.

Many of the organisms which make these diurnal movements must be extremely sensitive to light. During the day their level rises when clouds cross the sun. At night, moonlight has a noticeable effect, some creatures remaining further from the surface in bright moonlight than on dark nights, while others appear to be attracted by moonlight and reach the surface in greatest numbers at full moon on clear nights.

The majority of migrating forms are fairly active swimmers which regulate their depth by the rate and direction of swimming. In some cases, strong light may have an inhibitory effect on swimming, preventing these animals from reaching the surface during day-time even though they make no directional response to light. However, the movements of many zooplanktonts show an orientation to the direction of light, moving away from a source of strong illumination. A creature which is both negatively phototactic (moving away from light) and negatively geotactic (moving against gravity) presumably swims towards the surface in darkness, but in day-time moves down to the level where its responses to light and gravity are in balance, moving upwards or downwards as illumination changes. The movements may be partly regulated by an intrinsic 'physiological clock' operating in phase with changes of illumination. For instance, it has been shown that the copepod *Calanus*, if removed from the sea and kept in total darkness in a tank, continues for some days to exhibit periodic changes of activity having a 24 h rhythm which results in diurnal alterations of its level[51].

These mechanisms perhaps explain the movements of some organisms but do not account for the tendency of many species to disperse during total darkness. It seems as if many planktonts move towards a preferred 'optimum' intensity or spectral composition of illumination which differs according to the species. During day-time the animals tend to assemble at whatever level they find their preferred illumination, and move up or down as this level changes with rising and setting of the sun. During darkness, when there is no longer sufficient illumination to cause the population to concentrate at one level, they scatter. Experiments in which planktonic animals have been enclosed in long tubes exposed to an illumination gradient have demonstrated that some species do collect within particular ranges of light intensity.

To study the effects of illumination on speed and direction of swimming,

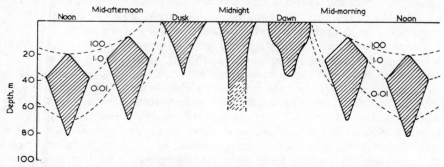

Figure 4.11. Generalized diagram illustrating diurnal changes of vertical distribution of many epipelagic zooplanktonts. The dotted lines show levels of approximately equal illumination, indicated in arbitrary units. Logarithmic decrease of illumination accounts for the kite-shaped pattern of distribution in daytime

Hardy and Bainbridge[48] devised an apparatus which has been termed a 'plankton wheel'. It consisted of a circular transparent tube, 4 ft in diameter, mounted vertically by spokes on a central axle on which it could be rotated. The tube was filled with sea-water, and small animals could be inserted. In effect, the animals were enclosed within an endless water column. By rotating the wheel one way or the other as the enclosed animals swam upwards or downwards, they could be kept at a fixed position relative to the observer. On the inside wall of the tube were mounted a series of small valves, automatically operated by floats and weights as the wheel was turned. These valves ensured that the water turned with the wheel, but did not interfere with the animals in the observation position. The apparatus was placed in a small glasshouse provided with movable screens to give variations in illumination. Movements of the wheel were recorded on a rotating smoked drum.

With this instrument it was possible to measure the speed and duration of upward and downward movement of small animals and to record the pattern of their activities under controlled conditions of lighting, although the apparatus did not provide the gradients of illumination, temperature or pressure that organisms encounter in the sea. Among the animals studied were copepods, euphausids, *Tomopteris*, *Sagitta elegans*, and nauplii of *Balanus* and *Calanus*. The experiments confirmed that the swimming capabilities of these animals are more than adequate to account for the extent of the migrations inferred from net hauls. Results were incomplete because of the difficulty of setting up the experiments, and the time needed for recording, but many of the animals tested showed a positive movement downwards in strong light. Their responses varied at different times of day, but during the evening in dim illumination the animals moved upwards towards the light. If kept in total darkness during day-time, they did not consistently move upwards, indicating that geotaxis was not automatically operating in the absence of strong light. The results generally suggested a directional movement towards an optimum level of illumination of low intensity.

Other factors have been suggested as contributing causes of vertical migration; for example, temperature changes in the surface layers of water may have some effect. At night the surface layers undergo slight cooling, and this might permit some of the inhabitants from colder water below to approach the surface, while in day-time the warmer surface temperature would discourage their rising so far. Apart from physiological effects, the physical effects of changes of temperature upon water density and viscosity must have some influence on buoyancy. Surface cooling at night might be expected to cause passively floating forms with neutral buoyancy to float at slightly higher levels than during the day, and the effect might be more marked on small upward-swimming creatures. It has also been suggested that the weight gained by organisms during their nightly feeding in the abundant food supplies of the surface layers may cause them to sink to lower levels until weight is subsequently lost through respiration, excretion and egestion. However, these processes cannot be closely correlated with either the extent or the periodicity of the migrations. The distances covered by most species are too great to be explained simply as passive rising or sinking due to alterations of buoyancy, but must involve active swimming upwards and downwards at considerable speed. Nor does the periodicity of the migrations correspond closely with the change of temperature of the surface water, which falls gradually throughout the evening and night until shortly before dawn, whereas the majority of vertically-migrating animals reach the summit of their ascent soon after sunset.

In the course of their upward and downward movements, organisms traverse a pressure gradient, and often a temperature gradient, and these must play some part in regulating the extent of the migrations. Increase of pressure causes some zooplanktonts to increase their upward swimming rate, and this presumably limits their depth of descent. Where there is a sharp thermocline between the surface and deeper levels, some migrating forms do not pass through the discontinuity layer. The thermocline then acts as a boundary between two groups of organisms, those of the warmer surface layers which descend during daylight only as far as the thermocline, but not through it, and those of deeper levels to which the thermocline is the limit of ascent during darkness.

It has been suggested that 'exclusion' (page 134) may have some connection with diurnal migration. If during photosynthesis phytoplankton liberates into the water metabolites which are distasteful or toxic to animals, the zooplankton would be expected to remain below the photosynthetic zone throughout the hours of daylight. Only during darkness, when photosynthesis ceases, would the surface water become suitable for the entry of animals. If the phytoplankton should be very abundant, the exclusion effect might persist throughout the night, upsetting the rhythm of the migration. However, it is uncertain to what extent exclusion operates in natural conditions. Some zooplanktonts become more photonegative when well-fed on phytoplankton and so would be expected to move downwards out of phytoplankton patches in daytime without any external metabolites being involved.

Vertical migrations involve the expenditure of so much time and energy by so many species that this behaviour must presumably be an adaptation of major importance, but the benefits are not altogether clear. Many of these migrating forms are herbivores grazing on the phytoplankton, and we may wonder what advantage it is to them to spend so much time away from the surface layers where they find their food. Probably the main benefit is safety from predation. Just as there are many terrestrial animals which are nocturnal, emerging from the safety of their hiding places only during darkness, innumerable pelagic creatures may also find safety during daytime in the darkness of deep water, ascending only at night to feed. For many of the herbivorous zooplanktonts sight is not essential for food capture. They gather their food by various processes of filtration and can feed effectively in darkness, but their chances of survival are greater if they avoid well-illuminated water because sight is far more important to most of the carnivores for the detection of their prey. Many of the migratory species which inhabit deep levels appear to make use of bioluminescence for camouflage (see page 108) and in these species it seems likely that their changes of level are adjustments of position to the appropriate intensities of dim illumination at which they are virtually invisible.

Hardy (page 33, reference 2) has pointed out that vertical migrations influence the horizontal distribution of planktonic organisms because the deeper layers of water usually move more slowly than the surface and often in a different direction. Planktonts which move downwards at dawn will therefore return to a different mass of surface water each night. This may be advantageous in several ways; for example, by enabling the organisms continually to sample fresh feeding grounds by simply moving up and down through the water, or, where surface and deep layers move in opposite directions, by stabilizing distribution within a particular locality, the population drifting one way at night and back again during the day. When the phytoplankton is sparse the surface water is likely to be clear and therefore strongly illuminated in daytime. This would lead to deeper descent and consequently quicker drift to beneath a different body of surface water. On reaching an area of richer phytoplankton, where more light would be absorbed near the surface, zooplanktonts would tend to rise into the better food supply. In localities where the deep levels move in opposite direction to the surface, descent is likely to carry organisms towards areas of upwelling (*see Figure 5.3*), which are always regions of high fertility and usually correspondingly rich in planktonic food.

There are many anomalies. The migratory behaviour sometimes becomes erratic or ceases completely. Animals which usually seem consistently to move up or down with changes of illumination may occasionally swarm at the surface in bright sunshine. The extent of the movements sometimes varies seasonally: for example, around the North Atlantic they are of more general occurrence during the summer months, and many planktonts cease this behaviour during winter when they remain at a deep level. Often the migrations are made only by particular stages of the life history. There are also records of reverse migration by some species, both phytoplankton and zooplankton, their movements being towards the surface in day-time and deeper water at night.

Wynne-Edwards[17] has suggested that swarming, patchy distribution and the vertical migrations of zooplanktonts may involve aspects of social behaviour related to control of population density. By congregating within thin layers and diurnally moving up and down, animals which would otherwise be widely distributed in a homogeneous environment can compete for food without much risk of too seriously depleting the food stocks of the whole volume of water through which they pass. In many species the quantity of egg production varies with the food supply and therefore the number of progeny is to some extent regulated by the availability of food for the adults. In this way competition for food provides a natural mechanism which may prevent the population reaching a size which would over-exploit the food resources. Also, concentration of populations within thin layers or patches facilitates display behaviour in sexual competition, whereby reproduction may be limited at supportable numbers.

Adaptations for Life in Darkness

Certain creatures living at levels down to about 1,000 m show modifications for vision in very dim illumination. The eyes of some deep-water fish are large and have exceptionally wide pupils; and, like many nocturnal land vertebrates, the light-sensitive cells of the retina are all rods. These rod cells are unusually long, and numerous rods supply each fibre of the optic nerve, an arrangement thought to form a retinal surface of extremely high sensitivity to light. The visual pigments[40,72] are also different from those of shallow water fish, giving maximum light absorption at the shorter wavelengths of the bluer light of deep ocean levels. In coastal fish the main

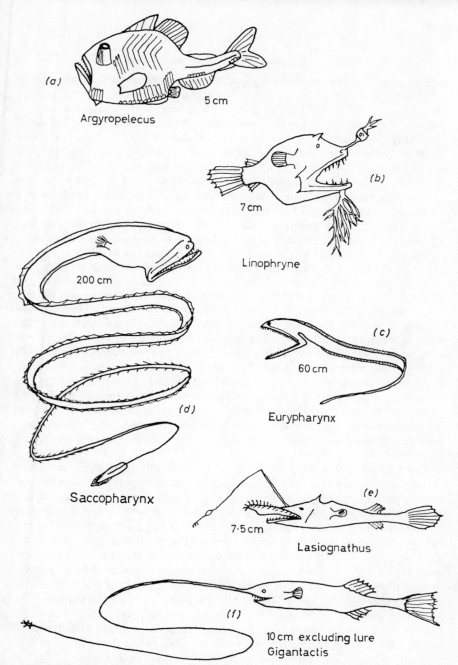

(a)

5 cm

Argyropelecus

(b)

7 cm

Linophryne

200 cm

(c)

60 cm

Eurypharynx

(d)

Saccopharynx

(e)

7·5 cm

Lasiognathus

(f)

10 cm excluding lure
Gigantactis

Figure 4.12 Some bathypelagic fishes. Approximate lengths in centimetres. Note the small size of most of them. (a) Argyropelecus—photophores and tubular eyes. (b) Linophryne—luminous lure and elaborate barbel. (c) Eurypharynx—large mouth and distensible stomach. (d) Saccopharynx—large mouth, distensible stomach and luminous tissue at end of elongate tail. (e) Lasiognathus—luminous lure with hooks. (f) Gigantactis—luminous lure

retinal pigments are the red-purple rhodopsins with maximum absorption at about 500 mμ; but the majority of deep-sea species of the mesopelagic zone have yellow or gold chrysopsins with a maximum absorption around 480 mμ. It is estimated that the eyes of some deep-sea fish are 60 to 120 times as sensitive as human eyes and are able to detect daylight to depths of about 1,150 m. In some deep-water species the eyes are tubular structures directed either forwards (*Gigantura*) or upwards (*Argyropelecus, Figure 4.12a*), and this probably provides a degree of binocular vision to assist perception of distance, a particular difficulty in very dim light. Below 1,000 m where darkness is virtually complete, there are fish with eyes which are very small (many ceratioids), degenerate (*Bathymicrops regis*) or even absent (*Ipnops*), although in larval stages living near the surface the eyes may be well formed.

The ears and lateral line organs[42] of fish are sensitive to vibration in the water, and enable fish to detect and locate objects in their vicinity. In some deep-level fish the lateral line system is especially well developed, for example, in myctophids and macrourids, and this must compensate to some extent for poor illumination and reduction or loss of vision. The olfactory organs of some species are exceptionally well developed and assist the detection and recognition of organisms and possibly also have a function in navigation. Very long tactile appendages are a frequent feature of animals that live in darkness, and are found in several forms among deep-sea creatures. Some fish have barbels of extraordinary length or elaboration (*Stomias boa, Linophryne, Ultimostomias*) (*Figure 4.12b*). Others have feeler-like fin rays (*Bathypterois*) or delicate, elongate tails (*Stylophorus*). The abundant family of deep-sea prawns, Sergestidae, have antennae of great length which are minutely hooked and possibly are used both for detecting and entangling their prey.

The coloration of marine creatures is obviously related to the illumination of their surroundings. Many fish of shallow water are protectively coloured, usually dark on the upper surface and whitish underneath. The upper parts are often mottled or patterned in a way which makes them very inconspicuous against their normal background, and in some cases they can rapidly change colour to match different surroundings. Some pelagic fish are difficult to see in water because their scales have a structure and arrangement which reflect an amount of light closely equivalent to the background illumination from almost any angle of view[30,37-39]. Deep water creatures which at times come close to the surface may have a reflecting surface or be almost transparent. At middle depths, where only very faint blue light penetrates, there are many highly pigmented species, usually black, red, or brown. By reducing reflection these colours must in the dim blue illumination give virtual invisibility below 500 m. In the total darkness of great depths there are numerous non-pigmented forms, light or buff in colour.

The marine fauna include a wide variety of bioluminescent species, with numerous examples known in almost all the major groups[22,24,30,49,70-73].

106

The common occurrence of bioluminescence in the sea, in contrast to its rarity on land, is probably the result of the much lower intensities of light in the sea. On land, even on the darkest nights, there is ample illumination for good vision by animals adapted for nocturnal life, but darkness in the sea is virtually total except for bioluminescence. It is especially common in warm surface waters, but evidently the deep levels are by no means uniformly dark because so large a part of the population carries light-producing organs (photophores). Light is produced by an oxidizing reaction, which in some

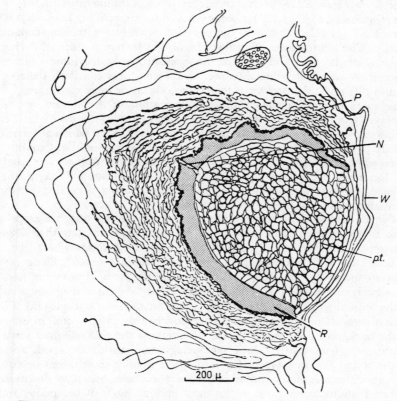

Figure 4.13   Transverse vertical section through a subocular light-organ of a stomiatoid fish, Astronesthes elucens. *N, nerve. P, pigment. p.t. photogenic tissue. R, reflector, W, clear window*

(From Nicol, J. A. C.[71], published by Cambridge University Press)

cases involves substances which have been separated and named luciferin, possibly a polypeptide, and an enzyme, luciferase. In many luminous organisms the reagents are not yet known. Some animals discharge luminous secretions into the water. In others the reaction is intracellular, the mass of photogenic cells often being backed by a reflecting layer (*Figure 4.13*) and sometimes covered by a lens. Masses of luminous bacteria within the tissues are responsible for the light production in some cases.

The function of light organs is not well understood, although they can obviously serve a variety of useful purposes. In some species, beams of light are emitted which evidently illuminate the field of vision. Light may also serve to attract prey, and a number of deep-water fish have remarkable luminous lures. These may take several forms, such as luminous barbels (*Linophryne, Figure 4.12b*) or structures resembling a fishing line with luminous bait (*Gigantactis, Figure 4.12f*) in some cases complete with hooks (*Lasiognathus, Figure 4.12e*). In *Galatheathauma* there is a luminous lure within the capacious mouth. The photophores often form highly distinctive patterns differing in closely related species and between the sexes, probably providing a means of recognition and facilitating shoaling. Some creatures can flash their lights on and off; and this may serve to confuse, alarm or dazzle attackers, or in some cases provide a means of communication or social and courtship display. The fish *Pachystomias* has light organs close to the eyes which emit flashes of red light[30]. In contrast to most deep-level fishes, which are sensitive mainly to blue light, the eyes of *Pachystomias* respond to red light enabling it to see without being seen. Luminous clouds squirted into the water, such as the luminous ink of the squid *Heteroteuthis*, are presumably a means of defence, perhaps by dazzling or possibly by making attacking creatures themselves visible to their own predators.

The photophores of many migratory species of the upper mesopelagic zone between about 200 and 700 m are positioned so that their light is apparently directed mainly downwards, even in animals whose eyes are directed upwards, for example *Argyropelecus, Opisthoproctus*. These photophores can hardly play any part in illuminating the field of vision, and the most likely explanation of their function is that they camouflage the shadow of the animal when viewed from below against the faint background of light from above. If these photophores emit downwards a light equal in intensity to the background illumination, the silhouette of the creature must virtually disappear (*Figure 4.14*). Many mesopelagic euphausids, decapods and fish of the upper 650 m have bodies which are mainly transparent or silvery, though with some pigmentation of the dorsal surfaces, and have downwardly directed photophores close to the most opaque parts of the body, and in many cases internally situated. Some of these animals may be able to adjust both reflectivity and light emission to suit changing illumination, and their vertical migrations may well serve to keep them within the light conditions under which they are least visible. Detailed examination of the reflectors associated with downwardly directed photophores in the hatchet fish *Argyropelecus aculeatus*[30] reveals that these reflectors direct part of the light laterally, the light intensity varying with direction in a way similar to the reflection of light from above by the scales of silvery fishes, whereby they are made inconspicuous when seen from the side. Animals which by day do not ascend above about 700 m are mostly more heavily pigmented and opaque, with photophores less often present, and when present more superficially placed.

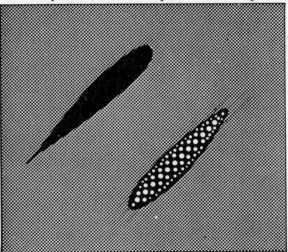

Figure 4.14. Diagram to illustrate how downwardly-directed photophores may reduce the conspicuousness of animals at deep levels when seen out of focus. The two objects represent the silhouettes of fish, one with and one without photophores, seen from below against a background of weak illumination from above. If the diagram is viewed with the eyes sufficiently closed to blur the outlines, the right-hand object becomes less visible than the left

In the sparse population of the dark levels of the sea, finding a mate must present a problem, and the possession of photophores may increase the chances of success. Another solution has been the evolution of dwarf parasitic males. For example, in some ceratioid angler fish the male is much smaller than the female, and relatively underdeveloped except that his eyes and olfactory organs are large and his teeth sharp and specialized for gripping the female. Once she has been found, perhaps partly by scent, the male apparently bites into her skin and remains permanently attached. His eyes and olfactory organs then degenerate and a partial fusion of tissues occurs between the two individuals, the male drawing nutriment from the female by an intimate, placenta-like association of the two blood systems. The testes develop and ripen so that sperms can be shed in the immediate vicinity of the eggs.

Certain fish of deep levels are equipped with relatively huge mouths and jaws and remarkably distensible stomachs, for example *Saccopharynx*, *Eurypharynx* (*Figure 4.12c* and *d*) and *Gigantura*. The jaws are often the only well-developed parts of the skeletal and muscular systems, and their swimming movements must be sluggish. In zones of darkness and scarce food it is evidently important to be able to take very large mouthfuls on the infrequent occasions when prey is found. These deep-level fish are mostly quite small, seldom more than about 10 cms in length. There is insufficient food to support many large creatures.

CURRENTS

The major ocean currents have been described earlier (page 6). Currents keep the water well mixed, influence the distribution of salinity and heat, bring to the surface the nutrients necessary for plant growth and carry down supplies of oxygen to deep levels. Because the bottom layers of water are in movement, it has been possible for a diverse benthic population to evolve which includes many forms living attached to or embedded in the sea bottom, and relying upon the flow of water to carry food and oxygen to them and waste products away. The benthic population is also influenced by the speed at which the bottom water moves because of effects on the nature of the sediment and the settlement of pelagic larvae.

Currents have a direct influence on the distribution of many species by transporting them from place to place, especially the smaller holoplanktonic forms. Many benthic and nektonic species also start life with a planktonic phase, and the direction of drift of their eggs and larvae must to some extent determine their areas of colonization. Currents probably also provide a means of navigation[5,52]. Some fish show a tendency to swim against the stream, at least during certain phases of life, and at the approach of the spawning period this behaviour is important in determining the position of spawning areas. Afterwards, the success of the brood depends upon the drift of larval stages to suitable nursery grounds.

How and to what extent can aquatic creatures detect water currents? Close to the bottom or any other fixed object they can obviously be aware of water movements through sight or touch; but where the water is flowing in a straight line at constant velocity, a floating organism out of sight or contact with fixed reference points can have little evidence of movement. Probably all animals are sensitive to acceleration and may therefore respond to velocity gradients or rotational flow of water. Because different water layers usually move at different velocities and often in different directions, there are possibilities of detecting water movement from effects of pressure or turbulence, or by observation of the movements of other floating objects above or below. There is also a theoretical possibility, though no firm evidence supports the idea, that organisms may be sensitive to the slight electromotive forces generated in the water or within themselves by movement through the earth's magnetic field.

Plankton Indicators

Because different communities of organisms are found in different parts of the sea, it is possible to distinguish particular bodies of water not only by their physical and chemical features, but also to some extent by their characteristic populations. Moving water carries with it an assortment of planktonts which can be regarded as natural drift-bottles, and by observing

110

the distribution and intermingling of different planktonic populations it may be possible to trace the movement and mixing of the water.

Large, easily-identifiable planktonts which are characteristic of particular bodies of water serve as convenient labels or 'indicators' of the water, and the study of the distribution of plankton indicators has several advantages as a method of investigating water movements. It is often simpler to obtain plankton samples than hydrographic data; and the biological evidence may be more informative because, where mixing occurs between different bodies of water, their characteristic populations may remain recognizable longer than any distinctive hydrographic features can be detected. Where conditions are changing, some species quickly succumb while hardier organisms survive longer. The distribution of a range of species of different tolerances may therefore give some indication of alterations in the quality of the water as it moves from place to place.

During the period 1921–1927, Meek[66] studied the distribution in the northern part of the North Sea of the two commonest chaetognaths of the area, *Sagitta setosa* and *Sagitta elegans*, which are also found in America's coastal waters. He observed that the two species are seldom found together. *S. setosa* occurs throughout the greater part of the North Sea, and is the only planktonic chaetognath normally found in the southern part of the North Sea. Inshore along the east Scottish coastline *S. setosa* is less common, and here *S. elegans* usually predominates. Further south along the Northumbrian coast, the abundance of the two species fluctuates. Meek concluded that when hydrographic conditions indicate a strong flow of Atlantic water into the northern part of the North Sea, *S. elegans* spreads further south, mainly in the inshore waters along the British coast. Alternatively, when the flow of Atlantic water into the North Sea is weak, the distribution of *S. elegans* retreats and *S. setosa* extends further north.

S. elegans and *S. setosa* both occur in the English Channel, and their distribution there has been investigated by Russell[78-80]. As in the North Sea, the two species have rather different distributions which vary from time to time. Prior to 1931, *S. elegans* was dominant to the west of Plymouth, *S. setosa* to the east. In the 'elegans' water Russell found several other planktonic animals often present, including the medusae of *Cosmetira pilosella*, the hydroid stage of which does not usually occur in the Channel but is found in deeper water along the Atlantic shores of the British Isles, also the trachymedusan *Aglantha digitale*, the pteropod *Clione limacina* and the euphausids *Meganytiphanes norvegica* and *Thysanoessa inermis*. These were absent from the 'setosa' water further east. This group of organisms found in 'elegans' water was thought to come from the west, flowing into the Channel from the Celtic Sea area to the south of Ireland and around the Scillies. Russell designated this water as 'western' water to distinguish it from the water of the main part of the English Channel, and also from 'south-western' water which occasionally flows into the Channel from the Bay of Biscay bringing a different characteristic collection of warmer-water forms, notably the copepod *Euchaeta hebes* and the medusa *Liriope tetraphylla*. It is now known that the species associated by Russell with 'western' water usually enter the Channel from the northwest rather than from the west, and consequently they are nowadays referred to as 'northwestern' forms.

After 1930, the boundary between Channel water and 'northwestern' water, as indicated by the distribution of the two species of chaetognath, lay further west than previously, approximately at the longitude of Lands End. This shift of the boundary was accompanied by changes in the quality of the water off Plymouth. Chemically, a lower concentration of phosphate during the winter months was observed. Biologically, it was noted that in subsequent years there were changes in the population, the plankton becoming sparser and less varied with a poorer survival of most types of fish larvae during the summer months. Herring shoals became so much reduced that the Plymouth fishery was eventually abandoned. The changed conditions, however, seem to have favoured the pilchard, which increased in numbers in the Channel as the herring declined[29]. Since 1966–7, there has been evidence of a return to the pre-1930 conditions in the English Channel, with *S. elegans* occurring in greater numbers off Plymouth, more herring and fewer pilchards in the area and much larger numbers of surviving fish larvae.

Russell also related the distribution of the two species of chaetognath to the rate of flow of the current through the Strait of Dover, normally a net flow of water eastwards from the Channel into the North Sea. Russell showed that the stronger this current through the Strait, the further eastwards *S. elegans* spreads up the English Channel and vice versa.

In both the western part of the Channel and the northern part of the North Sea, *S. elegans* and *S. setosa* are regarded as indicators of different qualities of water. Studies of the overall distribution of the two species have shown that, around the British Isles, *S. setosa* is restricted to the neritic water of lowish salinity found in the Channel, the North Sea, the Bristol Channel and Irish Sea. *S. elegans* is also a neritic form but occurs in slightly more saline water than *S. setosa*. It predominates in those regions where oceanic and coastal water become mixed, and is consequently limited around the British Isles chiefly to the western and northern areas and the northern part of the North Sea. Its distribution varies with the extent to which Atlantic oceanic water flows over the continental shelf. The boundary between the two species is by no means distinct. In the western English Channel they often exist together, one or the other being more numerous at different times or places.

Since these investigations by Meek and Russell, plankton surveys around the British Isles[21, 25, 26, 44, 47, 82–84] have provided further information about the distribution of organisms which can be related to particular areas of water, but it is beyond our scope to attempt any detailed account of this complicated subject. Very briefly, we may note that the water of the North Sea and

English Channel is derived from several sources, being a mixture of ocean water entering through the western mouth of the Channel and the northern mouth of the North Sea, and diluted by appreciable quantities of low-salinity water from the Baltic and fresh water from rivers. The plankton of the area therefore includes endemic neritic forms, to which may sometimes be added oceanic species carried in from deep water, and estuarine or brackish species carried out to sea.

The plankton of the eastern end of the English Channel, the southern part of the North Sea and much of the Irish Sea consists predominantly of organisms which prefer a slightly reduced salinity and are appreciably euryhaline and eurythermal to suit the somewhat fluctuating conditions of the region. This water is indicated by the presence of *S. setosa*. The copepods *Temora longicornis, Centropages hamatus, Isias claviceps, Labidocera wollastoni* and *Oithona nana* are common in the area, and the phytoplankton often includes *Biddulphia sinensis, Asterionella japonica, Nitzschia closterium, Eucampia zodiacus* and *Bellarochia malleus*. Because this water is shallow and close to extensive coastlines its plankton also contains many meroplanktonic forms, eggs, spores, medusae and larval stages of innumerable benthic and intertidal organisms which are not often found in more remote, deeper ocean water. These constituents vary seasonally according to the breeding habits of the different contributors and are a characteristic feature of neritic plankton.

In the northern part of the North Sea, the western part of the English Channel and off the west and north coasts of the British Isles, there is a higher proportion of water of oceanic origin than occurs in the southern North Sea and eastern Channel. The higher concentration of nutrients derived from deep water gives a rather richer and more varied plankton than that of 'setosa' water. We have already referred to this area of mixed oceanic-neritic water in connection with the distribution of *S. elegans*, and other organisms often found in this water include the trachymedusa *Aglantha digitale*, the copepods *Metridia lucens, Candacia armata* and *Centropages typicus*, the amphipod *Euthemisto gaudichaudi*, the euphausids *Meganyctiphanes norvegica, Nyctiphanes couchi, Thysanoessa inermis* and *T. longicaudata*, the polychaete *Tomopteris helgolandicus* and the pteropod *Spiratella retroversa*. All these animals are not confined to the same water as *S. elegans*. Most of them are also abundant in deep water further west and north. *T. inermis* and *N. couchi* are essentially shallow-water forms approximately limited to 'elegans' water. Another euphausid, *T. raschii*, occurs in 'elegans' water and also extends right across the northern part of the North Sea to the Skagerrak. The distribution of several species often alters seasonally in a somewhat irregular way. Along the east coast of Scotland and England there is a general tendency for the 'elegans' association to spread southwards during summer and autumn towards the southern part of the North Sea. But *T. inermis* does not usually penetrate far south in the North Sea, where it is typically found only in spring in the northern part. *N. couchi* is usually absent from the

North Sea in summer but appears in the north in autumn and spreads south during winter. The population of 'elegans' water also contains many meroplanktonts from the shore and sea bottom, which change seasonally. As in the English Channel, the 'elegans' water of the North Sea has larger numbers of surviving fish larvae than the 'setosa' water.

Ocean water flowing over the continental shelf towards the British Isles comes from three main sources (a) the North Atlantic, (b) the Arctic and (c) the Bay of Biscay and Mediterranean (*Figure 4.15*). The major influence is the North Atlantic Drift, driven north-eastwards from lower latitudes by the Westerly winds and bringing with it a population of oceanic species. These may reach the western end of the Channel and Irish Sea, or, when the

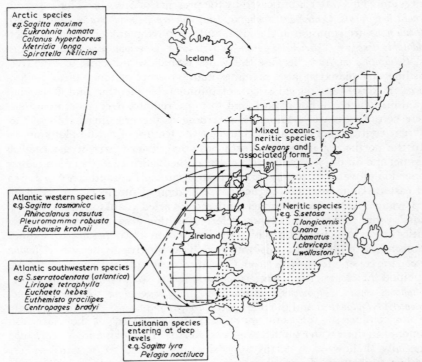

Figure 4.15. The main areas of distribution around the British Isles of neritic species and the species of mixed oceanic-neritic water, and the chief directions from which oceanic species may enter the region

flow of North Atlantic Drift water is strong, may be carried right round the north of Scotland and enter the northern part of the North Sea. The water derives from a vast expanse of ocean, and the population it carries varies somewhat with the direction of inflow. Distinctive species from the west of Ireland include the chaetognath *Sagitta tasmanica*, the copepods *Rhincalanus nasutus*, *Pleuromamma robusta* and *Mecynocera clausi*, and the euphausids *Nematoscelis megalops* and *Euphausia krohnii*. An inflow of water from further south (southwestern water) may bring with it the chaetognath *S. serratodentata* (*atlantica*), the trachymedusan *Liriope tetraphylla*, the siphonophoran *Muggiaea kochi*, the copepods *Euchaeta hebes* and *Centropages bradyi* and the amphipod *Euthemisto gracilipes*. Depending upon movements of the water, all these Atlantic forms may become very intermingled and sometimes carried far to the north. In the northern hemisphere, all salps, doliolids, heteropods and species of *Pyrosoma* are said by Russell to be of warm water origin, but *Ihlea asymmetrica* and *Salpa fusiformis* are carried well to the north of the British Isles.

If water from the Arctic spreads southwards around the British Isles, it may bring with it a cold-water population including the chaetognaths *S. maxima* and *Eukrohnia hamata*, the copepods *Calanus hyperboreus* and *Metridia longa*, and the pteropod *Spiratella helicina*. Being cold, this water tends to move under the surface layers.

Water of Mediterranean origin reaching the British Isles derives from the bottom outflow through the Straits of Gibraltar. Most of this water joins the Atlantic deep current, but a stream appears to diverge northwards and sometimes upwells along the continental slope to the north-west of the British Isles. It has been named 'Gulf of Gibraltar' water and carries a Lusitanian plankton which may include the chaetognath *S. lyra*, species of *Sapphirina*, the scyphomedusan *Pelagia noctiluca* and various siphonophorans, salps and doliolids.

It must of course be appreciated that many of the organisms mentioned here are widely distributed, and the occurrence of any one species in a tow-net sample does not alone indicate a particular source of water. Also, the distribution of certain species changes because of seasonal variations of water temperature or other parameters rather than because of any transport of water. Attempts to distinguish water masses on the basis of their plankton populations depend essentially on the recognition of particular associations of species, and require a wide knowledge and experience of planktology for correct interpretation.

REFERENCES AND FURTHER READING

Books

[1] Barnes, H. (1959). *Apparatus and Methods of Oceanography*. 'Part 1: Chemical.' London; Allen and Unwin
[2] Brown, M. (1957). *The Physiology of Fishes*. New York and London; Academic Press
[3] Ekman, S. (1953). *Zoogeography of the Sea*. London; Sidgwick and Jackson
[4] Gunther, K. and Deckert, K. (1956). *Creatures of the Deep Sea*. London; Allen and Unwin
[5] Harden Jones, F. R. (1968). *Fish Migration*. London; Edward Arnold
[6] Harvey, H. W. (1955). *The Chemistry and Fertility of Sea Water*. Cambridge U.P.

[7] Hill, M. N. (Editor) (1962–3). *The Sea.*
  'Vol. 1. (1962).
    Chap. 10. Light and Animal Life.
    Chap. 12. Sound in the Sea.
    Chap. 13. Sound Scattering by Marine Organisms.
    Chap. 14. Sound Production by Marine Animals.
  Vol. 2. (1963)
    Chap. 1. The Oceans as a Chemical System
    Chap. 2. The Influence of Organisms on the Composition of Sea Water
    Chap. 18. Biological Species, Water Masses and Currents'
[8] MacGinitie, G. E. and MacGinitie, N. (1949). *Natural History of Marine Animals.*
  New York; McGraw-Hill
[9] Marshall, N. B. (1954). *Aspects of Deep Sea Biology.* London; Hutchinsons
[10] Marshall, N. B. (1965). *The Life of Fishes.* London; Weidenfeld and Nicolson
[11] Moore, H. B. (1958). *Marine Ecology.* New York; Wiley
[12] Nicol, J. A. C. (1960). *The Biology of Marine Animals.* London; Pitman
[13] Norman, J. R. (1963). *A History of Fishes.* 2nd ed. by Greenwood, J. H. London;
  Benn
[14] Pickard, G. L. (1964). *Descriptive Physical Oceanography.* Oxford; Pergamon
[15] Smith, J. E. (1968). Editor. *Torrey Canyon Pollution and Marine Life.* Cambridge
  U.P.
[16] Sverdrup, H. U., Johnson, M. W. and Fleming, R. H. (1946). *The Oceans,* 'Chap. 3.
  Physical Properties of Sea Water. Chap. 4. General Distribution of Temperature, Salinity and Density. Chap. 6. Chemistry of Sea Water. Chap. 7.
  Organisms and the Composition of Sea Water.' New York; Prentice-Hall
[17] Wynne-Edwards, V. C. (1962). *Animal Dispersion in Relation to Social Behaviour,*
  'Chap. 16. Vertical Migration of the Plankton,' p. 366. Edinburgh and
  London; Oliver and Boyd

Papers

[18] Alvarino, A. (1965). 'Chaetognaths.' *Oceanogr. Mar. Biol. Ann. Rev.* **3**, 115
[19] Badcock, J. (1970). 'The Vertical Distribution of Mesopelagic Fishes Collected on
  the SOND Cruise.' *J. mar. biol. Ass. U.K.* **50**, 1001
[20] Bainbridge, R. (1952). 'Underwater Observations on the Swimming of Marine
  Zooplankton.' *J. mar. biol. Ass. U.K.* **31**, 107
[21] Bary, B. McK. (1963). 'Distribution of Atlantic Pelagic Organisms in Relation to
  Surface Water Bodies.' *Special publication. Royal Society of Canada*, **5**, 51
[22] Boden, B. P. and Kampa, E. M. (1964). 'Planktonic Bioluminescence.' *Oceanogr.
  Mar. Biol. Ann. Rev.* **2**, 341
[23] Bone, Q. and Roberts, B. L. (1969). 'The Density of Elasmobranchs.' *J. mar. biol.
  Ass. U.K.* **49**, 913
[24] Clarke, G. L., Conover, R. J., David, C. N. and Nicol, J. A. C. (1962). 'Comparative Studies of Luminescence in Copepods and other Pelagic Marine
  Animals.' *J. mar. biol. Ass. U.K.* **42**, 541
[25] Colebrook, J. M., John, D. E., and Brown, W. W. (1961). 'Contribution Towards
  a Plankton Atlas of the Northeast Atlantic and North Sea. Part 2. Copepoda.'
  Bull. Mar. Ecol. **5**, 90
[26] Colebrook, J. M. and Robinson, G. A. (1963). 'Ecological Differentiation in the
  Plankton of the Waters Around the British Isles.' Systematics Assoc., London.
  Publication No. 5: Speciation in the Sea, p. 157
[27] Corner, E. D. S., Denton, E. J. and Forster, G. R. (1969). 'On the Buoyancy of
  some Deep Sea Sharks.' *Proc. Roy. Soc. B* **171**, 415–29
[28] Cushing, D. H. (1951). 'The Vertical Migration of Planktonic Crustacea.' *Biol.
  Rev.* **26**, 158

REFERENCES AND FURTHER READING

[29] Cushing, D. H. (1961). 'On the Failure of the Plymouth Herring Fishery.' *J. mar. biol. Ass. U.K.* **41**, 799

[30] Denton, E. (1971). 'Reflectors in Fishes.' *Scient. Am.* January **224**, 65

[31] Denton, E. J. and Gilpin-Brown, J. B. (1961). 'The Buoyancy of the Cuttlefish *Sepia officinalis*'. *J. mar. biol. Ass. U.K.* **41**, 319

[32] Denton, E. J. and Gilpin-Brown, J. B. (1961). 'The Effect of Light on the Buoyancy of the Cuttlefish.' *J. mar. biol. Ass. U.K.* **41**, 343

[33] Denton, E. J. and Gilpin-Brown, J. B. (1961). 'The Distribution of Gas and Liquid within the Cuttlebone.' *J. mar. biol. Ass. U.K.* **41**, 365

[34] Denton, E. J. and Gilpin-Brown, J. B. (1966). 'On the Buoyancy of the Pearly Nautilus.' *J. mar. biol. Ass. U.K.* **46**, 723

[35] Denton, E. J., Gilpin-Brown, J. B. and Howarth, J. V. (1961). 'The Osmo Mechanism of the Cuttlebone.' *J. mar. biol. Ass. U.K.* **41**, 351

[36] Denton, E. J. and Marshall, N. B. (1958). 'The Buoyancy of Bathypelagic Fish without a Gas-filled Swimbladder.' *J. mar. biol. Ass. U.K.* **37**, 753

[37] Denton, E. J. and Nicol, J. A. C. (1965). 'Studies on Reflexion of Light from Silvery Surface of Fishes.' *J. mar. biol. Ass. U.K.* **45**, 683

[38] Denton, E. J. and Nicol, J. A. C. (1965). 'Reflexion of Light by External Surfaces of the Herring, *Clupea harengus*.' *J. mar. biol. Ass. U.K.* **45**, 711

[39] Denton, E. J. and Nicol, J. A. C. (1966). 'A Survey of Reflectivity in Silvery Teleosts.' *J. mar. biol. Ass. U.K.* **46**, 685

[40] Denton, E. J. and Warren, F. J. (1957). 'The Photosensitive Pigments in the Retinae of Deep-sea Fish.' *J. mar. biol. Ass. U.K.* **36**, 651–662

[41] Dietz, R. S. (1962). 'The Sea's Deep Scattering Layers.' *Scient. Am.* August

[42] Dijkgraaf, S. (1963). 'The Functioning and Significance of the Lateral-line Organs.' *Biol. Rev.* **38**, 51

[43] Foxton, P. (1970). 'The Vertical Distribution of Pelagic Decapods Collected on the SOND Cruise: 1965. II, The Penaeidae and General Discussion.' *J. mar. biol. Ass. U.K.* **50**, 961

[44] Glover, R. S. (1952). 'Continuous Plankton Records: The Euphausiacea of the Northeastern Atlantic and the North Sea, 1946–48.' *Hull Bulletin of Marine Ecology* **3** (23), 185

[45] Gray, J. (1953). 'The Locomotion of Fishes.' p. 1, *Essays in Marine Biology*. Edinburgh and London; Oliver and Boyd

[46] Hardy, A. C. (1953). 'Some Problems of Pelagic Life.' p. 101, *Essays in Marine Biology*. Edinburgh and London; Oliver and Boyd

[47] Hardy, A. C. (1960). 'Towards Prediction in the Sea.' p. 149, *Perspectives in Marine Biology*. Ed. by Buzzati-Traverso, A. Univ. of California Press

[48] Hardy, A. C. and Bainbridge, R. (1954). 'Experimental Observations on the Vertical Migrations of Plankton Animals.' *J. mar. biol. Ass. U.K.* **33**, 409

[49] Hardy, A. C. and Kay, R. H. (1964). 'Experimental Studies of Plankton Luminescence.' *J. mar. biol. Ass. U.K.* **44**, 435

[50] Harris, J. E. (1953). 'Fin Patterns and Mode of Life in Fishes.' p. 17, *Essays in Marine Biology*. Edinburgh and London; Oliver and Boyd

[51] Harris, J. E. (1963). 'The Role of Endogenous Rhythms in Vertical Migration.' *J. mar. biol. Ass. U.K.* **43**, 153

[52] Hasler, A. D. (1960). 'Perception of Pathways by Fishes in Migration.' p. 451, *Perspectives in Marine Biology*. Ed. by Buzzati-Traverso, A. Univ. of California Press

[53] Johnston, R. (1964). Recent Advances in the Estimation of Salinity.' *Oceanogr. Mar. Biol. Ann. Rev.* **2**, 97

[54] Johnston, R. (1969). 'On Salinity and its Estimation.' *Oceanogr. Mar. Biol. Ann. Rev.* **7**, 31

[55] Johnston, R. (1963–4). 'Sea-water. The Natural Medium of Phytoplankton. 1. General Features. 2. Trace Metals and Chelation.' *J. mar. biol. Ass. U.K.* **43**, 427–456 and **44**, 87–109

[56] Jones, F. R. H. and Marshall, N. B. (1953). 'The Structure and Function of the Teleostean Swimbladder.' *Biol. Rev.* **28**, 16

[57] Kinne, O. (1963). 'The Effects of Temperature and Salinity on Marine and Brackish Water Animals. 1. Temperature.' *Oceanogr. Mar. Biol. Ann. Rev.* **1**, 301

[58] Kinne, O. (1964). 'The Effects of Temperature and Salinity on Marine and Brackish Water Animals. 2. Salinity and Temperature–salinity Relations.' *Oceanogr. Mar. Biol. Ann. Rev.* **2**, 281

[59] Knight-Jones, E. W. and Morgan, E. (1966). 'Responses of Marine Animals to Changes in Hydrostatic Pressure.' *Oceanogr. Mar. Biol. Ann. Rev.* **4**, 267

[60] Korringa, P. (1947). 'Relation Between the Moon and Periodicity in the Breeding of Marine Animals.' *Ecol. Monogr.* **17**, 347

[61] Krauss, W. (1966). 'Internal Waves in the Sea.' *Oceanogr. Mar. Biol. Ann. Rev.* **4**, 11–32

[62] Lange, R. (1970). 'Isosmotic Intracellular Regulation and Euryhalinity in Marine Bivalves.' *J. exp. mar. Biol. Ecol.* **5**, 170–179

[63] Lucas, C. E. (1947). 'Ecological Effects of External Metabolites.' *Biol. Rev.* **22**, 270–95

[64] Lucas, C. E. (1961). 'External Metabolites in the Sea.' p. 139, vol. 3, suppl. *Pap. Mar. Biol. and Oceanogr., Deep-sea Research*

[65] McIntyre, F. (1970). 'Why the Sea is Salt.' *Scient. Am.* November. **223**, 5, 104

[66] Meek, A. (1928). 'On *Sagitta elegans*, and *Sagitta setosa* from the Northumbrian Plankton.' *Proc. Zool. Soc. Lond.*, 743

[67] Menzies, R. J. (1965). 'Conditions for the Existence of Life on the Abyssal Sea-floor.' *Oceanogr. Mar. Biol. Ann. Rev.* **3**, 195

[68] Moulton, J. M. (1964). 'Underwater Sound: Biological Aspects.' *Oceanogr. Mar. Biol. Ann. Rev.* **2**, 425

[69] Nelson-Smith, A. (1970). 'The Problem of Oil Pollution at Sea.' *Adv. Mar. Biol.* **8**, 215

[70] Nicol, J. A. C. (1958). 'Observations on Luminescence in Pelagic Animals.' *J. mar. biol. Ass. U.K.* **37**, 705

[71] Nicol, J. A. C. (1960). 'Studies on Luminescence. On the Subocular Light-organs of Stomiatoid Fishes.' *J. mar. biol. Ass. U.K.* **39**, 529

[72] Nicol, J. A. C. (1963). 'Some Aspects of Photoreception and Vision in Fishes.' *Adv. Mar. Biol.* **1**, 171

[73] Nicol, J. A. C. (1967). 'The Luminescence of Fishes.' *Symp. Zool. Soc. Lond.* **19**, 27

[74] Pytkowicz, R. M. (1968). 'The Carbon Dioxide-carbonate System at High Pressures in the Oceans.' *Oceanogr. Mar. Biol. Ann. Rev.* **6**, 83

[75] Rice, A. L. (1964). 'Observations on the Effects of Changes of Hydrostatic Pressure on the Behaviour of some Marine Animals.' *J. mar. biol. Ass. U.K.* **44**, 163

[76] Riley, G. A. (1970). 'Particulate Organic Matter in Seawater.' *Adv. Mar. Biol.* **8**, 1

[77] Russell, F. E. (1965). 'Marine Toxins and Venomous and Poisonous Marine Animals.' *Adv. Mar. Biol.* **3**, 255

[78] Russell, F. S. (1935). 'On the Value of Certain Plankton Animals as Indicators of Water Movements in the English Channel and North Sea.' *J. mar. biol. Ass. U.K.* **20**, 309

[79] Russell, F. S. (1939). 'Hydrographical and Biological Conditions in the North Sea as indicated by Plankton Organisms.' *J. Cons. perm. int. Explor. Mer.* **14**, 171

REFERENCES AND FURTHER READING

[80] Russell, F. S. (1952). 'The Relation of Plankton Research to Fisheries Hydrography.' *Rapp. P.-v. Réun. Cons. perm. int. Explor. Mer.* **131,** 28

[81] Shaw, E. (1962). 'The Schooling of Fishes.' *Scient. Am.* June

[82] Southward, A. J. (1961). 'The Distribution of some Plankton Animals in the English Channel and Western Approaches. 1. Samples taken with Stramin Nets in 1955 and 1957.' *J. mar. biol. Ass. U.K.* **41,** 17

[83] Southward, A. J. (1962). 'The Distribution of some Plankton Animals in the English Channel and Western Approaches. 2. Surveys with the Gulf 111 High-speed Sampler, 1958–60.' *J. mar. biol. Ass. U.K.* **42,** 275

[84] Southward, A. J. (1963). 'The Distribution of some Plankton Animals in the English Channel and Western Approaches. 3. Theories about Long-term Biological Changes, including Fish.' *J. mar. biol. Ass. U.K.* **43,** 1

[85] Waterman, T. H. (1960). 'Polarized Light and Plankton Navigation.' p. 429, *Perspectives in Marine Biology.* Ed. by Buzzati-Traverso, A. Univ. of California Press

[86] Wilson, D. P. (1951). 'A Biological Difference between Natural Seawaters.' *J. mar. biol. Ass. U.K.* **30,** 1

[87] Wilson, D. P. and Armstrong, F. A. J. (1952). 'Further Experiments on Biological Differences between Natural Seawaters.' *J. mar. biol. Ass. U.K.* **31,** 335

[88] Wilson, D. P. and Armstrong, F. A. J. (1954). 'Biological Differences between Seawaters: Experiments in 1953.' *J. mar. biol. Ass. U.K.* **33,** 347

[89] Wilson, D. P. and Armstrong, F. A. J. (1958). 'Biological Differences between Seawaters: Experiments in 1954 and 1955.' *J. mar. biol. Ass. U.K.* **37,** 331

[90] Wilson, D. P. and Armstrong, F. A. J. (1961). 'Biological Differences between Seawaters: Experiments in 1960.' *J. mar. biol. Ass. U.K.* **41,** 663

CHAPTER 5

ORGANIC PRODUCTION IN THE SEA

THE ORGANIC FOOD CYCLE

The synthesis of organic compounds from the inorganic constituents of sea-water by the activity of organisms is termed 'production'. It is effected almost entirely by the photosynthetic activity of marine plants, with traces of organic matter also formed by chemosynthesis. The raw materials are water, carbon dioxide and various other substances, the nutrients, mainly inorganic ions, principally nitrate and phosphate. Chlorophyll-containing plants, by making use of light energy, are able to combine these simple substances to synthesize complex organic molecules. This is termed 'Gross Primary Production'. The chief products are the three major categories of food materials, namely carbohydrates, proteins and fats.

Some of the organic material manufactured by plants is broken down again to inorganic state by the plants themselves in the course of their respiration. The remainder becomes plant tissue, and is referred to as 'Net Primary Production'. This is of major importance as the source of food for herbivorous animals. Upon the Net Primary Production therefore depends, directly or indirectly, the animal population of the sea.

By far the greater part of Primary Production in the sea is performed by the phytoplankton. Under favourable conditions this is capable of remarkably rapid growth, sometimes producing its own weight of new organic material within 24 h, a rate greater than that achievable by land plants. The large marine algae growing on the sea bottom in shallow water make only a relatively small contribution to the total production in the sea because they are of very restricted distribution.

The consumption of plants by herbivorous animals leads to the formation of animal tissue. This is 'Secondary Production', which in turn becomes food for the first rank of carnivorous animals (Tertiary production), and these may then fall prey to other carnivores, and so on (*Figure 5.1*). These successive stages of production of living tissue form a series of 'trophic levels', or links in the 'food chain'. The connections are complicated because many creatures take food from several trophic levels, the links of the food chain being interconnected and forming an intricate 'food web'. Between each trophic level there are large losses of organic material caused in several ways; for instance, a proportion of the organisms at each trophic level are not eaten by animals but simply die and decompose by autolysis and bacterial action, some of the food that animals consume is egested unassimilated and most of their assimilated food is broken down by respiration, leaving only a small proportion to form new tissue. The efficiency of transfer of organic matter from one trophic level to the next varies with the types of organisms, herbivores generally doing rather better than carnivores. In broad terms about 100 g of food are consumed for every 10 g of animal tissue formed, i.e. a gross conversion efficiency of 10% (*see* page 288). Herbivorous zooplanktonts sometimes exhibit efficiencies of about 30% and certain larval stages even better, but allowing for these higher efficiencies only a small part of the original plant production becomes incorporated in animal tissue.

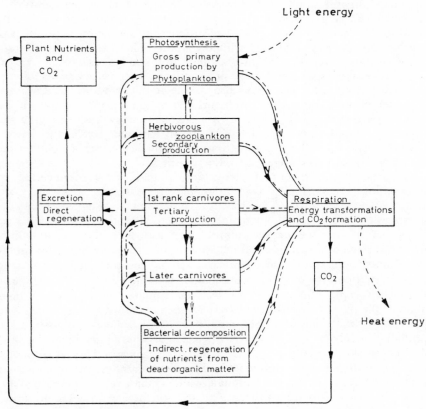

Figure 5.1. An outline of the main stages of the organic cycle in the sea

Eventually, as a result of respiration and excretion, death and decomposition, organic materials become broken down and returned to the water as simple substances which plants can utilize in primary production. In this way, matter is continually cycled from inorganic to organic forms and back to inorganic state. The initial synthesis of organic material involves the intake of energy to the system, and this is supplied by sunlight. The transference of organic matter from one trophic level to the next is part of the energy flow of the cycle (*Figure 5.1*), energy being continually lost from the system and in due course becoming dissipated as heat. This is discussed further in Chapter 9.

Regeneration

The processes of return of plant nutrients to the water following the degradation of organic compounds are termed regeneration. Regeneration is 'direct' when the products set free into the water by metabolism are directly utilizable by plants, which is the case with most of the excretory products of marine animals except for the trimethylamine oxide excreted by teleosts. Phosphorus is excreted mainly as phosphate, and in most cases nitrogen is excreted as ammonia, and both these products can be absorbed and assimilated by phytoplankton. Some animals also excrete amino acids, uric acid or urea, and each of these compounds can be utilized by some plants. However, the greater part of the plant nutrients are regenerated by 'indirect' processes involving bacterial activity. Bacteria are an essential part of the organic cycle, necessary for the decomposition of particulate organic matter from faecal pellets and the bodies of dead organisms.

After death, the tissues of plants and animals become converted by degrees into soluble form. Dissolution may be initiated by autolysis, the tissues being broken down by the dead organisms' own enzymes, but decomposition is brought about mainly by bacterial action. Although free-living bacteria are seldom very numerous in sea-water, they are abundant on the surface of organisms and detritus and are specially numerous in the uppermost layers of bottom deposits. Bacterial metabolism converts solid organic matter into organic solutes, and eventually into inorganic form. Phosphorus compounds are regenerated as phosphate. On the death of some animals the phosphorus in their tissues returns to the water very quickly as phosphate, indicating that decomposition of phosphorus compounds is probably mainly by autolysis[16]. Nitrogenous organic materials are broken down more slowly, mainly by bacterial activity, regenerated first as ammonia and then further oxidised to nitrite and finally nitrate. Sulphur-containing compounds are regenerated as sulphate, with sulphide often an intermediate stage. In the course of these processes, bacteria themselves grow and multiply, and constitute an important component of the food supply. Bacteria therefore perform two major functions in marine food cycles—(*a*) the breakdown of dead organic matter into soluble inorganic ions which can be utilized by plants, and (*b*) the transformation of dead organic matter into bacterial protoplasm which is directly utilizable as food by some animals.

There are continuous losses of organic material from the euphotic zone due to sinking and to movements by animals down to deeper levels after feeding. Some of this material may reach the bottom and become lost from the cycle by being permanently incorporated in the sediment. The greater part is regenerated in deep layers of water or on the bottom, and nutrients therefore accumulate below the euphotic zone. The continuation of production depends upon the restoration of nutrients from the deep to the surface layers by vertical water mixing (page 129).

MEASUREMENTS OF ORGANIC PRODUCTION

There are several ways in which estimates of the amount of primary production in the sea have been attempted, and certain of these methods are outlined below. The accuracy of any of these techniques is in doubt. Production rates are usually expressed as the weight of carbon fixed in organic compounds beneath unit area of sea surface in unit time; for example gC/m^2/day, or sometimes as weight of carbon fixed in unit volume of water in unit time, for example gC/m^3/day. Estimates of net primary production by phytoplankton often fall within the range 0·05–0·5 gC/m^2/day, with values as high as 5·0 gC/m^2/day in most productive sea areas. These are of similar order to values obtained for much of the land surface, but production rates in fertile agricultural lands or forests may exceed 10 gC/m^2/day, and the total production of the sea is almost certainly lower than that of the land surface.

Standing Stock Measurements

The earliest attempts to measure organic production in the sea were indirect, being based on estimates of the total amount of plant material in the water, i.e. the standing stock. This does not give a direct indication of the rate of production because account must be taken of the rate of turnover. If the plants are being very rapidly eaten, high production may maintain only a small standing stock. Alternatively, where the consumption rate is very low and the plants are long lived, a large standing stock is not necessarily the result of a rapid production rate. A large standing stock may itself limit production by reducing the penetration of light through the water and diminishing the supply of nutrients.

The size of the standing stock depends upon the balance between the rate of production of new plant cells and the rate at which they are lost by animal consumption and by sinking below the photosynthetic zone. To determine production rates from standing stock measurements it is therefore necessary to estimate both the rate of change in size of the population and also the rate of loss, the latter being particularly difficult to assess with any certainty. The following methods have been used for measuring the standing crop.

Direct counts—It is possible to count the number of plant cells in a measured volume of water. Because nets are not fine enough to filter the smallest plants, the phytoplankton is usually obtained by collecting water samples and removing the cells by centrifuging or sedimentation. Subsamples are made with a haemocytometer and counted under a microscope. When the number of plant cells in unit volume of water has been determined, the weight of plant protoplasm must be calculated. For this it is necessary to know the size and weight of the plant cells, and to make due allowance for the inorganic content of the cells.

Chlorophyll estimations—The quantity of chlorophyll that can be extracted from unit volume of sea-water depends upon the number of plant cells present, and it is possible to calibrate a scale of pigment concentration against quantities of plant tissue[12]. A measured volume of raw sea-water is filtered to remove all cells. These are then treated with a standard volume of acetone or alcohol to extract the chlorophyll. The intensity of colour in the extract is measured colorimetrically or absorptiometrically to determine the concentration of pigment, and the results are expressed as chlorophyll concentration or arbitrary units of plant pigment (U.P.P.).

Carbohydrate estimations—It has been shown that the tissues of most zooplankton contain little carbohydrate. The carbohydrate-content of material filtered from sea-water derives almost entirely from the plants present. Measurements of carbohydrate therefore provide a means of estimating the amount of plant material in a sample[17]. Carbohydrate can be determined absorptiometrically from measurements of the intensity of brown coloration developed by the action of phenol and concentrated sulphuric acid.

Zooplankton counts—To estimate the rate of loss of plant cells due to grazing by animals, quantitative zooplankton samples are needed to determine the number of herbivores present. It is also necessary to measure their feeding rates.

If the size of the animal population and its food requirements are known, it is possible also to make a calculation of the primary production necessary to support this number of animals.

Measurement of Nutrient Uptake

Where production is seasonal, estimates of production can be based on measurements of the decrease of nutrients in the water during the growing period. In temperate areas, concentrations of nitrate and phosphate in the surface layers reach a maximum during the winter months when photosynthesis is minimal and convectional mixing is occurring. The concentrations fall during the spring and summer due to the absorption of these nutrients by the phytoplankton. The N- or P- content of the phytoplankton being known, measurements of the reduction of nitrate and phosphate in the water enable estimates to be made of the quantity of new plant tissue formed. Allowances are necessary for the regeneration of nitrate and phosphate within the photosynthetic zone by decomposition of tissues, replenishment from deeper water, and utilization of other sources of nitrogen, for example nitrite, ammonium or organic nitrogen. Production estimates can also be based on measurements of change of oxygen, carbon dioxide or silicate content of the water.

Measurement of Photosynthesis

In photosynthesis, carbohydrate is formed according to the equation

$$nCO_2 + nH_2O \rightarrow [H_2CO]_n + nO_2.$$

The rate of photosynthesis may be determined by measuring either the evolution of oxygen or the absorption of carbon dioxide. These are not exactly equal because proteins and fats are also formed, but allowance can be made for this.

Oxygen-bottle experiments — Samples of sea-water are collected from several depths within the photosynthetic zone and their oxygen-contents measured. Pairs of bottles are then filled from each sample and sealed. The bottles are identical except that one of each pair is transparent — the light bottle — and the other is covered with black opaque material — the dark bottle.

The pairs of bottles are next suspended in the sea at the series of depths from which their contents were obtained, and left for a measured period. They are then hauled in and the amount of oxygen measured in each. The reduction in oxygen in the dark bottles with respect to the original measurements is due to the respiration of the plant, animal and bacterial cells contained. On the assumption that respiration is not influenced by light, the difference in oxygen-content between the light and dark bottles of each pair is regarded as due to the production of oxygen by photosynthesis. This assumption is probably not justified but the method has been widely used and can give tolerably consistent results. One complication which applies to all experiments in which sea-water is enclosed in bottles is the rapidity of bacterial growth in these conditions, and this may vary with the intensity of illumination.

Measurement of carbon dioxide uptake. The Steeman Nielsen ^{14}C *method* — This is a method of measuring carbon fixation by using the radioactive isotope of carbon, ^{14}C, as a tracer. Samples of sea-water are collected from a series of depths and the carbon dioxide content in each is measured. Bottles are filled from these samples and a small measured quantity of bicarbonate containing ^{14}C is added to each. The bottles are then sealed and suspended in the sea at appropriate depths for a measured period. Alternatively, if temperature and illumination at the sampling depths are known, the bottles can be immersed in tanks at corresponding temperatures and provided with artificial illumination at the correct intensity. This is advantageous at sea because the vessel does not need to remain hove-to at each station for the duration of the experiment. With bottles in tanks, it is necessary to keep them in sufficient motion to prevent settlement of the plant cells.

When the bottles are hauled in, the water is filtered to collect the phytoplankton. The cells are washed and their ^{14}C-content estimated by measurement of the β-radiation. The total carbon fixation is calculated from the known amounts of $^{14}CO_2$ and total CO_2 originally present in the water, making due allowances for the slight differences in rates of assimilation of $^{14}CO_2$ and $^{12}CO_2$. This method is widely thought to give the most accurate estimates of net production.

SOME FACTORS REGULATING PRODUCTION

1. Light

In the process of photosynthesis the energy of solar radiation becomes fixed as chemical energy in organic compounds. The efficiency of the ocean surface in this energy transformation must vary with locality and conditions, but is probably on average about 0·1–0·2 per cent overall, an efficiency a little lower than that of the land surface.

Photosynthesis is confined to the illuminated surface zone of the sea, and a useful measure of the extent of this productive layer is the Compensation Depth, i.e. the depth at which the rate of production of organic material by

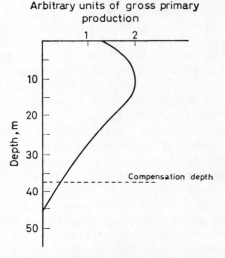

Figure 5.2. Generalized diagram relating primary production rate to depth in middle latitudes during bright sunshine.   Below the Compensation Depth there is no net production

photosynthesis exactly balances the rate of breakdown of organic material by plant respiration. Below the Compensation Depth there is no net production. The Compensation Depth obviously varies continually with changes of

126

illumination, and must be defined with respect to time and place. In clear water in the tropics, the noon Compensation Depth may be well below 100 m throughout the year. In high latitudes in summer, the noon Compensation Depth commonly lies somewhere between 10 and 60 m, reducing to zero during the winter months when virtually no production occurs.

Photosynthesis varies in proportion to the light intensity up to a limit at which plants become light-saturated, and further increase of illumination produces no further increase of photosynthesis. Exposure to strong light is harmful and depresses photosynthesis, the violet and ultra-violet end of the spectrum having the most unfavourable effects. In bright daylight the illumination at the sea surface seems often to be at or above the saturation level for most of the phytoplankton, and measurements of photosynthesis in these conditions show that maximum production occurs some distance below the surface, usually somewhere between 5–20 m depending upon light intensity, and falls off sharply above this level (*Figure 5.2*). Correspondingly, the maximum quantity of phytoplankton is seldom found very close to the surface, and except for a few species that seem to thrive in the uppermost few centimetres the greater part of the phytoplankton can be regarded as 'shade plants'. By absorbing light the plants themselves reduce light penetration through the water, and as the population increases the Compensation Depth tends to decrease.

Above the Compensation Depth the rate of photosynthesis exceeds the rate of respiration and there is a net gain of plant material; below it there is a net loss. At a particular level the total loss by algal respiration in the water column above will exactly equal the total gain by photosynthesis. This level is termed the Critical Depth[26]. The distance between Compensation Depth and Critical Depth depends upon the proportions of the phytoplankton stock above and below the Compensation Depth. This is determined mainly by vertical water movements.

Wind action induces vertical eddies at the sea surface, which distribute the phytoplankton through the wind-mixed layer. When the depth of wind-mixing exceeds the Compensation Depth some of the stock of phytoplankton is carried below the level at which it can grow. If the standing stock of phytoplankton is to increase, its total photosynthesis must be more than its total respiration. This is possible as long as the wind-mixed layer is not deeper than the Critical Depth. But when wind-mixing extends below the Critical Depth, total losses exceed total gains and the stock declines. This is a major cause of the cessation of net primary production in temperate seas in winter. Around the British Isles during the productive season the water column is stabilized by thermal stratification. The depth of wind-mixing is then seldom much more than the Compensation Depth, and the Critical Depth is estimated to lie some five to nine times deeper. In autumn the Critical Depth rises, vertical mixing soon reaches well below it and the standing stock of phytoplankton is sharply reduced.

2. Temperature

The rate of photosynthesis increases with rising temperature up to a maximum, but then diminishes sharply with further rise of temperature. Different species are suited to different ranges of temperature, and photosynthesis is probably performed as efficiently in cold water by the phytoplankton of high latitudes as it is in warmer water by the phytoplankton native to the tropics.

Seasonal variations of production-rate in temperate latitudes are related to changes of both temperature and illumination. Apart from its direct effect on rate of photosynthesis, temperature also influences production indirectly through its effects on movement and mixing of the water, and hence on the supply of nutrients to the euphotic levels.

3. Nutrients

In addition to dissolved carbon dioxide, which is present in sea-water in ample quantities to support the most prolific naturally-occurring plant growth, there are other substances, the nutrients, which plants also extract from the water and which are essential for their growth. Many of these are minor constituents of sea-water, present only in very low concentration, and their supply exerts a dominant control over production. Nitrate and phosphate are of special importance. Where the quantities of these ions are known, theoretical estimates of the potential productivity of the water generally accord well with observed values. Iron, manganese and copper are other essential nutrients, silicon is required by diatoms, and molybdenum and cobalt and probably other elements are necessary for some plants. Organic compounds dissolved in the water may be important in some cases.

The absorption of nutrients by the phytoplankton reduces the concentration of these substances in the surface layers, and this limits the extent to which the plant population can increase. A certain amount of the nutrients absorbed by phytoplankton may be regenerated and recycled within the euphotic zone, but plants are continually being lost from the surface layers through sinking and by consumption by zooplankton which moves to deeper levels during the day-time. Much of the nutrients absorbed from the surface layers are therefore regenerated in the deeper and darker layers of water where plants cannot grow. Consequently, nutrients accumulate at deep levels due to the continuous transfer of material from the surface. This loss of nutrients from the productive layer of the sea to deep levels contrasts with the nutrient cycle of the land surface. In soil the breakdown of organic compounds releases nutrients where they are quickly available for reabsorption by plant roots, thereby maintaining the fertility of the land. In the sea the continuance of plant growth depends to a great extent upon the rate at which nutrients are restored to the euphotic zone by mixing with the nutrient-rich water from below. The lower overall productivity of the sea compared with the land is largely a consequence of the regeneration of

128

nutrients in the sea far below the zone of plant growth, with recycling dependent upon relatively slow processes of water movement. Some of the vertical mixing processes which restore fertility to the surface layer of the sea are summarized below.

Upwelling[23]

Offshore winds, by setting the surface water in motion, may cause water from deeper levels to be drawn up to the surface (*Figure 5.3*). We have mentioned earlier the upwelling which occurs in low latitudes along the western coasts of the continents to replace the westward-flowing surface water in

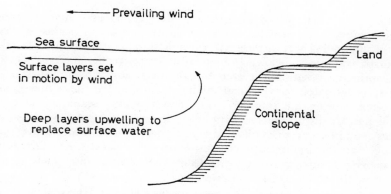

Figure 5.3. Upwelling due to wind action at the surface

the equatorial currents (page 10). Although this upwelling water probably does not rise from depths greater than some 100–200 m, this is deep enough to supply nutrients to the Canaries current, Benguela current, Peru current, California current and West Australia currents, and these are all areas of high fertility. In the Southern Ocean, continuous upwelling ensures that, even during the highly productive period of the Antarctic summer, plant growth is probably never limited by shortage of nutrients. In the Arctic, where upwelling is less, there is some depletion of surface nutrients during the summer months and production is correspondingly reduced.

Upwelling also occurs at divergences, i.e. areas where adjacent surface currents move in different directions. In low latitudes, divergences between the equatorial currents and countercurrents cause upwelling close to the Equator and along the northern boundary of the equatorial counter-currents, and these are important in maintaining the productivity of tropical waters.

Turbulence (Eddy diffusion)

Turbulence is a term loosely applied to various complex and irregular movements of the water in which different layers become mixed by vertical eddies. Turbulence has various effects on production depending upon circumstances. It may promote production by bringing nutrients to the surface, or it may sometimes reduce production by carrying down a considerable part of the plant population below the Compensation Depth. Broadly, high production is likely to follow a limited period of turbulence once the water column becomes sufficiently stabilized for the phytoplankton to flourish in the replenished surface layer without undue loss of plants by downward water movements (*see* page 127).

The following are examples of how turbulent water movements may arise—

Convection—When surface water cools, its density increases and convectional mixing commences once the density of the surface layers begins to exceed that of underlying water, the surface water sinking and being replaced by less dense water from below. In high latitudes, convectional mixing is virtually continuous because heat is continually being lost from the surface. In temperate latitudes, convectional mixing occurs during the winter months but ceases during the summer. The corresponding seasonal changes in production are discussed later (page 137). In low latitudes where the surface waters remain warm throughout the year there is little if any convectional mixing, and the concentration of nutrients at the surface is generally low unless vertical mixing is occurring from some other cause, such as wind action on the surface causing upwelling.

Currents—Vertical eddies may arise where adjacent layers of water are moving at different speeds, or where currents flow over irregularities on the sea-bed. On the continental shelf, especially where the bottom is uneven, strong tidal currents may cause severe turbulence and keep the water well mixed throughout its depth. Tidal flow in the eastern part of the English Channel is representative of a condition which produces sufficient turbulence to mix almost the full depth of water, and this helps in maintaining the fertility of the area throughout the summer months.

Internal waves—The depth to which surface waves appreciably move the water is rather less than their wave-length. Although swell waves in deep water occasionally exceed 100 m in wave-length, most surface waves are considerably shorter than this and do not mix the water column to any great depth. However, where the water column is not homogeneous, it is evident that very large internal waves can exist far below the surface. Over a great part of the oceans the water is stratified in layers of different density—density increasing with depth—and these layers do not remain still but oscillate about a mean level. Internal wave movements of at least 200 m in height have been detected in the deep ocean.

Internal waves may be produced in several ways. For example, strong onshore winds driving light surface water towards the continental slope will bend the equal-density layers downwards. When the wind ceases, the low density water which has been forced down the continental slope will return to the surface, and the displaced density layers may not return simply to the horizontal but will probably oscillate up and down, transmitting their motion over great distances as waves in the deep layers.

Oscillations in the deep layers of the North Atlantic may also arise from irregularities in the rate of southward flow of deep water from the Arctic over the North Atlantic ridges, due perhaps to changes of Arctic climate influencing the rate of sinking of surface water. Cooper[8] has suggested that this flow is intermittent, and that sometimes enormous 'boluses' of cold water flood down the south side of the ridge, displacing the deep density layers and setting up internal waves.

Cooper[7] considers that internal waves can cause vertical mixing where they impinge upon the continental slope, their motion here becoming translated in a manner comparable with that of waves breaking on the shore, carrying deep water up the continental slope much as surface waves run up a sloping beach. In this way, oceanic deep layers rich in nutrients may sometimes spill over the continental edge, mixing with and increasing the fertility of shelf water.

If onshore winds are sufficiently strong and continuous, the displacement of the density layers may become so severe that the water column becomes unstable. A profound disturbance may then ensue which Cooper has termed 'capsizing' or 'culbute mixing'. There is no exact English equivalent for the French word 'culbuter', meaning to upset violently resulting in a confused heap or jumble, which well describes this process.

Capsizing or 'culbution' is a cataclysmic disturbance of the water column which is believed to occur if low density water is forced so far down the continental slope that it eventually comes to lie beneath water of greater density (*Figure 5.4*). This unstable, topsy-turvy condition is thought to resolve by the low density water, which has been forced downwards, bursting up towards the surface, i.e. the water column capsizes. The ensuing upheaval must produce a homogeneous mass of water from surface to bottom. This water mass, being a mixture of surface and deep water, is denser than adjacent surface water and will subsequently subside to its appropriate density level. The process must be continuous as long as onshore winds of sufficient strength persist, the line of capsizing gradually receding seawards and involving progressively deeper water. The water blown over the continental shelf will be capsized water containing a component of deeper water, and therefore richer in plant nutrients than ordinary oceanic surface water. Its nutrient content may be expected to increase as the sequence proceeds and extends to deeper levels.

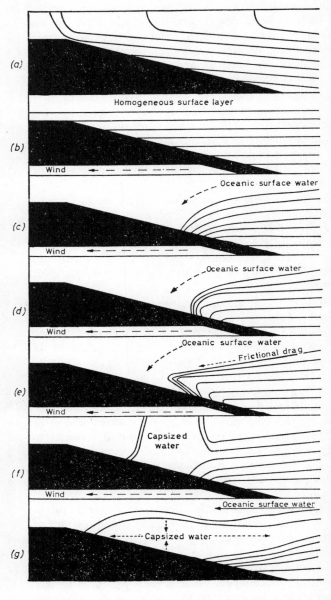

(a)

Homogeneous surface layer

(b)

Wind

Oceanic surface water

(c)

Wind

Oceanic surface water

(d)

Wind

Oceanic surface water
Frictional drag

(e)

Wind

Capsized water

(f)

Wind

Oceanic surface water

Capsized water

(g)

(From Cooper, L. H. N.[7] By courtesy of Cambridge University Press)

Figure 5.4. (a) a frequent pattern of isosteres south of the Celtic Sea in winter when the uppermost 75– 100 m of water in the ocean is homogeneous and is overlying water with density and content of nutrient salts increasing downwards. (b) a pattern of isosteres in winter over the continental slope, drawn to a scale of 1 : 4, with no forces operative. The amount of light water which will later lie to windward is considered unlimited. (c) and (d) cushioning of light oceanic surface water against the continental slope brought about by on-slope gales. The resistance of the solid slope to further progress of the foot of the light water is, however, absolute causing the isosteres to curl and ultimately to become vertical. Restoring Archimedian forces in (d) have then become zero. (e) the drag of the surface wind current will draw the upper strata of stratified water with it leading to an unstable density inversion as illustrated. (f) the unstable tongue of heavier water will capsize violently, leading to a homogeneous mass of mixed water extending from the surface to the depth of the bottom of the capsizing water mass. New isosteres, bracketing the capsized water mass, are so created. (g) the newly formed surface water will be heavier than the surface water inshore and to seaward and will subside to form a lens of homogeneous water at its appropriate density level and will be replaced by fresh oceanic surface water blown in from seaward. Processes (d)– (g) must be considered as simultaneous parts of a continuous process, with the line of capsizing receding from the continental slope and the maximum depth of the phenomenon increasing as the gale proceeds. Consequently, the water blown on to the continental shelf of the Celtic Sea by a strong southerly gale should be entirely capsized water and richer in nutrients than it would be if it were solely oceanic surface water. Moreover, the nutrient content of the water passing on to the shelf should increase as the gale proceeds, and the depth of capsizing becomes greater

132

4. The Grazing Rate

Although the interactions between plant and animal populations are difficult to elucidate, the grazing rate of the herbivorous zooplankton is certainly one of the factors which regulates the size of the standing stock of phytoplankton, and therefore influences the production rate. The quantity of epipelagic zooplankton generally correlates more closely with the quantity of plant nutrients in the surface layer than with the size of stock of phytoplankton, indicating how greatly grazing reduces the number of plants in fertile water. In the long term, the primary productivity of an area must determine the size of the animal population it supports, but in the short term there are often wide, and sometimes rapid, changes in both numbers and composition of populations due to a variety of causes. Interactions between species often involve a time lag, and there is consequently a tendency for numbers to fluctuate about mean levels. Although some natural populations show homeostatic mechanisms which control reproduction within limits which do not exhaust the food resources of their environment, there is obviously a general trend for animal numbers to increase as long as there is sufficient food until rising food consumption diminishes the food supply. Food shortage may then cause a decline in the feeding population, and eventually the reduction in food consumption may allow the food supply again to increase, and these oscillations may involve many links of the food web. If the inorganic environment were to remain uniform the system would in due course settle to a steady state, but in nature the physical conditions fluctuate and the equilibrium is forever being disturbed.

A dominant cause of short term fluctuations in the plankton of middle and high latitudes is seasonal variation of climate which influences both the production rate and the sequence of species which predominate. These changes are discussed later (page 137) but in the present connection we should note that the sharp reduction in numbers of diatoms which follows their period of rapid multiplication in the spring occurs before the nutrients are fully exhausted, but coincides with the growth in quantity of zooplankton. There can be little doubt that the increasing rate of grazing is one cause of the decline of the standing stock.

A striking feature of the distribution of marine plankton is its unevenness, with localized patches in nearby areas differing in both quantity and composition. One aspect of this patchiness is the inverse relationship of quantities of phytoplankton and zooplankton which has often been reported. Where phytoplankton is especially plentiful, herbivorous zooplanktonts are sometimes few in number; and where herbivores abound, the phytoplankton may be sparse.

This appearance of an inverse relationship may be due simply to the different reproductive rates of phytoplankton and zooplankton, and the effects of grazing on the size of the standing stock. In favourable conditions phytoplankton can multiply rapidly and produce a dense stock. Zooplankton populations increase more slowly, but as the number of herbivores rise the phytoplankton will be increasingly grazed and the stock correspondingly diminished. Measurements of feeding in various herbivorous planktonts indicate that filtering rates and food intake are not much influenced by food requirements. In high concentrations of phytoplankton, large numbers of plant cells are rapidly ingested, often far in excess of the animals' needs. Many of these cells pass through the gut virtually undigested but are killed in the process. It may therefore be impossible for any abundance of phytoplankton and zooplankton to coexist for any length of time in natural conditions because of the rapidity with which plant cells can be removed from the water by herbivorous animals.

Another explanation for an inverse phytoplankton/zooplankton relationship involves the concept of 'animal exclusion'. According to this hypothesis, animals avoid water rich in phytoplankton because the plants have some effect on the quality of the water which animals find unpleasant. The nature of the excluding influence is uncertain, but it might perhaps be due to secretion of external metabolites by the plants. Small zooplanktonts could avoid this water by controlling their depth so as to remain at deeper levels until the relative movement of the different layers of water carries them to areas where the surface water is less objectionable. The exclusion effect is also observed with some pelagic fish. For instance, the occurrence from time to time of dense patches of *Rhizosolenia*, *Biddulphia* or *Phaeocystis* on the North Sea herring fishing grounds are usually associated with poor catches (*see* Chap. 8[11]), the shoals seeming to avoid phytoplankton-laden water. In certain mackerel fishing areas, it is known that poor catches are obtained in areas which the fisherman recognize as 'stinking water', and this has also been shown to contain a large amount of phytoplankton[25]. The absence of fish from this water may be due, however, merely to the absence of the zooplankton on which they feed.

To discover if any evidence of exclusion could be demonstrated experimentally, Bainbridge[4] devised laboratory apparatus in which the behaviour of zooplankton could be observed in the presence of high concentrations of phytoplankton. He constructed a horizontal, circular Perspex tube, 4 ft in diameter, divided into three equal compartments by sliding watertight doors (*Figure 5.5*). One compartment was filled with sea-water enriched with cultures of phytoplankton. The other compartments were filled with filtered sea-water. The doors were opened for a period to allow the phytoplankton to spread around the tube until a distinct gradient of phytoplankton concentration was seen to exist. At this stage, equal numbers of small planktonic animals were introduced into each of the three compartments, and their distribution around the tube was counted at intervals. Pure and mixed cultures of phytoplankton were tested, including the diatoms *Skeletonema*, *Thalassiosira*, *Biddulphia*, *Coscinodiscus*, *Lauderia*, *Eucampia* and *Nitzschia*, the flagellates *Chlamydomonas*, *Dicrateria*, *Rhodomonas*, *Syracosphaera*,

Oxyrrhis, Exuviella, Peridinium and *Gymnodinium*, and some bacterial cultures. The animals studied included the mysids *Hemimysis lamornae, Praunus neglectus, P. flexuosus, Neomysis integer* and *Mesopodopsis slabberi*, also *Artemia salina, Calanus finmarchicus* and various other small copepods, and some decapod larvae.

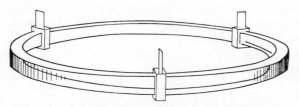

Figure 5.5. The Perspex horizontal circular apparatus used by Bainbridge[4] to study the interrelations of zooplankton and phytoplankton, showing the three sliding doors. Openings at these three points allow for filling, and introduction of animals

(From Bainbridge, R.[4] published by Cambridge University Press)

The results showed definite migrations by the mysids towards concentrations of certain diatoms, notably *Skeletonema, Thalassiosira, Biddulphia, Nitzschia* and mixed cultures, and towards the flagellates *Chlamydomonas, Peridinium, Dicrateria* and *Oxyrrhis*. *Artemia salina* and some small copepods moved into cultures of *Nitzschia, Biddulphia* and *Thalassiosira*. Cultures of *Lauderia, Coscinodiscus, Eucampia, Syracosphaera* and *Exuviella* showed no attractive effect. The mysids showed a definite movement away from the flagellates *Rhodomonas* and *Gymnodinium II*, and the bacterial cultures. No repellant effect was demonstrated for any diatoms except *Nitzschia* at highest concentration.

In this circular tube the results with *Calanus finmarchicus* were inconsistent. Observations on this animal were also made in another apparatus consisting of a pair of straight, parallel, vertical tubes. One tube was filled with normal sea-water, the other with sea-water enriched with phytoplankton, and equal numbers of animals were inserted in each. The tubes were suspended in an aquarium tank under lighting of moderate intensity, and the number of animals at different levels of the tubes was counted at intervals. The experiment demonstrated significantly greater numbers of *Calanus* swimming upwards in cultures of *Coscinodiscus, Skeletonema, Ditylium, Chlamydomonas, Gymnodinium, Oxyrrhis* and mixed phytoplankton. *Chlorella* appeared to depress the numbers swimming up.

Bainbridge's experiments were not designed to elucidate the nature of attractive or repellent substances, but these did not appear to be associated with changes in the concentration of carbon dioxide or oxygen or pH. Positive migrations into concentrations of ammonia were observed. As this is the usual excretory product of marine invertebrates, its attractive effect might be partly accountable for the tendency of many small pelagic organisms to collect in swarms.

From his observations, Bainbridge concluded that 'exclusion' is of quite restricted occurrence in natural conditions, although it may operate during intense blooms of some toxic flagellates. It seems that, in general, natural concentrations of diatoms are likely to be attractive to grazing animals, and Bainbridge suggests that the inverse phytoplankton/zooplankton relationship may be explained in terms of a dynamic cycle of growth, grazing and migration as follows (*Figure 5.6*).

(*a*) Where conditions are favourable, rapid growth produces a dense patch of phytoplankton.

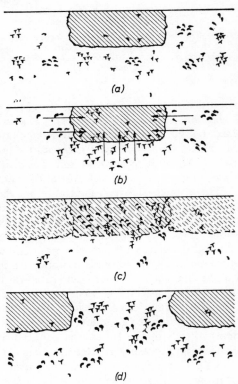

Figure 5.6. Scheme of grazing and migration cycle. (a) initial state with inverse relationship; (b) start of migration with some grazing; (c) completion of migration and heavy grazing; (d) reversal of initial state and return to inverse relationship. Oblique hatching represents concentrations of phytoplankton

(From Bainbridge, R.[4] published by Cambridge University **Press**)

(b) Herbivorous animals are attracted horizontally and vertically into the phytoplankton patch. The grazing rate increases but the concentration of plants will not decline appreciably until the rate of removal by grazing exceeds the rate of increase by cell division.

(c) As more animals are attracted to the area, heavy grazing rapidly diminishes the number of plants until the phytoplankton patch is virtually eliminated.

(d) Meanwhile, in adjacent water, rapid growth of phytoplankton can now occur because these areas have become denuded of their grazing population by migration into the original phytoplankton patch, and conditions are set for a repetition of the cycle.

Hardy and Gunther noted three species of herbivore commonly present in dense phytoplankton in the Antarctic, namely, the copepods *Calanus simillimus* and *Drepanopus pectinatus* and the mysid *Antarctomysis maxima*. These three forms are strong swimmers, and Bainbridge suggests that their presence in high concentrations of phytoplankton is due to the rapidity with which they move. They arrive before the other, slower-moving herbivores become numerous, but their grazing effect is insufficient to make much reduction in the number of plants.

Bainbridge's experiments were conducted using fresh cultures of phytoplankton. Others have pointed out that there is evidence that the production of antibiotic substances by algae occurs mainly in ageing cells, and this may invalidate some of Bainbridge's conclusions. His experiments at least demonstrate that certain species of phytoplankton exert an influence, either attractive or repellent, to which some zooplanktonts react.

OCEAN SEASONS

Seasonal variations in temperature, illumination and availability of nutrients in the surface layers of the sea give rise to fluctuations in production and composition of the plankton (*Figure 5.7*). The four seasons of the sea have the following general features[6]—

Winter

A period when the surface layers of water have a low temperature and poor or absent illumination, but a high concentration of nutrients due to convectional mixing. The quantity and growth of both phytoplankton and zooplankton are at a minimum.

Spring

A period when the surface water temperature is rising, the water column is becoming stabilized by thermal stratification, illumination is increasing and the Critical Depth becomes lower than the zone of wind mixing. The concentration of nutrients in the surface layer is initially high, but begins to decrease due to absorption by the rapidly growing phytoplankton. The rate of primary production soon becomes very high and there is an enormous increase in the quantity of phytoplankton, especially diatoms, which soon reach their greatest abundance for the year (the spring diatom peak). The zooplankton undergoes a more gradual increase, but during late winter and early spring it becomes augmented by great numbers of eggs and larvae, which by late spring have developed to more advanced larval or juvenile stages. As the zooplankton increases in amount, the quantity of phytoplankton declines rapidly.

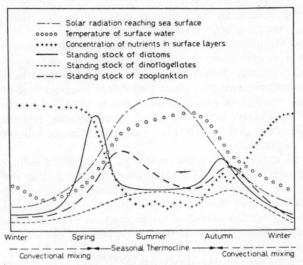

Figure 5.7. Generalized diagram illustrating seasonal changes of temperature, nutrients, phytoplankton and zooplankton in the surface layers of temperate seas

Summer

A period during which the surface water is warm and well illuminated. The concentration of nutrients at the surface is now low because they have been taken up by the phytoplankton, and there is little replenishment from deeper water because vertical mixing is prevented by a sharp thermocline. The dinoflagellates are at their greatest numbers, but the phytoplankton as a whole has declined in amount and primary production is reduced due to grazing by zooplankton and shortage of nutrients. Diatoms are often quite scarce at this time. The zooplankton, mainly holoplankton, now reaches its greatest amount for the year, and after that diminishes.

Autumn

During this season the surface water is cooling and illumination is becoming less. The deeper layers are still getting slightly warmer until eventually the thermocline breaks and convectional mixing is re-established. This leads to rapid replenishment of nutrients in the surface layer, and a consequent increase in primary production, both diatoms and dinoflagellates becoming more numerous. This autumn increase of phytoplankton is always less than the spring peak. It is often followed by a slight increase in zooplankton. These increases are short-lived. Vertical mixing disperses much of the phytoplankton below the Critical Depth and the size of stock falls quickly. As temperature and illumination decrease further, the quantities of both phytoplankton and zooplankton gradually reduce to their winter levels, and their over-wintering stages appear.

This sequence of four ocean seasons occurs in middle latitudes where the temperature of the surface water undergoes the greatest seasonal change. In high latitudes the surface temperature does not vary much with season. Here, illumination is the dominant factor regulating productivity, and only two ocean seasons are apparent, a long winter period of poor or absent illumination and virtually no primary production, followed by a short period of very high production when the light becomes sufficiently good to enable the phytoplankton to grow. This productive period lasts only a few weeks, but during part of this time daylight is continuous throughout the 24 h period. This makes possible a very rapid growth of a large quantity of phytoplankton, and this abundance of food allows a great increase in zooplankton. Because the rich food supply lasts only a short time, the developmental stages of zooplankton at high latitudes must be passed through rapidly. After that, illumination declines and primary production falls to zero. The phytoplankton virtually disappears, probably over-wintering mainly as spores locked within sea ice, and the zooplankton population decreases to its over-wintering level. This single short season of rapid growth soon followed by decline is a merging of the spring and autumn seasons with elimination of the intervening summer period, the biological winter being correspondingly prolonged.

In low latitudes, conditions are mostly those of continual summer. The surface water is consistently warm and well illuminated but production is limited by shortage of nutrients, there being little vertical mixing across a strongly developed thermocline. However, production continues throughout the year and extends to a greater depth than in high latitudes, and the rate of turnover is probably rapid. The result in some tropical areas is a total annual production some five to ten times greater than in temperate seas[27]. Generally the rate of production in warm seas remains fairly uniform but there are parts where seasonal changes of wind, for example the monsoons, cause variations of water circulation and seasonal improvements in nutrient supply which are quickly followed by periods of increased production. In the Mediterranean the main season for the production of algae is November to April when there is vertical mixing. Except for a few areas, seasonal peaks in warm seas seldom exceed a tenfold increase of production, compared with fluctuations sometimes as great as fiftyfold in temperate waters.

Seasonal changes in plankton of temperate seas

Eggs, larvae and spores of benthic plants and animals are often a conspicuous part of the plankton of neritic water, and some of the most obvious seasonal changes are related to the reproductive seasons of the benthos. We can summarize certain seasonal features of the plankton of shallow water in temperate seas as follows (*Figures 5.8, 5.9, 5.10 and 5.11*).

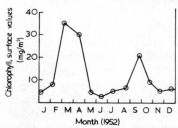

Figure 5.8. Changes in the standing stock of phytoplankton in western part of English Channel as indicated by measurements of chlorophyll concentration (see *page 124*)

(Data from Atkins, W. R. G. and Jenkins, P. G., *J. mar. biol. Ass. U.K.*, **31**, 495–508 (1953) published by Cambridge University Press, and Jenkins, P. G., *Deep Sea Research*, suppl. to vol. 3, 58–67 (1961), published by Pergamon Press)

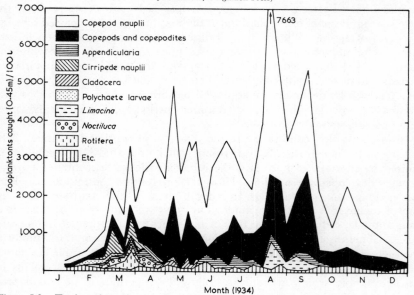

Figure 5.9. Total numbers of zooplanktonts per 100 litres caught between the surface and 45 m off Plymouth through 1934

(From Harvey, H. W. *et al. J. mar. biol. Ass. U.K.*, **20**, 407–441 (1935), published by Cambridge University Press)

140

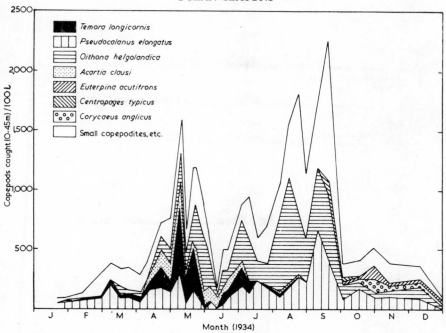

Figure 5.10. *Numbers of copepods caught per 100 litres between surface and 45 m off Plymouth through 1934*

(From Harvey, H. W. *et al.*, *J. mar. biol. Ass. U.K.*, **20**, 407–441 (1935) published by Cambridge University Press)

Winter, November–March—This is the season of minimum quantity of holoplankton. There is very little phytoplankton in the water, although certain diatoms are often present, mainly species of *Coscinodiscus* and *Biddulphia*. Dinoflagellates are very scarce except for *Ceratium tripos*. Copepods are few in number, over-wintering mainly as the pre-adult copepodite stage V, and much of the holoplankton lies at a deeper level during winter than in summer. Many invertebrates and the majority of fishes in this area spawn during the latter part of the winter, setting free into the plankton a great number and variety of eggs and early larval stages. Usually the most abundant larvae in tow-net samples taken in late winter are nauplii of *Balanus balanoides*. There may also be nauplii of *Verruca stroemia* and *B. crenatus*, together with numerous plutei, trochophores, bivalve and gastropod veligers, zoeas and fish eggs of many species.

Spring and early Summer, March–June—Diatoms rapidly increase in numbers to their annual maximum which usually occurs during late March or early April. The diatom population is now very mixed, *Skeletonema*, *Chaetoceros*, *Lauderia*, *Thalassiosira*, *Coscinodiscus* and *Biddulphia* becoming

141

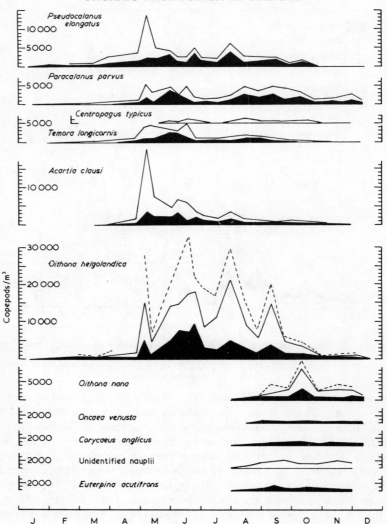

Figure 5.11. *Numbers of smaller copepods per cubic metre through the year 1947 off Plymouth, all to same scale. Solid black: copepodites and adults. Continuous line: all stages including nauplii. Dotted line: total plus eggs*

(From Digby, P. S. B., *J. mar. biol. Ass. U.K.*, **29**, 393–438 (1950) published by Cambridge University Press)

very numerous. Different species tend to predominate in succession, although the order is not constant from year to year. Photoperiodism may play some part in setting off the rapid growth of diatoms during the spring, and also the effects of external metabolites may be partly accountable for the successive rise and decline of different species. Following their spring peak, the diatom population declines fairly rapidly.

During early spring the plankton contains many eggs and larvae, but by the early summer months many of these have metamorphosed and disappeared from the plankton. The nauplii of *B. balanoides*, most abundant in March, develop to cypris stages in April and have mostly settled by mid-May. Over the same period, many crab zoeas advance to megalopas, and fish larvae to young fish stages. Overwintering copepodites mostly become adult in February and produce a brood of eggs in March, hatching to nauplii by April. The individuals from this brood usually grow to larger size than any others during the year. The cladocerans *Podon* and *Evadne* often reach their greatest numbers between April and May.

Summer, June–August—Diatoms continue to decrease in number and are often scarce between July and August, species of *Rhizosolenia* often being the commonest at this period. Dinoflagellates are now more numerous, and usually reach their maximum abundance between June and July. The majority of larvae produced in the spring have now completed their metamorphosis and disappeared from the plankton, but other species require warm water for breeding and shed their eggs and larvae at this time. For example, the nauplii of *Chthamalus stellatus*, *B. perforatus* and *Elminius modestus* appear in plankton samples taken off our south west coastline. However, it is the holoplanktonic species, copepods, chaetognaths, ctenophores and larvaceans, which become very numerous and form the greater part of the zooplankton during the summer months.

Autumn, September–October—After the thermocline breaks, diatoms show a brief but definite increase in numbers and again produce a very mixed population in which species of *Rhizosolenia*, *Coscinodiscus*, *Biddulphia* and *Chaetoceros* are often found together. Autumn plankton samples are sometimes rich in the larvae of benthic invertebrates, mainly plutei and bivalve veligers. An autumn brood of copepods appears which in some cases does not complete its development to the adult stage but survives the winter as stage V copepodites. Note the copepods *Oncaea venusta*, *Oithona nana*, *Corycaeus anglicus* and *Euterpina acutifrons*, which are not numerous earlier in the year, but have their main abundance from August to the end of the year (*Figure 5.11*).

SOME MATHEMATICAL MODELS

The aim of ecological studies is to achieve sufficient understanding of the interactions between organisms and their environments to be able to express these relationships in precise numerical terms. If this could be done, mathematical models could be constructed from which predictions might be made.

The processes regulating phytoplankton production can be formulated in several ways. For example, the rate of change of a phytoplankton stock, P, obviously depends on the rates of addition and loss of cells. Fleming[11] expressed this relationship as follows—

$$\frac{dP}{dt} = P[a - (b + ct)]$$

where a = rate of cell division of phytoplankton, b = initial rate of loss of cells by grazing, and c = rate of increase of grazing intensity. Taking observed values for diatom populations at the beginning and the peak of the spring bloom in the English Channel, he computed curves which fitted observed values well and indicated that, following the peak, the production rate continued to increase despite the fall in phytoplankton stock due to heavy grazing.

Cushing[10] formulated the rate of change of phytoplankton as—

$$\frac{dP}{dt} = P(r - M - G)$$

where P = number of algae (or weight of carbon) per unit volume or beneath unit surface, r = instantaneous reproductive rate of algae, G = instantaneous mortality rate of algae due to grazing, and M = instantaneous mortality rate from other causes. r was calculated for observed rates of algal cell division at various light intensities, and account taken of light penetration, Compensation Depth and depth of wind-mixed layer. In computing grazing rates for observed populations of herbivores, adjustments were made to allow for their reproductive and mortality rates. Applying the equation to the North Sea during the spring bloom, in an area where nutrient depletion was thought not greatly to depress production, Cushing showed fairly good agreement between observed and calculated values for standing stocks of phytoplankton and herbivores.

An alternative to these equations, based on rates of cell reproduction and loss, is a general equation advanced by Riley which takes account of energy considerations, as follows:

$$\frac{dP}{dt} = P(Ph - R - G)$$

where P = phytoplankton population, Ph = rate of photosynthesis, R = plant respiration rate, and G = grazing rate.

Riley derived expressions for the coefficients Ph, R and G which take account of illumination, concentration of nutrients, temperature and the number of herbivores, and combined these into the following expanded equation:

$$\frac{dP}{dt} = P\left[\frac{pI_0}{kz_1}(1 - e^{-kz_1})(1 - N)(1 - V) - R_0 e^{rT} - gZ\right],$$

where

P = Phytoplankton population in gC/m^2.

p = Photosynthetic constant, 2·5. (gC produced per gramme of phytoplankton C per day per average rate of solar radiation.)

I_0 = Average intensity of surface illumination in g cal/min.

k = Extinction coefficient (1·7 ÷ depth of Secchi disc reading in metres).

z_1 = Depth of euphotic zone (depth at which light intensity has a value of 0·0015 g cal/cm^2/min).

N = Reduction in photosynthetic rate due to nutrient depletion, $\left(\dfrac{mg\ At\ P/m^3\ \text{at surface}}{0·55}\ \text{when concentration is less than } 0·55\right)$.

V = Reduction in photosynthetic rate due to vertical turbulence.

R_0 = Respiratory rate of phytoplankton at 0°C.

r = Rate of change of R with temperature, r being chosen so that R is doubled by a 10°C increase in temperature.

T = Temperature in °C.

g = Rate of reduction of phytoplankton by grazing of 1 g C of zooplankton.

Z = Quantity of zooplankton in g C/m^2.

From this equation the likely fluctuations of the phytoplankton population over a 12 month period were predicted and showed good agreement with the observed values (*Figure 5.12*).

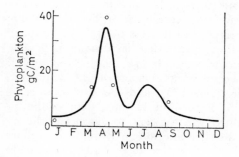

Figure 5.12. Curve of calculated values of phytoplankton standing crop on Georges Bank. Circles are observed values

(From Riley, G. A.[21] by courtesy of Sears Foundation for Marine Research)

An equation was also derived for changes in the population of herbivores, H.

$$\frac{dH}{dt} = H(A - R - C - D)$$

A = Coefficient of assimilation of food

R = Coefficient of respiration

C = Coefficient of predation by carnivores

D = Coefficient of death from other causes.

Despite the many difficulties of evaluation of this equation, predicted and observed values were in substantial agreement (*Figure 5.13*).

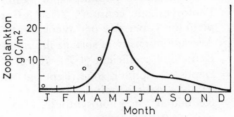

Figure 5.13. Curve of calculated values of zooplankton population on Georges Bank. Circles are observed values
(From Riley, G. A.[22] by courtesy of Sears Foundation for Marine Research)

GEOGRAPHICAL DIFFERENCES OF FERTILITY

Differences in the fertility of the seas in different localities, and at different times, depend upon the availability of plant nutrients in the surface layers. Certain areas are of consistently high fertility, others are ocean deserts, and in many regions the fertility fluctuates seasonally.

The areas of good fertility include most of the seas overlying wide continental shelves. There are several reasons why this shallow water is relatively rich in nutrients. Waves erode the coastline and stir up the sediments, releasing nutrients into the water. Fresh water running off the land may carry additional nutrients, including trace elements such as iron and manganese which are often scarce in deep water due to precipitation. Where there are centres of human population, sewage is usually poured into the sea and provides nutrients after decomposition. The tidal flow of the water above the shelf may cause sufficient turbulence to keep the water column well mixed, ensuring that nutrients regenerated below the euphotic zone are quickly restored to the surface, and there is unlikely to be much loss of nutrients to deep levels of the ocean unless shelf water is flowing down the continental slope, as occurs for example in cascading (page 147). Processes leading to enrichment of oceanic surface water near the continental slope have been mentioned earlier (page 130), and much of this fertile water may be blown over the shelf. From time to time, some of the innumerable deep water pelagic creatures which ascend to the surface during darkness may also be carried over the shelf and augment the food supplies of neritic water. The very dense growth of benthic algae that exists on some coasts contributes greatly to the primary production of these areas.

The fertility of deep water depends largely upon the extent to which water from deep levels is brought to the surface. In temperate areas, the surface layers are well provided with nutrients by convection during the winter and early spring, but the supply of nutrients diminishes during the summer when the formation of a thermocline prevents replenishment by vertical mixing. Upwelling of deep water around Antarctica produces in the Southern Ocean the world's widest expanse of highly fertile open sea. Fertile areas are also produced by upwelling in the currents along the

146

eastern part of the Atlantic, Indian and Pacific Oceans at low latitudes (page 10 and *Figure 1.5*).

Poor fertility occurs where vertical water mixing is minimal. Throughout the tropics, wherever a permanent thermocline is present, production rates are mainly low despite rapid regeneration. However, because production continues through the year without much seasonal decline, it is probable that the total annual production in most tropical areas substantially exceeds that of temperate seas. The Sargasso is a semi-tropical marine desert area where even horizontal mixing by surface currents is slight. Here the production rate is very poor except during the brief winter period when some convectional mixing may take place.

In the Arctic there is relatively little upwelling compared with the Southern Ocean, and the Arctic is correspondingly less fertile. In the Mediterranean, fertility is low because nutrients are continually lost in the deep outflow which forms the bottom current through the Straits of Gibraltar. The inflowing surface current is derived from surface levels of the Atlantic which are relatively poor in nutrients.

Cascading

A process termed 'cascading' is believed sometimes to cause losses of both nutrients and planktonic organisms from neritic water. During autumn and winter, loss of heat from the sea surface may cause the shallow shelf water to become appreciably colder and denser than water at similar levels beyond the continental edge. Consequently, the heavier shelf water will tend to flow down the continental slope, i.e. to cascade, to its appropriate density level. The loss of cascading water from above the shelf must be compensated by a corresponding inflow of water elsewhere.

The effect of cascading on the fertility of shelf water must depend on the quality of the compensation water. Cooper and Vaux[9] studied cascading in the Celtic Sea (south of Ireland and west of the Bristol and English Channels), and concluded that surface layers of the Atlantic and Bay of Biscay formed the main sources of compensation water. Because the phosphate content of this water was lower than that of the Celtic Sea at the time, cascading presumably led to a net loss of nutrients from the area. Iron and other trace elements which tend to concentrate in particles in the lower layers are probably specially liable to be swept off the shelf in the cascade. Annual fluctuations in the fertility of shelf water could be caused by variations in the volume of cascading.

The distribution of organisms could be influenced by cascading in several ways. Small planktonts carried down in the cascade would be lost in deep water. Stronger swimming forms might ascend out of the cascade as it flows down the continental slope, and this could lead to patches of large, active zooplanktonts over deep water near the continental edge. Where submarine valleys cut into the slope, cascade streams are likely to be strongest in these channels. Cascading might also occur on the sides of submarine banks where a flat top lies near the surface. The preference of certain animals for the sides of submarine valleys, troughs and mounds during winter may be due to the relative abundance of food carried by cascading to these parts of the sea floor.

Turbidity Currents

When water contains a large quantity of suspended particles its overall density is increased. On occasions the excess density of highly turbid water above the continental shelf may cause it to flow as a coherent fluid down the continental slope. This downslope motion due to turbidity is termed a turbidity current. It is a major mechanism of transport of material from the shelf to the deep-sea floor. The current mainly follows the course of valleys in the slope. The velocity of turbidity currents is exceptionally great for deep levels and this is undoubtedly important in the erosion of submarine canyons. It is possible that turbidity currents may at times cause changes in fertility of shelf water and in distribution of organisms comparable with the effects of cascading.

REFERENCES AND FURTHER READING

Books
[1] Sverdrup, H. L., Johnson, M. W. and Flemming, R. H. (1946). *The Oceans.* 'Chap. 16. Phytoplankton in Relation to Physical-Chemical Properties of the Environment.
Chap. 19. Organic Production in the Sea.'
New York; Prentice-Hall
[2] Raymont, J. E. G. (1963). *Plankton and Productivity in the Oceans.* Oxford; Pergamon
[3] Hill, M. N., (Editor) (1963). *The Sea.* Vol. 2,
'Chap. 7. Productivity, Definition and Measurement.
Chap. 8. Organic Regulation of Phytoplankton Fertility.
Chap. 17. Geographic Variations in Productivity.
Chap. 20. Theory of Food-chain Relations in the Ocean.'
London; Allen and Unwin.

Papers
[4] Bainbridge, R. (1953). 'Studies of the Interrelationships of Zooplankton and Phytoplankton.' *J. mar. biol. Ass. U.K.* **32,** 385
[5] Bainbridge, R. (1957). 'The Size, Shape and Density of Marine Phytoplankton Concentrations'. *Biol. Rev.* **32,** 91
[6] Bogorov, B. G. (1960). 'Perspectives in the Study of Seasonal Changes of Plankton and of the Number of Generations at Different Latitudes.' *Perspectives in Marine Biology.* Ed. by Buzzati-Traverso. Univ. of California Press
[7] Cooper, L. H. N. (1952). Processes of Enrichment of Surface Water with Nutrients due to Winds Blowing on to a Continental Slope. *J. mar. biol. Ass. U.K.* **30,** 453
[8] Cooper, L. H. N. (1961). 'Hypotheses connecting Fluctuations in Arctic Climate

REFERENCES AND FURTHER READING

with Biological Productivity of the English Channel.' Pap. Mar. Biol. and Oceanogr., *Deep-Sea Research*, suppl. to Vol. 3.

[9] Cooper, L. H. N. and Vaux, D. (1949). 'Cascading over the Continental Slope from the Celtic Sea.' *J. mar. biol. Ass. U.K.* **28,** 719

[10] Cushing, D. H. (1959). 'On the Nature of Production in the Sea.' *Fish. Investig. Lond., Ser. II,* **22,** No. 6

[11] Fleming, R. H. (1939). 'The Control of Diatom Populations by Grazing.' *J. Cons. Int. Explor. Mer.* **14,** 210

[12] Harvey, H. W. (1950). 'On the Production of Living Matter in the Sea off Plymouth.' *J. mar. biol. Ass. U.K.* **29,** 97

[13] Harvey, H. W., Cooper, L. H. N., Lebour, M. V. and Russell, F. S. (1935). 'Plankton Production and its Control.' *J. mar. biol. Ass. U.K.* **20,** 407

[14] Margalef, R. (1967). 'Some Concepts Relative to the Organisation of Plankton.' *Oceanogr. Mar. Biol. Ann. Rev.* **5,** 257

[15] Marshall, S. M. and Orr, A. P. (1953). 'The Production of Animal Plankton in the Sea.' *Essays in Marine Biology*, p. 122. Edinburgh and London; Oliver and Boyd

[16] Marshall, S. M. and Orr, A. P. (1961). 'On the Biology of *Calanus finmarchicus.* XII. The Phosphorus Cycle.' *J. mar. biol. Ass. U.K.* **41,** 463

[17] Marshall, S. M. and Orr, A. P. (1962). 'Carbohydrate as a Measure of Phytoplankton.' *J. mar. biol. Ass. U.K.* **42,** 511

[18] Mullin, M. M. (1969). 'Production of Zooplankton in the Ocean: the Present Status and Problems.' *Oceanogr. Mar. Biol. Ann. Rev.* **7,** 293

[19] Murphy, R. C. (1962). 'The Oceanic Life of the Antarctic.' *Scient. Am.* Sept.

[20] Raymont, J. E. G. (1966). 'The Production of Marine Plankton.' *Adv. Ecol. Res.* **3,** 117

[21] Riley, G. A. (1946). 'Factors Controlling Phytoplankton Populations on Georges Bank.' *J. mar. Res.* **6,** 54

[22] Riley, G. A. (1947). 'A Theoretical Analysis of the Zooplankton Population of Georges Bank.' *J. mar. Res.* **6,** 104

[23] Smith, R. L. (1968). 'Upwelling.' *Oceanogr. Mar. Biol. Ann. Rev.* **6,** 11

[24] Steele, J. H. (1959). 'Quantitative Ecology of Marine Phytoplankton.' *Biol. Rev.* **34,** 129

[25] Steven, G. A. (1949). 'A Study of the Fishery in the South-West of England with Special Reference to Spawning, Feeding and Fishermen's Signs.' *J. mar. biol. Ass. U.K.* **28,** 555

[26] Sverdrup, H. U. (1953). 'On Vernal Blooming of Phytoplankton.' *J. Cons. Int. Explor. Mer.* **18,** 287

[27] Wickstead, J. H. (1968). 'Temperate and Tropical Plankton; a Quantitative Comparison.' *J. zool. Lond.,* **155,** 253

CHAPTER 6

THE SEA BOTTOM

The sea-floor provides several conditions favourable for animal life. There is food available, mainly in the form of fragments of organic matter sinking from the overlying water, and in some areas this food supply is sufficient to support a large population. Many bottom-dwelling creatures are able to live and grow to large size with relatively little expenditure of energy in hunting and collecting food because they can obtain adequate nourishment simply by gathering the particles that fall within their reach or are carried to them by the currents, or by digesting the organic matter and associated bacteria contained within the sediment. Most of the sea bottom is covered with soft deposits which give concealment and protection to burrowing creatures. Where the substrate is hard it provides a secure surface for the attachment of sessile forms and affords protection for creatures which hide in crevices or burrow in rock. Compared with the pelagic division of the marine environment the sea bottom provides a far wider variety of habitats because the nature of the bottom differs greatly from place to place. The benthic population of the sea is correspondingly more diverse than the pelagic population.

Except in very shallow depths, the temperature, salinity, illumination and movements of the water at the bottom are less variable than in the surface layers. Below 500 m seasonal changes are negligible, and the deeper the water the more constant are the conditions.

THE SUBSTRATE

Conditions which determine the types of materials forming the sea-floor include (a) the speed of the bottom current, (b) the depth, (c) the proximity of land, and the geographical and geological features of the coastline (for example, the rock formations and the outflows of rivers or glaciers), (d) the types of suspended matter in the overlying water, including the pelagic organisms, and (e) the type of benthic population.

Sediments cover the sea-floor except where the bottom current is strong enough to sweep away particles, or where the gradient is too steep for them to lodge. The scouring action of tidal currents may expose rocks beneath shallow water, and in deep water uncovered rock occurs on steep sides of submarine peaks or valleys.

In sediments, the type of deposit varies with the speed of the bottom current and the size and density of suspended particles. The faster the water moves, the coarser is the texture of the substrate because finely divided material is more easily held in suspension than larger particles of the same density. Stokes' equation for the settling velocity, W, applies fairly well to small particles in sea water.

$$W = \frac{2}{9} g \, \frac{D - d}{\mu} r^2$$

g = gravitational acceleration
D = density of particle
d = density of liquid
μ = dynamic viscosity of liquid
r = radius of spherical particles

Marine sediments are classified in two main groups, terrigenous and pelagic deposits.

Terrigenous Deposits

Terrigenous deposits are found near land, covering the continental shelf and upper parts of the continental slope. Much of this material is derived from weathering and erosion of exposed land surfaces, and consists largely of particles worn from the coast by wave action or carried into the sea by rivers or glaciers. Terrigenous deposits contain some organic material, often some 0·01–0·5 per cent of the dry weight, the finer-texture deposits usually having the greater proportion of organic matter. Microscopic examination sometimes reveals recognizable traces of various materials of biological origin, both terrestrial and marine. The former are mainly fragments of leaf and wood from land plants. The latter are very diverse, deriving from both benthic and pelagic sources, and often include small particles of seaweed, diatom cell walls, sponge spicules, polychaete chaetae, and fragments of the shells and skeletons of foraminifera, hydroids and corals, polyzoa, crustacea, echinoderms and molluscs. Minute animals may be seen in samples from the surface layers of the deposit; for example, flagellates, ciliates, foraminifera, nematodes and copepods. Superficial deposits from shallow water may also contain a microflora of benthic diatoms. The mud from shallow creeks is sometimes rich in fragments of marine angiosperms, for example *Spartina*.

Terrigenous deposits show considerable differences of composition from place to place, varying with the nature of the adjacent coastline, the movements of the water and the contours of the sea-bed. They range from large boulders close to rocky shores where they have been dislodged by violent wave action, through all grades and mixtures of pebbles, gravel and sands down to fine clay. The continental shelf is nowadays an important source of sand and gravel for building operations.

The particles in the sediment can be classified according to size (Table 6.1), but sediments are seldom uniform in composition and usually contain particles of many grades and types. The term 'mud' is loosely applied to mixed deposits containing a large proportion of silt and clay.

TABLE 6.1
Simple Classification of Sediment Particles

Description of particles	Maximum diameter of particles in mm	Settling velocity in water at 20°C cm/sec
Coarse sand	2·0	347
Fine sand	0·2	3·47
Silt	0·02	0·0347
Clay	0·002	0·000347

Pelagic Deposits

Pelagic deposits occur beneath deep water beyond the continental edge, carpeting the deep ocean basins. Much of this material is of fine texture, and its nature varies with the depth and with the types of organisms that abound in the overlying water. At depths less than about 6,000 m, pelagic deposits contain a considerable proportion of material of biological origin, commonly some 30 per cent or more by weight. Although these deposits are termed 'organic', they seldom contain much decomposable carbon but consist almost entirely of skeletal fragments of planktonic organisms. Down to about 4,000 m, organic deposits are rich in calcareous matter except in high latitudes where great areas of sea-bed are covered by siliceous diatom ooze. Below 4,000 m, hydrostatic pressure causes some forms of calcium carbonate to dissolve, and organic deposits below this depth therefore contain less calcific material and a larger proportion of silica.

The organic deposits are classified as follows:

1. Calcareous Oozes

(a) *Globigerina ooze*—This is the most widespread of the pelagic deposits over the greater part of the deep Atlantic and much of the Indian and Pacific Oceans, covering nearly 50 per cent of the deep sea bottom and extending to depths of 6,000 m. It contains up to 95 per cent calcium carbonate mainly in the form of foraminiferan shells.

(b) *Pteropod ooze*—This contains many pteropod shells and occurs below subtropical parts of the Atlantic at depths down to 3,500 m.

(c) *Coccolith ooze*—A high proportion of coccolith material, sometimes amounting to 25 per cent or more of the total weight, is occasionally found in samples of globigerina ooze, chiefly beneath areas of warm surface water.

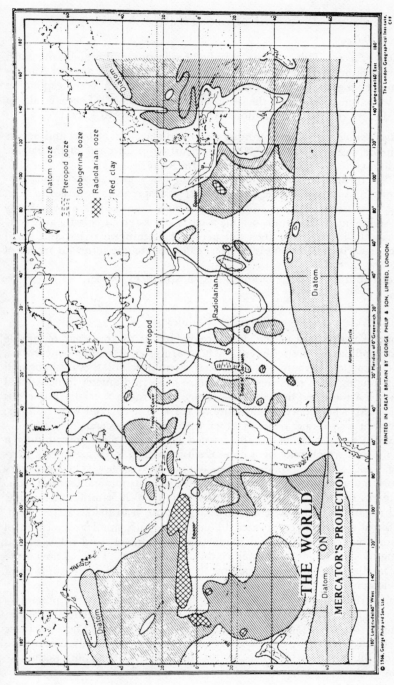

Figure 6.1. Ocean bottom deposits

(Based on a Map of the World by courtesy of G. Philip & Son Ltd.)

153

2. *Siliceous Oozes*

(*a*) *Diatom ooze*—This consists mainly of siliceous material in the form of diatom fragments. It occurs as an almost continuous belt around Antarctica beneath the Southern Ocean, its northernmost limit corresponding closely with the position of the Antarctic convergence. There is also a strip of diatom ooze across the northern part of the North Pacific.

(*b*) *Radiolarian ooze*—This contains many radiolarian skeletons and occurs at depths between 4,000 and 8,000 m beneath tropical parts of the Pacific and Indian Oceans and also recorded in the Atlantic.

At 6,000 m and below, sediments contain less material of obvious biological origin, and at these depths the most widespread deposit is *Red Clay*, covering nearly 40 per cent of the deep ocean floor. It is a very finely divided sediment of uncertain origin, usually brick-red in colour, and containing about 50 per cent silica, 20 per cent aluminium oxide and small amounts of various compounds of iron, calcium, magnesium and traces of many other metals.

The distribution of deep sediments is shown in *Figure 6.1.*

Deep Sea Nodules

A feature of wide areas of the deep ocean floor is the presence of sizable lumps or nodules in the sediment. These are often irregular concretions, several centimeters in diameter, somewhat resembling potatoes in shape and size, but ranging from particles a millimetre or so in diameter to occasional huge lumps a metre or more across. They seem to be specially numerous in the Pacific, where parts of the bottom are almost covered with nodules, but they are also found beneath other areas of deep ocean.

In structure the nodules usually have a lamellated form surrounding a central nucleus which may be a core of silty material, or sometimes a fragment of rock, a fish tooth, or even a whale ear-ossicle. They are very variable in composition[17], but are commonly rich in manganese and iron. Mero[13] gives the average composition by weight of 30 samples of nodules from all oceans as manganese dioxide 32 per cent, iron oxides 22 per cent, silicon dioxide 19 per cent, water 14 per cent, with smaller quantities of aluminium oxides, calcium and magnesium carbonates, and other metals including nickel, copper, cobalt, zinc and molybdenum. There is now some interest in the possibilities of dredging the deep ocean floor for supplies of scarce elements.

Nodule formation in sea-water is thought to be initiated by the precipitation of colloidal particles, for example, manganese dioxide or ferric hydroxide, which tend to attract ions of other metals from the water. These particles are electrically charged, and may be attracted to electrically conductive objects on the bottom which become the nucleus of a developing nodule.

BENTHIC POPULATIONS

Communities of the Sea-floor

Using the term 'community' simply to refer to groups of species consistently occurring together in broadly similar environmental conditions, we find that different parts of the sea-floor are populated by characteristic communities. The differences of environment accountable for the association of particular communities with particular parts of the sea bottom can be related to features of both the water and the substrate.

We have already outlined the major hydrographic parameters which control the distribution of marine organisms—the temperature of the water, its composition, movements, pressure and illumination. Except on the shore and in shallow water, these vary less at the bottom than they do in the upper levels of water. They are none the less important in relation to the distribution of benthic populations, restricting certain species to particular localities and often having a major effect during the early stages of life when the majority of benthic creatures pass through pelagic phases. The planktonic eggs and larvae of many species are highly susceptible to the quality of the water in which they float, and the period prior to completion of metamorphosis is always the time of heaviest mortality. Predation and shortage of suitable food exact a heavy toll, and losses are specially severe if water temperature or salinity are unfavourable.

At abyssal depths, where temperature and salinity are uniform over great areas, hydrostatic pressure may well be the chief factor which accounts for differences between the populations within the ocean trenches and those of other parts of the deep-sea bottom. The turbidity of the water is another factor to which some benthic organisms are very sensitive. The quantity of suspended matter is often considerably greater in water close to the bottom than in layers nearer the surface. In shallow water this reduces illumination and may therefore restrict the distribution of benthic plants. High turbidity may also have adverse effects on animals by clogging the feeding apparatus or smothering the respiratory surfaces. Many benthic creatures are filter-feeders, notably the lamellibranchs which form the major part of the population of many sediments. These obtain food by drawing in a current of water from which they filter suspended food particles, and special adaptations are required to cope with the problem of separating food from large quantities of silt.

The benthic population is also influenced by the speed of the bottom current because this controls the particle size of the substrate, its oxygenation and organic content, and also affects the dispersal of pelagic larvae and the ease with which they can settle on the bottom. The bottom current is also important in the transport of food particles, sweeping them away from some areas and concentrating them in others, especially in depressions in the sea bed.

155

The substrate material exerts a dominant influence over the distribution of organisms on the sea-floor. Where the bottom is rock or large stones, the population consists chiefly of forms which live on the surface of the substrate, i.e. it is an epifauna and epiflora. The animal population usually consists of sessile and encrusting coelenterates, sponges, polyzoa, barnacles, tubicolous worms, mussels and sea squirts, and crawling among them a variety of errant polychaetes, starfish, echinoids, gastropods and large crustacea such as crabs, lobsters and crawfish. In shallow water where sufficient illumination reaches the bottom, large algae grow attached to rock or to stones heavy enough to give secure anchorage. There is usually a wide diversity of forms among the population of a rocky bottom because the irregularities of the rock surface provide a great variety of microhabitats, with innumerable differences of living space, water movement, food supply, illumination and temperature. A rocky bottom does not support a numerous infauna, i.e. animals dwelling within the substrate, but burrowing creatures occur in accumulations of silt in rock crevices, and there are also a few forms capable of boring into rock, mainly bivalve molluscs (*Hiatella*, *Pholas*, *Lithophaga*), a few annelids (*Polydora*, *Dodecaceria*), the sponge *Cliona* and certain barnacles and sea urchins. In some areas there are also species of red algae which bore superficially in calcareous rock.

Where the bottom is covered with sediment, most of the inhabitants live within the deposit. Local conditions are generally more uniform and the populations more homogeneous than on a rocky bottom. The infauna includes burrowing sea anemones, polychaetes, bivalves, gastropods, echinoderms and crustacea, and certain fish also burrow superficially in the deposit. The particle-size of a sediment is an important factor regulating the distribution of the infauna because the mode of burrowing of many creatures is specialized, and suitable only for a certain grade of substrate[15,18]. Burrowing can be done by forcing or digging through the sediment, pushing the particles aside, or eating through it, or often by a combination of methods. Large particles are more difficult to displace or ingest than small ones, and the mechanical difficulty of burrowing in coarse deposits may be one reason why these are usually less populated than finer ones. On the other hand, very fine sediment can compact into a dense, unyielding mass in which it is not easy to burrow and which requires adaptations for dealing with silt. Certain combinations of particles form thixotropic deposits which are readily reduced to a semi-fluid consistency by repeated, intermittent pressures and yield easily to burrowing. Where the deposit is exceptionally soft, such as occurs beneath some areas of very deep water, animals may simply sink into it, and here we find adaptations in the way of stalks or extremely long appendages to lift the main body clear of the bottom.

Although differences of population can often be correlated with differences of particle-size of sediments, other factors are also involved. The grade of a deposit depends upon the speed of bottom current, and this also controls several other features of the substrate. Slow-moving water allows organic matter to settle, giving a sediment that may be not only fine in texture but also rich in organic content. Poor or absent circulation of the contained water leads to deficient oxygenation of the subsurface layers, and high concentrations of sulphide. Beneath shallow water these conditions often support a large population because there is a good food supply for creatures which feed on the surface or digest organic matter from the sediment, but the infauna must be able to cope with silt and a deoxygenated medium. Where the bottom water moves more swiftly there is likely to be less settlement of food and a lower organic content in coarser sediments, but better oxygenation of the interstitial water. The poorer food supply supports a smaller population, but these conditions favour animals which can burrow in coarse material and capture floating food suspended in the water. Therefore, several interrelated factors must operate to limit certain species to particular substrates.

During the second decade of this century, pioneer studies of marine benthos were made by Petersen, who carried out detailed investigations by grab samples of the larger animals (the macrofauna) of soft deposits in shallow water off the Danish coast. He found that different areas supported characteristic associations of animals, and he distinguished nine communities, naming each after the most conspicuous components of the population as follows:

(1) *Macoma* communities, widespread in shallow muds. (*M. baltica, Mya arenaria, Cardium edule*, etc.)

(2) *Syndosmya* communities in shallow, muddy sands, often in sheltered creeks. (*S. alba, S. prismatica, Macoma calcarea, Astarte spp*, etc. sometimes with *Echinocardium cordation*.) (*Syndosmya = Abra*, pages 158, 159.)

(3) *Venus* communities, widespread on shallow sandy bottoms on open coasts. (*V. striatula, Tellina fabula, Montacuta ferruginosa*, etc., often with *E. cordatum*.)

(4) *Echinocardium-filiformis* communities, in sandy mud at intermediate depths. (*E. cordatum, Amphiura filiformis, Turritella communis, Nephtys spp.*, etc.)

(5) *Brissopsis-chiajei* communities on deeper, soft mud. (*B. lyrifera, Amphiura chiajei, Abra nitida*, etc.)

(6) *Brissopsis-sarsi* communities on soft mud below *Brissopsis-chiajei* depths (*B. lyrifera, Ophiura sarsi, A. nitida*, etc.)

(7) *Amphilepis-Pecten* communities on deep mud in Skagerrak (*Amphilepis norvegica, Chlamys (Pecten) vitrea*, etc.)

(8) *Haploops* communities on deep, firm mud in Kattegat (*Haploops tubicola, Chlamys septemradiata*, etc.)

157

(9) Deep *Venus* communities, widespread on coarse sands (*V. striatula, Spatangus purpureus, Echinocardium flavescens, Spisula spp.*, etc.)

Following Petersen's work, investigations over wider areas have shown that in middle latitudes certain dominant genera occur in deposits of similar grade and depth in widely separated parts of the sea, with differences of species associated with temperature and salinity. N. S. Jones[10] has concluded that the north-east Atlantic area contains the following categories of benthic communities, limited to particular substrates and depths, and ranges of temperature and salinity.

(A) Shallow water and brackish communities with upper limits of distribution extending on to the shore. Eurythermal and euryhaline within wide limits. Temperature range 3–16°C, salinity 7–34 per mille, but always exposed to periodical diminution below 23 per mille.

(I) Shallow Soft Bottom Community

(*a*) *Boreal shallow sand association*—occurring on relatively exposed coasts in north-west Europe. Important species include *Arenicola marina, Tellina tenuis, Donax spp., Nephtys caeca, Bathyporeia pelagica, Yoldia spp.*

(*b*) *Boreal shallow mud association*—equivalent to Petersen's *Macoma* community, occurring on more sheltered coasts of north-west Europe, and in estuaries and in the Baltic.
Arenicola marina, Macoma baltica, Mya arenaria, Cardium sp., Corophium volutator.

(II) Shallow Hard Bottom Community

(*a*) *Boreal shallow rock association*—shores and shallow water with rocky substrates.
Balanus balanoides, Mytilus edulis, Littorina spp., (Patella vulgata, Nucella lapillus —Europe).

(*b*) *Boreal shallow vegetation association*—algae of the shore and shallow water with associated fauna.
Hyale prevosti. Idotea spp., Hippolyte varians, Littorina spp., Rissoa spp., Lacuna vincta.

(B) Offshore communities having upper limits of distribution below extreme low-water spring tidal level. Eurythermal and euryhaline, but between narrower limits than A. Temperature range 5–15°C, salinity 23–35·5 per mille.

(I) Offshore Soft Bottom Community

(*a*) *Boreal offshore sand association*—equivalent to Petersen's *Venus* community, occurring offshore on sandy bottom in the north-east Atlantic.
Sthenelais limicola, Nephtys spp., Ampelisca brevicornis, Bathyporeia guilliamsoniana, Dosinia lupinus, Venus striatula, Tellina fabula, Gari fervensis, Abra prismatica, Ensis ensis, Echinocardium cordatum.

(*b*) *Boreal offshore muddy sand association*—equivalent to Petersen's *Echino-cardium-filiformis* community, occurring on muddy sand offshore and in modified form in sheltered and estuarine situations around Europe.
Nephtys incisa, Goniada maculata, Lumbriconereis impatiens, Pectinaria auricoma, Diplocirrus glaucus, Eumenia crassa, Scalibregma inflatum, Notomastus latericeus, Owenia fusiformis, Ampelisca spinipes, A. tenuicornis, Nucula turgida, Dosinia lupinus, Abra alba, A. prismatica, A. nitida, Cyprina islandica, Cardium echinatum, C. ovale, Cultellus pellucidus, Spisula subtruncata, Corbula gibba, Dentalium entalis, Turritella communis, Aporrhais pes-pelicani, Philine aperta, Ophiura texturata, Amphiura filiformis, Echinocardium cordatum, E. flavescens, Leptosynapta inhaerens.

(*c*) *Boreal offshore mud association*—equivalent to Petersen's *Brissopsis-chiajei* community, occurring on soft mud, usually outside the muddy sand association but gradually grading into it.
Leanira tetragona, Nephtys incisa, Glycera rouxi, Lumbriconereis impatiens, Maldane sarsi, Notomastus latericeus, Eudorella emarginata, Calocaris macandreae, Nucula sulcata; N. tenuis, Abra nitida, A. prismatica, Ampiura chiajei, Brissopsis lyrifera.

(II) Offshore Hard Bottom Community

(*a*) *Boreal offshore gravel association*—occurring in Europe at moderate depth wherever the deposit is very coarse, whether sand, gravel, stones or shells. Masses of *Modiolus modiolus* sometimes form a numerous epifauna.
Polygordius lacteus, Glycera lapidum, Potamilla spp., Serpula vermicularis, Crania anomala, Balanus porcatus, Galathea spp., Eupagurus spp., Hyas coarctatus, Nucula hanleyi, Glycimeris glycimeris, Lima loscombi, Venus casina, V. fasciata, V. ovata, Vencrupis rhomboides, Gari tellinella, Spisula elliptica, Modiolus modiolus, Buccinum undatum, Asterias rubens, Ophiothrix fragilis, Ophiopholis aculeata, Echinus spp., Echinocyamus pusillus, Echinocardium flavescens, Spatangus purpureus.

(**C**) **Deep communities** with upper limits of distribution not above 70 m depth and usually much lower. Stenothermal and stenohaline. Temperature range 3–7°C, salinity 34–35·5 per mille.

(I) Deep Soft Bottom Community

(*a*) *Boreal deep mud association*—equivalent to Petersen's *Brissopsis-sarsii* and *Amphilepis-pecten* communities, occurring below the offshore mud association.
Glycera alba, Spiophanes kroyeri, Chaetozone setosa, Maldane sarsi, Clymene praeter-missa, Sternaspis scutata, Notomastus latericeus, Melinna cristata, Proclea graffi, Eriopisa elongata, Nucula tenuis, Nuculana pernula, Chlamys vitrea, Thyasira flexuosa, Abra nitida, Portlandia lucida, Cardium minimum, Ophiura sarsi, Amphilepis norvegica, Brissopsis lyrifera.

(II) Deep Hard Bottom Community

(*a*) *Boreal deep coral association*—a little known, deep-water epifauna of corals and associated animals.
Lophohelia prolifera, Paragorgia arborea, and *Gorgonocephalus caput-medusae.*

The dominance of particular species, which is so evident in marine benthos in the temperate oceans, is less apparent at low latitudes where the composition of communities generally shows a greater diversity. This diversification may be a feature of more mature communities which have evolved in stable conditions over a long period, permitting the survival of species specialised for narrow ecological niches. The communities of the temperate oceans may be regarded as relatively immature, having evolved in fluctuating conditions since the extremes of the Pleistocene period. This favours the evolution of polymorphic populations which survive by virtue of their wide variability, with dominant species occupying broad ecological niches.

Studies of benthic communities have mostly been confined to examination of the larger animals, or *macrobenthos*. These communities also contain many smaller forms, the *meiobenthos*[12] and *microbenthos*, about which relatively little is known. Numerous small organisms just large enough to be seen by the unaided eye comprise the meiobenthos, including foraminifera, turbellarians, nematodes and various small polychaetes, bivalves and crustacea. The smallest organisms, visible only with a microscope, comprise the microbenthos, which includes bacteria, a great variety of protozoa, mainly flagellates, ciliates and amoebae, and often other small organisms such as rotifers, crustacean larvae and the smallest nematodes. In shallow water there is often a microflora of diatoms and coloured flagellates. Organisms small enough to live in the interstices between the grains of sediment are described as an *interstitial* fauna or flora.

Although inorganic factors exert a major control, there are also biological factors which influence the distribution and composition of benthic communities. The physical and chemical features of the environment determine a range of species which compete, but success or failure in a particular habitat depends ultimately on qualities inherent in the organisms themselves. For example, they are able to some extent to choose their position. Free-living forms can move about to find areas that suit them, but the majority of adult benthic animals remain more or less stationary, confined within burrows or attached to the bottom. Most of these start life as pelagic larvae dispersed by the water[14,15,19]. Once they settle and metamorphose they stay in place, and die if conditions are unsuitable. Undoubtedly there are great losses, but the larvae of many species show behavioural features which influence dispersal and favour their chances of reaching situations where survival is possible. Many species of larvae have some control over the depth at which they float, often by virtue of their response to light. Larvae of shallow water species are usually photopositive for a time, collecting near the surface, while those of deeper-dwelling forms mostly prefer dim illumination or darkness, and therefore occupy deeper levels. The depth at which the larvae float must obviously influence the depth at which they settle.

It has also been demonstrated that larvae of certain species can discriminate between different substrates, and have some powers of selection;

for example the larvae of the small polychaete *Ophelia bicornis*, studied by Wilson[19]. This worm has a patchy distribution in various bays and estuaries around the British coast, living in a particular type of clean, loose sand. Pelagic larvae are produced which are ready to metamorphose when about five days old. At this stage they begin to enter the deposit, and Wilson discovered that the larvae are able to distinguish between sands from different areas, preferring to complete their metamorphosis in contact with certain 'attractive' sand samples, and avoiding contact with other sands which have a 'repellent' effect. When the larvae settle, they appear to explore the deposit. If the sand is of the 'repellent' type, they leave it and swim away, shortly settling again and repeating their exploration. This behaviour continues over a period of several days, with metamorphosis delayed until a suitable substrate is found. If the larvae find nothing suitable, they eventually attempt to metamorphose none the less, but then usually die. During an early series of experiments, the particle size of the sand seemed to be the main factor to which the larvae were sensitive, but further experiments with artificially-constituted sands showed the importance of other factors, notably the coating of organic materials and bacteria on the surface of the sand grains. If sand samples were washed in hot concentrated sulphuric acid, they became neutral, losing their attractive or repellent qualities.

Selective examination of the substrate, with metamorphosis delayed as long as possible until a suitable substrate is discovered, has now been demonstrated for a fairly wide range of organisms, including annelids, barnacles, molluscs[4], polyzoa and echinoderms. A number of different properties of the bottom have been shown in particular cases to influence the choice, including slope, contour, texture, colour, particle size and chemical nature. The bacterial covering of the substrate often has an important influence on settlement, although it remains uncertain why. Probably in some cases the bacteria are the first food of the newly-settled larvae. In certain species, for example *Mytilus*, the presence of adults or even old byssus on the substrate stimulates attachment in the immediate vicinity. In some cases the successful settlement of a few larvae makes the substrate attractive to others. Gregarious settlement obviously facilitates the maintenance of the population in suitable locations. The attractive influences appear generally to be due to chemical agencies which stimulate settlement and metamorphosis. Although in some cases this may operate by aqueous diffusion, it has been shown for several species that the larvae must first make actual contact with a particular surface such as the shell, cuticle or cementing substance of their own kind before settlement is attempted, the stimulus for settlement being dependent upon a 'tactile chemical sense.'

Once the larvae have settled, many other biological factors begin to influence their chances of survival. A community is a society of organisms with many interactions between the individuals. There is often competition for living space, and the outcome of this aspect of the struggle for existence

161

is influenced by the breeding periods and reproductive capacities of different species, and by relative growth rates. One species may gain advantage by early settlement following a seasonal decline in numbers of the community, or a fast-growing species may oust a slower-growing competitor by over-growing it, or by claiming an increasing proportion of a shared food-source. Interaction between predator and prey must also regulate the numbers of both. Mortality due to predation is usually highest during the period following settlement while the individuals are still small. Certain predators, notably ophiuroids, are sometimes so numerous that they virtually carpet the bottom and it seems surprising that any small creatures suitable for food can escape. In these conditions, survival may depend upon the time of settlement. It has been observed[16] that some benthic carnivores have phases when feeding diminishes or ceases, usually in association with breeding. This passive period may last several weeks, and some species which settle during this time may be able to reach a sufficient size to become relatively safe from predation before their enemies start active feeding again. The composition of a community may therefore reflect coincidences between passive periods of predators and settlement periods of other species.

The individuals of a community are in various ways interdependent, and some organisms thrive only in the presence of particular associated forms. Each type of animal is dependent upon other organisms for food, and the quantity and quality of food sources obviously exerts a profound control over numbers and composition of communities. Certain organisms depend upon others to provide surfaces of attachment upon which they can grow, and in some cases are seldom found elsewhere; for example, *Pyrgoma anglicum* on *Caryophyllia smithii*, *Adamsia palliata* on *Pagurus prideauxi*. Some animals share the burrows formed by others; for example, *Montacuta ferruginosa* and *M. substriata* in the burrows of heart urchins, and *Lepidasthenia argus* with *Amphitrite edwardsi*. *Harmothoe lunulata* may be free-living but often shares the burrows of *Arenicola marina*, *Amphitrite johnstoni*, *A. edwardsi*, *Acrocnida brachiata*, *Leptosynapta inhaerens*, *Labidoplax digitata* and others. One species may even live inside another; for example, the shrimp *Typton spongicola* within the sponge *Desmacidon*, and the barnacle *Acasta spongites* embedded in the sponge *Spongelia fragilis*.

Relationships between species certainly involve numerous associations of an epizoic, commensal, symbiotic or parasitic nature. A striking example of a group of species commonly found living in close proximity is associated with the hermit crab *Pagurus spp.*, widely distributed on gravelly deposits in shallow water around the British Isles and the U.S. The adult crab inhabits shells of a whelk or small snail. Several animals are commonly found on the surface of the shell, the most conspicuous in the U.K. being the anemone *Calliactis parasitica*. Other inhabitants of the outer shell surface, sometimes extending into the opening, include saddle oysters or slipper shells, the hydroid *Hydractinia echinata*, the serpulid worms *Spirorbis spirillum*, and *Hydroides spp.*, the

BENTHIC POPULATIONS

barnacle *Balanus spp.* and the sponge *Suberites spp.* In Europe the barnacle *Trypetesa lampas* apparently lives only in *Buccinum* shells inhabited by hermit crabs, where it burrows into the shell substance just inside the shell aperture. The shell is also sometimes bored by the worm *Polydora*. Living within the shell alongside the hermit crab there is frequently a large worm, *Nereis fucata*, which feeds on fragments of food dropped by the crab. Also inside the shell the tiny crab *Porcellana longicornis* is sometimes found. The hermit crab may carry parasites; for example, the isopod *Athelges paguri* in the branchial chamber, and the parasitic barnacle *Peltogaster paguri* extruding from the abdomen.

Beneath shallow water, benthic communities show seasonal and annual fluctuations, but over the long term they usually remain fairly constant in numbers and composition, indicating that the major factors moulding the community are stable. On occasions when rapid, permanent changes of population occur, these are usually associated with alterations of the environment, often the result of human activity; for example, a change of water circulation or temperature connected with industrial installations, or a change of the sea-floor due to dredging or the dumping of waste at sea[9]. Major changes may also follow the introduction of a new species to an area. Otherwise, the continual minor fluctuations are largely self-cancelling. Any tendency for one species to increase is eventually counteracted by increasing competition and predation so that a dynamic equilibrium is maintained and the natural balance of the community is preserved.

Over very long periods, climatic or geological changes may slowly alter the environment, and there are also gradual and permanent modifications brought about by the activities of the organisms themselves; for example, the substrate becomes changed by accumulations of shells or skeletons or through the erosion of rocks and stones by the boring of various plants and animals, and the composition of the water is influenced by biological processes of extraction, precipitation and secretion. The continual interactions between habitat and community lead very gradually to changes in both, the ecosystem comprising environment and population being a composite evolving unit.

The Food Supply

The food supply of benthic animals derives, directly or indirectly, almost entirely from particulate matter sinking from the overlying water. There is very little primary production of food on the sea-bed because green plants can grow only where there is sufficient light for photosynthesis. Vegetation on the sea bottom is therefore limited to shallow water. Large algae produce a lush growth on and near the shore, especially in middle latitudes, and form a primary food source which supports many omnivorous and vegetarian animals, and contributes quantities of organic debris to the local sediments. This vegetation seldom extends deeper than some 40–60 m and is confined to areas of rocky bottom, or stones large enough to provide secure attachment for the plants. Rock and sediment in shallow water may also be covered with a thin microfloral film, mainly diatoms and other unicellular algae. Surface layers of sediments contain large numbers of bacteria, and traces of food are also produced by chemosynthetic species capable of metabolizing inorganic compounds.

Particulate food sinking through the water reaches the bottom in great variety. There is sometimes an appreciable amount of vegetable matter derived from the land, and even in very deep water dredging has disclosed a surprising quantity of terrestrial material in the form of fragments of wood and leaf. Much of this may be carried into the deep ocean basins by turbidity currents flowing down the continental slope. However, in most areas the greater part of the food supply consists of the remains of pelagic organisms. In shallow water, the major component is usually planktonic diatoms and other microscopic plants, and the abundance of this food depends upon the rate of surface production. Seasonal changes in the quantity of phytoplankton produce fluctuations in the supply of food to the benthos. In temperate latitudes the numbers of planktonic diatoms reaching the bottom may be over a hundred times greater during the summer months than in winter, and there are associated changes in the weight of benthic populations.

There are broadly four ways in which benthic animals gather food; they filter suspended particles from the water, they collect food particles which settle on the surface of the sediment, they obtain nutriment from the organic material which has become incorporated in the deposit, or they prey upon other animals. Many of course take food from several sources. For example, *Crangon* feeds largely on surface debris but also preys on small animals; *Cancer* is omnivorous, taking a wide variety of plant and animal foods, *Galathea* takes large pieces of animal and vegetable matter and also uses the setae on its maxillipeds to filter micro-organisms and debris from the bottom deposit. However, in many other cases the feeding mechanism is adapted to deal with a particular source of food. Suspension feeders include sponges, ascidians, polyzoa and certain polychaetes (*Chaetopterus*), crustacea (*Balanus*), gastropods (*Crepidula*), many lamellibranchs (*Cardium, Venus, Dosinia, Gari*), and some holothurians (*Cucumaria*). Surface deposit feeders include polychaetes (*Terebella, Amphitrite*), and lamellibranchs (*Tellina, Abra*). Feeding on organic matter within the sediment are polychaetes (*Arenicola*), heart urchins (*Spatangus, Echinocardium*) and holothurians (*Leptosynapta*). Benthic predators include errant polychaetes (*Nephtys, Glycera*), the majority of actinians, many crabs, some gastropods (*Natica, Scaphander, Buccinum*), starfish and ophiuroids (*Asterias, Amphiura*) and many fish, including most of the commercially-fished species.

Depending on the availability of food, the biomass (weight of living material per unit area) of the sea-bed varies from place to place. Biomass may be expressed as 'rough weight' of fresh material, with or without shells, or more accurately as a 'dry weight' of organic material obtained after the removal of shells and drying of the remaining tissue to constant weight, allowance being made for any inorganic material present in the residue. Petersen found dry weights between 1 g and 35 g/m² beneath shallow water off the Danish coast. Holme[6,8] found a mean dry weight of about 11 g/m² for the biomass of the macrobenthos at 20 stations in the English Channel off Plymouth, the percentages of different animal groups being as follows:

Group	Percentage
Lamellibranchs	35·46
Polychaetes and Nemertines	25·79
Echinoderms	17·8
Crustacea	10·51
Coelenterates	7·00
Polyzoa	2·46
Protochordates	0·88
Gastropods	0·02

In areas of exceptional productivity, biomass dry weights as high as 100–150 g/m² occur.

Whereas in shallow water much of the food supply of the sea bottom comes directly from the surface layers, it is unlikely that much surface plankton reaches the bottom in very deep water because most of it is consumed by pelagic organisms on the way down. Between the productive surface layers and the deeper parts of the ocean there is a food pyramid with many links in the food chain, and the quantity of food available to support the population at deep levels can be only a very small fraction of the surface production. Knowledge of the distribution of deep sea benthos is scanty, but population studies in deep water generally indicate that the benthos is sparse and that the biomass decreases with depth. Broadly, the deeper the water and the further from land, the smaller the weight of animals on the sea bottom.

Occasionally, even at the greatest depths, large pieces of meat, the remnants of bodies of whales or very large fish, must reach the sea bottom. But almost all the organic material of surface origin that eventually settles on the bottom in very deep water must consist of structures which cannot be digested by pelagic animals during descent; for example, cell walls, shells and skeletons. Pieces of wood and other materials from the land also find their way to the deep sea-floor, sometimes carried off the continental shelf by turbidity currents (*see* page 148).

Further decomposition of these materials depends upon the activities of bacteria. These occur in the superficial layers of deposit even in the deepest water, and constitute an essential link in the organic cycle. Their food is drawn from many sources. Most of them are heterotrophic, obtaining energy from oxidation of organic compounds, but they can act upon a far wider range of materials than animals are able to digest, including chitin, keratin, cellulose and lignin. Even at the deepest levels these substances are likely to reach the sea-floor, and here they are digested and assimilated by bacteria. In this way, organic materials which animals cannot make use of directly are transformed into bacterial protoplasm, and in this form become assimilable by numerous animals which feed on bacteria.

Although populations of deep sediments are small compared with those of most shallow areas, they appear none the less to have a somewhat greater biomass than would be expected from food chains originating only at the surface. Dry weights of some 1–2 g/m² of animal material occur on the bottom at depths of 5,000–8,000 m. These values are not so much less than mean values in shallow areas, and raise the question of whether some sources of food are available in deep water in addition to particulate matter sinking from above. One possibility is the utilization of organic materials dissolved in the water. Although very low in concentration, the total amount of these solutes in the sea is extremely large (pages 83–87). The phytoplankton includes some obviously saprophytic forms, mainly colourless flagellates, and many chlorophyll-containing plants may also live by saprophytic feeding in some circumstances. Certain groups of plants, notably coccoliths and blue-green algae, have been found far below the illuminated layers, sometimes as deep as 4,000 m, in such large numbers as to suggest that they can live and multiply there, presumably feeding as saprophytes. Bacteria, too, may draw some of their nutriment from organic constituents of the water, which may well comprise a significant part of the food supply of deep levels. The bacterial population also includes chemosynthetic autotrophs which do not require organic material, but are able to obtain energy by oxidizing inorganic compounds such as ammonia, nitrites and sulphides. These forms can function as primary producers of organic matter below the photosynthetic zone. There is also some addition to the biomass of deep water from larval and juvenile stages which develop nearer the surface, drawing on the more abundant food available there before descending to deeper levels.

The food relationships of benthic communities are complex and not well known, but it is evident that bacteria are an extremely important source of food for the animal population at all depths[11]. The microbenthos includes many flagellates. amoebae and ciliates which feed mainly on bacteria; some also ingest minute particles of organic debris, and the larger flagellates and amoebae can probably take whole diatoms. For the meiobenthos and macrobenthosm, bacteria comprise an appreciable part of the food of many suspension and surface-deposit feeders. Animals which obtain food from the sediment presumably derive a significant amount of their nutriment from the bacteria and protozoa it contains. Some interconnections of the marine food-web are illustrated in *Figure 6.2*.

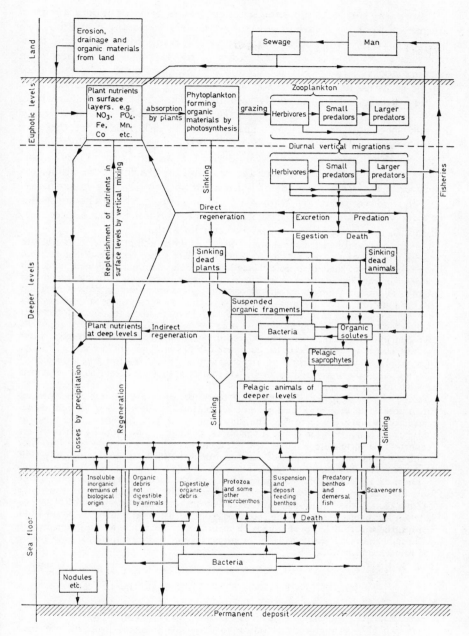

Figure 6.2. *Some connections of the marine food web*

167

Suspension, surface deposit and sediment feeding animals of the sea bottom are preyed upon by a variety of carnivores, including many of the food fish highly prized by man. Studies of biomass and composition of benthic communities therefore give some indication of the extent to which areas can support stocks of fish. The production of animal material on the sea-floor can be regarded as 'useful' or 'wasteful' according to whether it contributes to the formation of commercially-valuable species of fish or other creatures of no value to man. For example, starfish and brittle-stars are not used as human food and are relatively unimportant as fish food; but they are carnivores and compete with fish for the same sort of prey, especially molluscs. The growth of these predatory echinoderms is a wasteful type of production.

The invertebrate predators of the sea bottom are far more numerous, comprise a far greater weight of living material, and have a much higher rate of food consumption than the bottom-feeding fish. Thorson[16] has reviewed what is known of feeding habits and food requirements of predatory benthos in the north Atlantic area, and concludes that:

'it seems reasonable to assume that most bottom-dwelling fishes in temperate waters will consume an average portion of food per day corresponding to from 5 to 6 per cent of their own living weight during the summer half-year, and will reduce this rate significantly when the temperature decreases. During the coldest months of the year they may almost completely cease to feed . . . Summarizing the data for invertebrate predators, we find that at very young stages they are extremely voracious, taking an amount of food corresponding to about 25 per cent of their own living weight per day. Those species which remain active as adults and continue to increase in size do require somewhat less food although their average consumption is about 15 per cent of their living weight per day. Finally, those predators which nearly or totally stop growing when mature and are sluggish will in their adult phase, reduce their food claims to such an extent as to be cheap or fairly cheap to run.

'On an average, growing invertebrate predators seem to consume four times as much food per day and unit weight as bottom-dwelling fishes, which seems reasonable when we realize that the life cycle, or at least the time from birth to maturity, is much shorter in the invertebrate predators than in most of the fishes. A species of invertebrate predator will often produce three or more generations (i.e. build up three or more units of meat) while a flounder is producing only one generation (i.e. build up one unit).

'On the basis of our information that invertebrate predators consume food about four times as fast as the flounders, we must recognize the amazing fact that only 1–2 per cent of the "fish food" on the sea bottom is actually eaten by fish; the rest is taken by invertebrates. If the standing crop of "fish food" is increased for some reason, the invertebrate predators with their shorter life cycles and quicker growth will furthermore be ready to take the advantage of this circumstance long before the fishes are able to do so.'

We shall refer to these estimates of Thorson's later (pages 274 and 292).

REFERENCES AND FURTHER READING

Books

[1] Sverdrup, H. U., Johnson, M. W. and Fleming, R. H. (1946). *The Oceans.*
    'Chap. 17. Animals in Relation to Physical-Chemical Properties of the Environment.

Rocky Beach

This ecological zone contains many crevices and crannies in which various animals and plants may find their homes. They are thus protected from the full force of the surf and whether they are found on a man-made breakwater, or a natural shore front, the same types of plants and animals are likely to be discovered. Their presence is determined by the tides, a hard substrate, and the protection from the surf. These are the main features defining their world.

The herring gull (Larus argentatus) often colonizes the upper beach during their early summer breeding season.

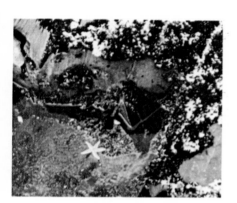

Above: Intertidal rocks give a hard substrate for barnacles and mussels to attach, fed on by starfish.

Above right: A simple nest of shore-strewn debris holds two speckled gull eggs.

Right: Rock weed (Fucus sp.) rims an intertidal pool where temperature and salinity vary as the tide shifts.

The Marsh

This area is a rich nursery for many kinds of animals which grow and then leave for open coastal waters. The protected and quiet marsh is supplied with large amounts of plants that help support the animal life. Marshes are ecological resources particularly sensitive to man.

Above: Well defined parallel zones of vegetation types are found in tidal marshes. Zones are caused by the tide and minor differences in land elevations. Upland trees and shrubs lead down to salt meadow grass (Spartina patens) which leads down to salt marsh grass (Spartina alterniflora).

Left: The mud snail (Nassa obsoleta) lives in the top silty layer of the marsh. Equipped with a snorkle, he pumps clean water from above the silt in order to breathe.

Sandy Beach

Beach sands present an ever shifting substrate relatively free of larger organisms. Smaller animals generally bury themselves when the tide passes, but when covered with water, they dig out and feed on the food carried to them by the tide. Rocks get worn by the slurry of water and sand cast over them.

Currents along the shore carve scalloped beaches between resistant head-land rocks.

Left: Waves break on a beach and batter any exposed animal or plant with a slurry of sand.

Right: Seabeach Sandwort (Arenaria peploides) is found above the high tide mark.

Far right: Cactus-like fleshy leaves serve Sandwort to store water.

The fiddler crab, Uca pugilator, is a scavenger seen perched on its burrow.

The ribbed mussel, Modiolus demissus, lives embedded in the marsh mud and feeds when the tide covers it. It filters microscopic food organisms from the water and its pumping prevents silt from burying them.

Right: Mud flats within the marsh are rich with microscopic plants and animals growing on their surface.

The red mangrove (Rhizophora mangle) has seeds which germinate on the tree as seen here.

Anemones often are on the roots.

Callinectes danae, the blue crab.

The microscopic hydroid Pennaria.

The Mangrove

Mangroves are found only in warm climates and are tidal, forming an interface between sea and land. Like the marsh, they contain vegetation zones and the red mangrove is nearest the sea level. Its roots harbor many organisms (Right).

Above: Pencil-sized growths, called pneumatophores, rise from the roots of the black mangrove by which it "breathes."

Right: The tangled growth of mangrove is home to many types of birds and other organisms.

REFERENCES AND FURTHER READING

Chap. 18. Interrelations of Marine Organisms.
Chap. 20. Marine Sedimentation.' New York; Prentice-Hall
[2] Hill, M. N. (Editor) (1963). *The Sea.* 'Vol. 2.
Chap. 19. Communities of Organisms.' London; Allen and Unwin
[3] Hardy, A. C. (1959). *The Open Sea*, 'Part 2. Fish and Fisheries.'
Chap. 5. Benthos—Life on the Ocean Floor.
Chap. 6. More about the Benthos, and its Relation to the Fisheries. London; Collins

Papers

[4] Bayne, B. L. (1969). 'The Gregarious Behaviour of the Larvae of Ostrea Edulis at Settlement.' *J. mar. biol. Ass. U.K.* **49,** 327

[5] Buchanan, J. B. (1966). 'The Biology of Echinocardium Cordatum from Different Habitats.' *J. mar. biol. Ass. U.K.* **46,** 97

[6] Holme, N. A. (1953). 'The Biomass of the Bottom Fauna of the English Channel.' *J. mar. biol. Ass. U.K.* **32,** 1

[7] Holme, N. A. (1961). 'The Bottom Fauna of the English Channel.' *J. mar. biol. Ass. U.K.* **41,** 397

[8] Holme, N. A. (1966). 'The Bottom Fauna of the English Channel.' Part II. *J. mar. biol. Ass. U.K.* **46,** 401

[9] Howell, B. R. and Shelton, R. G. J. (1970). 'The Effect of China Clay on the Bottom Fauna of St. Austell and Mevagissey Bays.' *J. mar. biol. Ass. U.K.* **50,** 593

[10] Jones, N. S. (1950). 'Marine Bottom Communities.' *Biol. Rev.* **25,** 283

[11] Mare, M. F. (1942). 'A Study of Marine Benthic Communities with Special Reference to the Micro-organisms.' *J. mar. biol. Ass. U.K.* **25,** 517

[12] McIntyre, A. D. (1964). 'Meiobenthos of Sublittoral Muds.' *J. mar. biol. Ass. U.K.* **44,** 665

[13] Mero, J. L. (1960). 'Minerals on the Ocean Floor.' *Scient. Am.* October

[14] Reese, E. S. (1964). 'Ethology and Marine Zoology.' *Oceanogr. Mar. Biol. Ann. Rev.* **2,** 455

[15] Thorson, G. (1950). 'Reproductive and Larval Ecology of Marine Bottom Invertebrates.' *Biol. Rev.* **25,** 1

[16] Thorson, G. (1960). 'Parallel Level-bottom Communities, their Temperature Adaptation, and their "Balance" between Predators and Food Animals.' p. 67, *Perspectives in Marine Biology.* Ed. by Buzzati-Traverso, A. Univ. of California Press

[17] Tooms, J. S. and Summerhayes, C. P. (1969). 'Geochemistry of Marine Phosphate and Manganese Deposits.' *Oceanogr. Mar. Biol. Ann. Rev.* **7,** 49

[18] Trueman, E. R. and Ansell, A. D. (1969). 'The Mechanisms of Burrowing into Soft Substrate by Marine Animals.' *Oceanogr. Mar. Biol. Ann. Rev.* **7,** 315

[19] Wilson, D. P. (1956). 'Some Problems in Larval Ecology Related to the Localized Distribution of Bottom Animals.' p. 87, *Perspectives in Marine Biology.* Ed. by Buzzati-Traverso, A. Univ. of California Press

169

CHAPTER 7

THE SEA-SHORE

Populations of the sea-shore face a number of special difficulties. There are problems caused by alternate submergence by water and exposure to air. There are usually additional difficulties due to the breaking of waves. In the shallow water along the coast the physical and chemical conditions are less stable than in deep water; in particular, there are wider and more rapid changes of temperature, and fluctuations of salinity associated with evaporation or fresh water dilution. Often the inshore water is also very turbid because quantities of suspended matter are churned up from the bottom by waves or carried into the sea by rivers.

On most shores the ecological conditions are dominated by the tides and the waves.

TIDES

Although the behaviour of tides is very complex, the underlying cause of their motion has been understood since Newton (1647–1727) accounted for tides as due to the gravitational attraction of the moon and the sun upon the oceans.

Considering the lunar effect alone, the tide-generating forces are the resultants of the moon's gravitational pull on the water and the centrifugal force due to the revolution of the earth–moon system. The centrifugal force on any small mass of water is the same at any point on the earth, and is equal to the moon's pull on an equal small mass at the earth's centre of gravity. Gravitational force varies inversely with the square of the distance at which it acts, and is therefore greater than the centrifugal force on the side of the earth facing the moon, and less than the centrifugal force on the side away from the moon. The horizontal components of the tide-generating forces therefore tend to move the water towards two points, one immediately below the moon and one in the same line on the opposite side of the earth (*Figure 7.1*). If the earth were covered with water, this would distort the water-layer to produce two tidal bulges at A and B, where the tide would be high. Low tide would be at positions half-way between A and B, such as C and D, where the moon would be on the horizon. During the earth's rotation the tidal bulges would remain stationary relative to the moon, and at any point on the earth's surface there would be a diurnal cycle of alternate high tide, low tide, high tide and low tide within the period of the lunar day, i.e. 24 h 50 min. This corresponds fairly closely with what is observed in many parts of the seas. The lunar effect on the tides varies slightly from day to day with changes in the declination of the moon and its position in its elliptical orbit.

The sun's effect on the tides is less than the moon's because of its greater distance from the earth, but is sufficient to exert an appreciable modifying influence. When the moon is new or full, the pull of the sun on the water is in nearly the same line as that of the moon. The combined pull of sun and moon then causes the specially high and low tides known as *spring tides*. At the moon's first and last quarters the sun pulls at right-angles to the moon, reducing the height of the tidal bulges. There is then less difference in the levels of high and low water, and these tides of reduced range are termed *neap tides*. The height of the tides therefore varies daily with the phases of the moon, spring and neap tides each recurring twice in every 28-day lunar cycle. The solar effect on the tides also varies with the sun's declination, being greatest at the equinoxes. The spring tides then have their maximum range and the neaps their minimum.

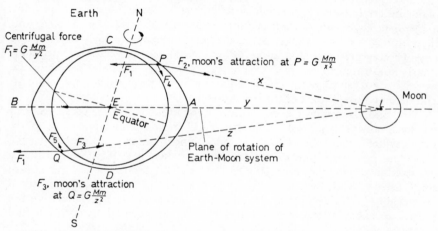

Figure 7.1. Tide-generating forces at P and Q are F_4 and F_5, the resultants of F_1, F_2 and F_1, F_3; M = mass of moon. m = point mass of water at P or Q. G = gravitational constant. E = earth's centre of gravity. L = moon's centre of gravity. x = distance PL. y = distance EL. z = distance QL

Many of the complexities of tidal behaviour arise because the oceans do not completely cover the earth, but are broken up by land. The oceans comprise many interconnected bodies of water, each with a natural period of oscillation, and the effect of tide-generating forces is to set a number of oceanic areas into complex tidal oscillations. The number and dimensions of these oscillating systems is not known with certainty. Within each system the oscillations tend to be deflected by the earth's rotation, causing them to swing around a centre known as the amphidromic point. Tidal oscillations are believed to occur most strongly in areas where the natural oscillation period is closely in phase with the rhythm of the tide-generating forces. Areas of little tidal movement, for example the Mediterranean, are presumably out of phase with the tidal period. A detailed discussion can be found in *The Tides—Pulse of the Earth* by E. P. Clancy[3].

Details of tides in the open ocean are not well known because of the difficulties of measurement. Along the shore the extent of tidal movement is determined partly by the shape of the coastline. In tapering channels, where the tide enters a wide mouth and moves forwards between converging coastlines, the height of the tide is increased by the constriction of the water between opposite shores. An example is the Bay of Fundy, where the tidal range sometimes exceeds 12 m. The average range of tides around the United States of America is about 3 m.

The coastal waters around the land masses of the world are set in oscillation by the rise and fall of the oceanic tides. The specific height to which a tide rises along a particular coast is, in large part, dependent on the configuration of the underlying continental shelf and shape of the shore line. The submarine and subaerial geology of the coast may cause the tide to be constricted, and thereby to amplify its range. Even geographic locations fairly close to one another may present remarkably different tidal ranges, and such variations are demonstrated in *Figure 7.2*. The range at any particular sector of the world's coast represents a physical parameter of the natural environment, and defines a major part of the sea-shore ecology.

Wind and atmospheric pressure have some influence on sea level, and can produce unexpected tidal anomalies. Of special importance are storm surges which occur when strong on-shore winds pile up the water along the coast and cause the tide to rise to abnormal heights. A striking instance occurred in recent years along the east coast of England, when a great storm surge in the North Sea during the night of 31st January 1953, resulted in remarkably high tides, up to 3 m above predicted levels, overwhelming the coastal defences in many areas and causing widespread devastation and flooding.

Tidal Levels on the Shore

The intertidal zone is that part of the sea bottom over which the margin of the water normally moves between the lowest and highest levels reached during spring tides of greatest range, i.e. between Extreme Low Water Spring level (ELWS) and Extreme High Water Spring level (EHWS). It is sometimes convenient to refer to various levels of the shore in terms of the tidal cycle, as indicated in Table 7.1. The strip of shore between average high tide level (AHTL) and average low tide level (ALTL) has been termed the middle shore, above AHTL the upper shore and below ALTL the lower shore.

The distribution of plants and animals on the shore is sometimes described in terms of the standard tidal levels but this can be misleading because, as discussed later (page 184), many other factors besides the tidal cycle influence

172

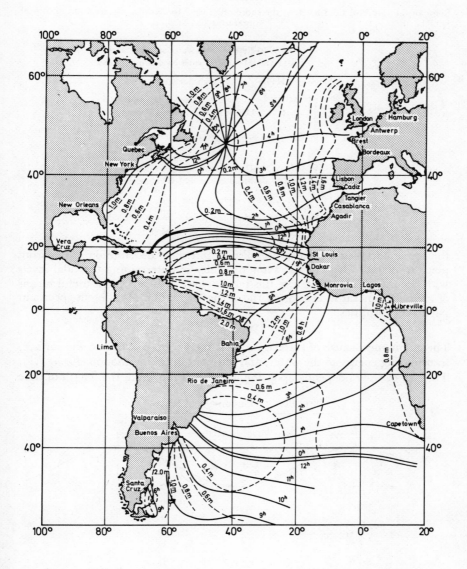

Figure 7.2. Co-tidal and co-range lines of the M_2 Atlantic tide (after W. Hansen). The times of high water refer to the moon's transit through the Greenwich meridian, the co-range lines are in meters

173

THE SEA-SHORE

TABLE 7.1

Some of the standard tidal levels of the shore often used to describe the zonation of littoral organisms

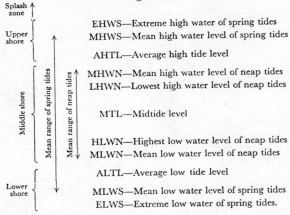

EHWS—Extreme high water of spring tides
MHWS—Mean high water level of spring tides

AHTL—Average high tide level

MHWN—Mean high water level of neap tides
LHWN—Lowest high water level of neap tides

MTL—Midtide level

HLWN—Highest low water level of neap tides
MLWN—Mean low water level of neap tides

ALTL—Average low tide level

MLWS—Mean low water level of spring tides
ELWS—Extreme low water of spring tides.

the levels occupied by shore organisms, and their distribution varies from place to place. There is also some difficulty in determining the tidal levels with accuracy. None the less, this terminology does provide a useful means of giving a general indication of the vertical zonation of different species in particular localities.

WAVES

The commonest cause of surface waves is the action of wind on the surface, transmitting energy to the water, and setting it in orbital motion (*Figure 7.3*). The size of waves depends upon the velocity of the wind, the length of time

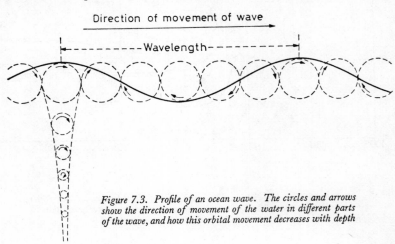

Figure 7.3. Profile of an ocean wave. The circles and arrows show the direction of movement of the water in different parts of the wave, and how this orbital movement decreases with depth

174

WAVES

during which it blows and the distance over which it acts. Surface waves do not mix the water to any great depth. Their motion falls off sharply with depth, and at a depth equal to the wave-length of the waves the water is virtually still.

When waves move into shallow water, their advance becomes slowed and the waveform changes when the depth becomes less than half the wave-length. The retarding effect of a shallow bottom causes *wave refraction*. Waves which approach the coast obliquely change direction in shallow water and reach the shore nearly parallel to the beach (*Figure 7.4*). Consequently, waves converge upon headlands, but diverge and spread out within bays. The energy transmitted by the waves is therefore concentrated on promontories, and is correspondingly reduced along equivalent lengths of coastline between the headlands (*Figure 7.5*).

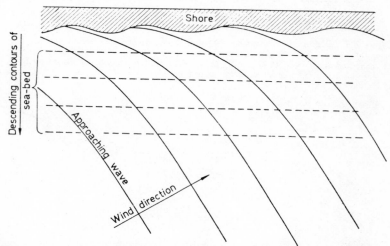

Figure 7.4.   Diagram of wave refraction as waves approach shallow water obliquely

The slowing of wave advance in shallow water reduces the wave-length. As the wave crests become closer together the wave height increases and the fronts of the waves become steeper. Where the water gets progressively shallower near the shore, the waveform becomes increasingly distorted until eventually the waves become unstable, their crests topple forwards as they break, and the energy of the wave motion is translated into the energy of a forward-moving mass of water (*Figure 7.6*).

The Effects of Waves on Beaches[8]

The up-rush of water formed by a wave breaking on the beach is known as the 'swash' or 'send' of the wave. Part of this water percolates down through the beach and the remainder flows back over the surface as the 'backwash'.

175

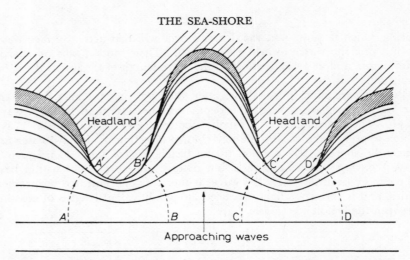

Figure 7.5. Diagram illustrating the concentration of wave energy on promontories. The energy of the wave-fronts between A–B and C–D become concentrated by wave refraction onto the short stretches of shore A'–B 'and C'–D'. Within the bay the energy of wave-front B–C becomes spread out around the shoreline B'–C'

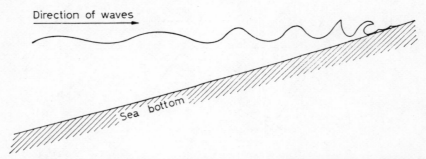

Figure 7.6. Changes of waveform on entering shallow water. At a depth of about half the wavelength, the waves become closer and higher. With decreasing water depth, the fronts of the waves become steeper until they are unstable and their crests topple forward

The effect of breakers on the shore may be destructive or constructive. Destructive breakers are usually formed by high waves of short wave-length; for example, those arising from gales close to the coast. As these waves break, they tend to plunge vertically or even curl seawards slightly, imparting little power to the swash, pounding the beach, loosening beach material and carrying some of it seawards in the backwash. Constructive waves are more likely to be low waves of long wave-length, sometimes the swell from distant storms. These waves move rapidly shorewards and plunge forwards as they break, transmitting much power to the swash and tending to carry material up the beach, leaving it stranded.

176

The continual transmission of energy from waves to shore gradually modifies the coastline, either eroding the beach by carrying away the beach material, or adding to the beach by deposition. In any sequence of breakers, there may be waves derived from many different sources which combine to form many different heights and wave-lengths, some destructive and some constructive. The condition of the shore is, therefore, a somewhat unstable equilibrium between the two processes of erosion and deposition. The balance varies from time to time, and differs greatly in different regions. Where the shore is exposed to very violent wave action, erosion usually predominates. The waves break up the shore, fragmenting the softer materials and carrying them away. Harder rocks are left exposed and are gradually fractured into boulders, making the coastline rocky and irregular. Strong currents may assist the process of erosion by carrying away finely-divided material. But materials carried away from the shore in one place may be deposited as beach material in another, and where the major effect of the waves is deposition, the beach is made up largely of pebbles, sand or mud.

Beach Construction

Beaches consist of a veneer of beach material covering a beach platform of underlying rock. In very sheltered situations the beach material may rest on a gentle slope of rock virtually unmodified by wave action, but in wave-washed localities the beach platform has usually been formed by wave erosion. Where the land is gradually cut back, both cliff and beach platform are formed concurrently (*Figure 7.7*). As the beach platform becomes wider,

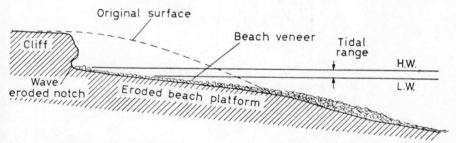

Figure 7.7. Beach section to illustrate erosion of the cliff to form the beach platform and veneer

waves crossing it lose power, and erosion of the cliff base is reduced. The seaward margin of the beach platform is itself sometimes subject to erosion by large waves breaking further out, so the cutting back of the beach platform and the cliff base may proceed together. Cliff erosion is caused largely by the abrasive action of stones, sand and silt churned up by the water and hurled against the base of the cliff, undercutting it until the overhanging rock collapses. Where the rocks are very hard, they may not be appreciably worn away, but can be cracked along lines of weakness by sudden air compression in holes and crevices when waves strike the cliff. This leads to falls of large pieces of rock, and produces a boulder strewn coastline on which the beach platform is often quite narrow. The range of tidal movement has a considerable bearing on the rate of erosion because the greater the depth of water covering the beach platform at high tide, the more powerful the waves that can cross it to erode the cliff base.

Various sources may contribute to the materials covering beach platforms; for example, fragments derived from erosion of adjacent cliffs, or churned up from the sea-bed, or eroded from the edge of the beach platform, or carried along the coast from other places by currents or beach drifting (*see* below). In sheltered regions, finely-divided material carried in suspension in the water may be deposited on the beach as sand, mud or silt.

Beach Drifting and Grading

When a wave breaks, the swash may carry stones or smaller particles up the beach. Waves which break obliquely on the shore carry materials up the beach at an angle, while the backwash and any particles it contains run directly down the slope of the beach. Consequently, each time a wave breaks obliquely, some of the beach material may be carried a short distance sideways (*Figure 7.8*). This process, known as beach drifting, can move huge quantities of material over great distances.

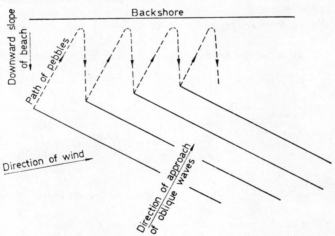

Figure 7.8. The movement of pebbles along a beach under the influence of oblique waves

Beach drifting has a sorting effect on the distribution of beach materials. Where cliffs are exposed to strong wave action, the shore is usually strewn with boulders too large to be moved by the waves. Boulders gradually become

178

broken up, and the fragments may then be carried along the shore in a series of sideways hops. A short distance from the original site of erosion, the beach is likely to consist mainly of large stones which only the more powerful waves can move. Further along the shore the particle-size of the deposit becomes progressively smaller, partly because the pebbles gradually wear away as they rub one against another, and partly because smaller particles are more easily transported.

Waves also exert a sorting action on the grade of material deposited at different levels of the beach, due to the difference in energy of swash and backwash. The swash has the full force of the wave behind it, but the backwash merely flows back down the beach under the influence of gravity, and contains less water than the swash because some is lost by percolation through the beach. The swash may move large stones up the beach but the energy of the backwash may be insufficient to carry them down again. This often leads to an accumulation of large pebbles at the back of the beach, causing the slope of the beach to become steeper towards the land until a gradient is reached at which stones begin to roll down under their own weight. The smaller particles of sand and gravel are more easily carried down by the backwash, and this has a grading effect, depositing coarse material at the back of the shore and progressively smaller particles lower down. Around the British Isles there are many beaches with steeply-sloping shingle at the higher levels, and flatter sand or mud on the middle and lower shore.

Drifted beach material accumulates alongside obstructions crossing the shore; for example, headlands, large rocks or groins. The purpose of erecting groins is to limit beach drifting by causing pebbles and sand to become trapped between them. Wave action then builds the beach into high banks between the groins, preventing the beach material from being carried away and protecting the land against erosion.

THE EVOLUTION OF COASTLINES

Erosion and deposition continued over long periods gradually change the configuration of a coastline, tending eventually to straighten it by wearing away the headlands. Over many millions of years these processes would have reduced long stretches of shore to virtual uniformity were it not for the changes in relative levels of land and sea which have occurred from time to time throughout the earth's history[34]. The causes of these changes are incompletely understood, but variation of world climate has certainly been one of the major factors during the last million years by altering the volume of water in the oceans. During this period there have been a series of 'ice ages' when the world climate has become colder than at present, polar ice caps have extended to much lower latitudes, and more snow has remained on the mountains instead of melting and flowing into the sea. Because a greater proportion of the earth's water has been locked up in frozen form,

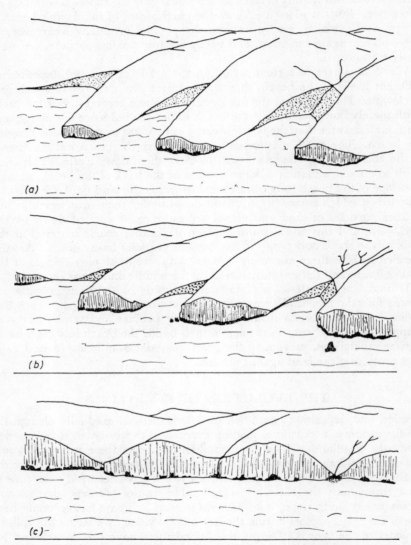

Figure 7.9. Stages of evolution of a submerged coast. (a) early phase. Flooded valleys, erosion of headlands and deposition of sand and silt within the inlets. (b) intermediate phase. Headlands cut back further, lengthening the cliff line and exposing the bays to stronger wave action. Beach material becomes coarser and less stable. (c) mature phase. Promontories eliminated and cliffs virtually continuous

sea level has fallen. During the warmer interglacial periods, the melting of ice and snow has increased the volume of water in the oceans, and raised sea level.

Changes in ocean volume do not produce equal relative changes of land and sea level in all parts of the world. The enormous weight of an ice cap can depress the level of the underlying land. When the ice cap melts, although the sea becomes deeper, that part of the land which has been relieved from the huge load of ice may rise considerably more than the sea around it, so that here sea level falls relative to the rising land. Depending upon the change of relative levels of land and sea, the changes of the coastline may be submergent or emergent.

Submergent Coasts

Where sea level rises relative to the land, deep inlets are formed by the sea flooding the lower part of river valleys. The headlands between the valleys are exposed to wave action, and are gradually eroded to form cliffs. Within the valleys, sheltered conditions permit the deposition of eroded material, so the inlets gradually fill with sand or silt (*Figure 7.9*). This submergent type of coastline is evident around the south-west peninsulas of the British Isles and the north-east of America.

If land and sea level become stable for any long period, the headlands are progressively cut back reducing the depth of inlets until they become bays containing beaches between extensive stretches of high cliff. Eventually, when the protecting headlands have been completely obliterated, the bays themselves become exposed to erosion and the coastline then tends to become fairly uniform, consisting mainly of cliffs with little if any beach, i.e. a mature coastline.

Emergent Coasts

Where the sea is receding from the land, the margin of the water meets a gentle slope that was originally the sea bottom. Waves tend to break a long way out because the water is shallow for a considerable distance from the shore. In the line of wavebreak the sea-bed becomes churned up, and loosened material may be thrown ahead of the breaking waves. This sometimes leads to the formation of ridges or bars of sand or gravel, known as offshore bars. These are often of transient duration, their shape and position changing from tide to tide, but occasionally the process is cumulative so that a bar is eventually built up above sea level. It may then become stabilized and consolidated by plant growth along its crest. The seaward side of the bar is now the new coastline. The lagoon between the earlier shore line and the newly formed bar becomes silted-up, forming an area of salt marsh which may later become converted into sand dunes, and finally into ordinary soil.

If sea level continues to fall, this sequence may be repeated several times. Once the sea level becomes constant over a long period, wave erosion is likely to encroach gradually upon the land, leading at length to the formation of cliffs and the development of a mature coastline.

Due probably to a rise of sea level since the last glaciation, much of our coast of the Atlantic and Pacific is the submerged type in various stages of evolution. However, there are signs of old cliffs and beaches, some now many feet above sea level and far behind the present coastline, in several places along the Atlantic Coast.

SOME PROBLEMS OF SHORE LIFE

(1) Water Loss

The majority of inhabitants of the sea-shore are essentially aquatic organisms which have evolved directly from fully marine forms. If left uncovered to the air, they lose water by evaporation and will eventually die from dehydration. To make the transition from sea to shore successfully, forms which live in exposed positions on the surface of the shore must have some means of retarding the rate of water loss sufficiently to permit survival during the periods when they are left uncovered by the receding tide. The danger of drying is most severe where exposure to air also involves exposure to sun or wind.

Water loss can be fatal in several ways. Death may be due to disturbances of metabolism resulting from the increasing concentration of internal fluids. In many cases the immediate hazard is asphyxia. Some organisms require a continuous current of water over the gills for adequate gaseous exchange. Others can survive for a time in air, but all must preserve at least a film of water over their respiratory surfaces. In a fish out of water the weight of the gill lamellae unsupported by water causes them to collapse against each other and adhere by surface tension. So little area of gill is then left exposed for respiratory exchange that the fish asphyxiates despite the high oxygen content of the air.

(2) Wave Action

Enormous forces are transmitted to the shore by the breaking of waves. The destructive impact of a great weight of water together with stones and other suspended matter hurled by the waves presents a major hazard to the inhabitants of the shore surface. Beneath the surface, burrowing forms are in danger of being crushed when waves churn the beach deposits. There is also the danger of dislodgement by waves, which may carry creatures up or down the shore to levels unsuitable for their survival. When masses of stones, sand, or organic debris are washed up and stranded by the waves, the shore population is in danger of becoming smothered. Continuous rapid movement of the water presents difficulties for the settlement and attachment of spores and larvae, and may prevent colonization of the shore except in crevices and sheltered parts.

(3) Temperature Fluctuations

During low tide, wide and rapid changes of temperature are sometimes encountered on the shore. Strong sunshine can produce high temperatures on exposed shore surfaces, and the temperature of water standing on the shore when the tide recedes may be raised well above the normal limits of sea temperature. Shallow pools on the coasts sometimes reach temperatures of 25–30°C on hot summer days. In the tropics, temperatures of 50°C or higher have been recorded in shore water.

Intertidal organisms may also be exposed to severe frosts. When shore water freezes they face the additional dangers of moving ice which may scrape them off the rocks or crush them within their hiding places. In high latitudes the shore may be kept virtually barren by the effects of moving ice.

(4) Salinity Fluctuations

Shore organisms often encounter water of much reduced salinity due to dilution of the shore water by rain, or fresh water flowing off the land. On the other hand, evaporation may raise the salinity of shore water above that of normal sea-water. Increased salinity and raised temperature usually occur together.

(5) Fluctuations of Oxygen, Carbon Dioxide and pH

These occur in shore water in association with photosynthesis and bacterial activity. In bright light, photosynthesis by dense algal vegetation in small pools sometimes raises the oxygen-content appreciably, and the withdrawal of carbon dioxide from the water raises the pH. Diminished oxygen, increased carbon dioxide and reduced pH may result from rapid bacterial decomposition in stranded detritus.

(6) The Range of Illumination

The illumination of the shore varies widely with rise and fall of the tide. When the tide recedes, the shore is directly exposed to light. When the shore is covered, the illumination is much reduced, especially where the water is very turbid. Inadequate light is probably one of the factors limiting the downward spread of some of the algae of the shore. Direct exposure to light may favour the growth of some plants but high intensities of light are probably lethal to many red algae. Strong sunlight is detrimental to many organisms due to the combined effects of radiation, heating and drying, but it is difficult to dissociate their separate influences.

(7) Predation

The inhabitants of the sea-shore are exposed to a double set of predators. During submergence they are preyed upon by other marine creatures. When uncovered they encounter enemies from the land and air. On some parts of our coastal areas, sea birds exact a very heavy toll from shore populations during low tide; for example, studies in an English estuary[30] indicate that oystercatchers (*Haematopus ostralegus*) feeding on cockles may sometimes take about 500 specimens per bird per day, and in certain conditions consume up to 70% of the stock of two-year-old cockles.

(8) Immersion

The majority of shore creatures utilize the oxygen dissolved in sea-water for their respiration, but a few are air-breathing. Some of these have evolved from land forms which have spread to the sea-shore, for example *Petrobius*, *Lipura*, *Otina*, *Leucophytia*. Others have evolved directly from marine forms, their respiratory organs having become adapted to absorb atmospheric oxygen, for example *Ligia*, *Orchestia*, *Talitrus*, *Littorina neritoides* and *L. saxatilis*. In either case, prolonged immersion is fatal by cutting off the essential supply of air. During immersion the shore population experiences varying water pressure, with very rapid fluctuations where there are waves. Sensitivity to water pressure must be a factor having some influence on distribution within the intertidal zone.

(9) Pollution

An additional hazard of shore life which has appeared in recent years is pollution from accidents at sea, especially by stranding of oil. Even more toxic than oil itself are some of the chemicals used to clean beaches (*see* Chapter 4, References 15 and 69).

FOOD SOURCES

Despite its dangers, the shore is often densely populated by a variety of organisms excellently adapted to the difficult conditions. So great are the numbers on some shores that every available surface is colonized, and there is severe competition for living space. A numerous population indicates an abundant supply of food, and this is derived from several sources. On rocky coasts there is often a copious vegetation of seaweed. The rapid growth of these plants is favoured by the excellent lighting conditions and by the good supply of nutrients, well distributed by the movements of the water and continually released by wave disturbance of the sediments, weathering of the coastline or in fresh water flowing off the land. Where there are no large plants, the surface of the beach may be covered by a film of microscopic plants. In addition to primary production on the shore, large quantities of food are contributed by the sea, inshore water often containing a rich plankton on which innumerable shore creatures feed. Also, pieces of plant material torn from the sea-bed below low tide level become deposited on the shore. The land, too, makes a contribution, various organic substances of terrestrial origin being stranded on the beach.

ZONATION

A gradient of environmental conditions extends across the shore, due mainly to the different durations of submergence at each level. The lowest parts of the shore are uncovered only during the lowest spring tides, and then only for brief periods; the highest levels are seldom fully submerged and are mainly wetted by wave splash; intermediate levels experience intermediate durations of alternating exposure and submersion. Even on non-tidal shores there are gradients between permanently submerged and fully terrestrial conditions, with intermediate levels wetted by splash or irregularly covered and uncovered as wind action alters the water level.

The requirements for life in air and water are so different that no organism is equally well suited to every level of the shore. Different levels are therefore occupied by different assemblages of plants and animals, each species having its main abundance within a particular zone where conditions are most favourable for it. Above and below this zone it occurs in reduced numbers, or is absent, because physical conditions are too difficult to allow its survival, or because it competes less successfully with other forms better suited to the different environment.

Alternations of uncovering and submergence present somewhat different problems to different organisms, depending on the type and level of shore they inhabit, and their mode of life. For surface-living forms, drying, wave action, illumination and temperature are major factors influencing zonation, but their relative importance must vary with level. The occupants of the upper shore are exposed to air for long periods between spring tides, and must therefore be able to withstand conditions of prolonged drying, extremes of temperature and strong illumination. Direct wave impact is a relatively brief and infrequent hazard, and strong wave action favours the upward spread of this population by increasing the height to which spray regularly wets the shore.

Lower on the shore, surface populations are never uncovered for more than a few hours at a time, and here the problems of desiccation, temperature fluctuations and excessive illumination are less severe. These organisms experience longer and more frequent periods of wave action with the attendant risks of damage or dislodgement, and during submergence the illumination may be inadequate for certain species of plants. Other factors which vary with shore level, and influence distribution, include the duration of periods during which feeding is possible, the maximum depth of water, and often the type of substrate. These factors may be largely accountable for the zonation of many of the forms which are not ordinarily left exposed, but hide in crevices or under the algae, or burrow into the rock or sediment.

Although the tides exert a major influence on zonation, the distribution of littoral (shore-dwelling) species relative to particular tidal levels is by no means constant. Wide variations occur from place to place due to differences of geography, geology, and climate. Factors which modify zonation and vary with locality include the intensity of wave action, the range of temperatures and humidities, the aspect of the shore with respect to the sun and prevailing wind, the type of rock or sediment, the amount of rainfall and fresh water run-off, and the period of day or night when extreme low tides occur. Lewis[9] has emphasized that the *littoral zone* cannot be satisfactorily defined solely in relation to sea level, but is better described in biological terms as the strip between sea and land inhabited by characteristic communities of organisms which thrive where the shore surface undergoes alternations of air and sea-water. These communities comprise two main groups occupying different levels, and the littoral zone can be correspondingly subdivided (*see* page 198). The uppermost communities are those requiring mainly aerial conditions, inhabiting the *littoral fringe*, a zone submerged only at spring tides or wetted only by wave splash. On very sheltered shores this zone is a narrow belt below EHWS level, but it becomes higher and wider with increasing exposure to wave action until, on the most wavebeaten rocky coasts, it lies entirely above EHWS level and may extend upwards for 20 m or more. Below the littoral fringe is the *eulittoral zone*, occupied by communities tolerant of short periods of exposure to air between tides but requiring regular submersion, or at least thorough wetting, at each tidal cycle. At their lowest levels the populations of the eulittoral zone overlap those of the *sublittoral fringe*, i.e. the communities of shallowest water (page 158) contiguous to the shore.

FITTING THE SHORE ENVIRONMENT[11]

The detrimental effects of exposure to air restrict much of the littoral population to sheltered parts of the shore which are not left completely uncovered at low tide. Many algae, anemones, hydroids, polyzoa, prawns and fish occur only in rock pools where they are safe from drying, and relatively protected from wide temperature fluctuations. Shelter can also be found between the fronds and holdfasts of the shore algae, under stones and boulders or in rock crevices. Some of the most numerous shore animals have a flattened shape well suited for hiding in narrow spaces, for example *Porcellana platycheles*, *Nebalia*, amphipods, isopods and chitons. Others burrow into the shore deposits for protection, and these comprise virtually the entire population of sandy and muddy shores.

Organisms which live on the shore surface in situations where they are frequently uncovered to air must have a protective covering to prevent excessive drying. The exposed algae have a thick cuticle and mucilagenous secretions to reduce evaporation. The beadlet anemone, *Actinia equina*, often inhabits uncovered rocks and is coated with slimy exudations which presumably help to conserve water. Most of the surface-dwelling animals have a strong shell, the orifice of which can be kept closed while they are uncovered. The limpet, *Patella*, draws down the rim of its shell so close to the rock surface that only a very narrow gap is left, sufficient for oxygen and carbon dioxide to diffuse through without allowing much evaporation. Winkles, top shells, dog whelks and serpulid worms close the shell aperture with an operculum.

In barnacles, the movable plates of the shell, the terga and scuta, are kept shut most of the time the animals are uncovered, occasionally opening momentarily for renewal of the air enclosed within the shell.

In addition to losses by evaporation, animals also lose water by excretion. The majority of marine creatures excrete ammonia as their chief nitrogenous waste product. This is a highly toxic substance which has to be eliminated in a very dilute urine, involving the passage out of the body of a copious amount of water. On the sea-shore, this must present a difficulty to animals already in danger of desiccation, and some of the littoral gastropods reduce their excretory water-loss by excreting appreciable amounts of uric acid, a less soluble and less toxic substance than ammonia which can be excreted as a semi-solid sludge, thereby conserving water. The different species of *Littorina* form a series, those which live highest on the shore (page 195) excreting the greatest amount of uric acid in the most concentrated urine (Table 7.2).

TABLE 7.2

Uric acid content of nephridia of *Littorina Spp.*

(From Nicol, J. A. C.[12] by courtesy of Pitman)

	mg/g dry weight
Littorina neritoides	25
L. saxatilis	5
L. obtusata	2·5
L. littorea	1·5

The surface population is also exposed to the dangers of wave impact and dislodgment, and shells are again the chief form of protection. The heavy wear sometimes visible on the shells of shore molluscs indicates the severity of the abrasion to which they are subjected. To resist dislodgment, some of the surface forms have great powers of adhesion. The large algae are anchored to the rock by strong holdfasts. Barnacles and serpulid worms have shells firmly cemented to the rock. The common mussel, *Mytilus edulis*, attaches itself to rocks and stones by strong byssus threads. Four different genera of fish found in the Atlantic, *Gobius*, *Liparis*, *Cyclopterus* and *Lepadogaster*, have the pelvic fins specialized to form a ventral sucker by which they can cling to a firm surface. The remarkable adhesion of the foot of *Patella* or *Acmaea* has given rise to the expression 'sticking like a limpet'.

Intermittent submergence presents problems in connection with respiration because no respiratory organs function equally well in both air and water. The majority of shore-dwelling animals perform aquatic respiration, but within the littoral fringe the infrequency of immersion calls for the ability to breathe air. Some of the inhabitants of this fringe zone are essentially marine forms which have become adapted for aerial respiration.

Talitrus saltator and *Orchestia gammarella* can live in moist air. In *Littorina neritoides* and *L. saxatilis* the ctenidium is reduced and the mantle cavity is modified to function as a lung. There are also a few animals which have colonized the shore from the land, and several of these show adaptations for storing air during the periods they are submerged. The collembolan *Anurida maritima*, widespread among rocks above mid-tide level, carries a layer of air among its surface bristles. The intertidal beetle, *Aepus marinus*, has internal air-sacs for air storage.

The difficulties of survival on the shore have their effects on all phases of life, including reproductive processes and larval and juvenile stages. The majority of benthic organisms start life as floating or swimming forms in the plankton, and may become widely dispersed in the water before they settle on the sea bottom. Shore creatures face special risks of great losses of pelagic eggs and larvae during this phase if they drift far from the shore and settle outside the zone in which survival is possible. For some inhabitants of the shore the chances of successful settlement in suitable areas are enhanced by certain aspects of the behaviour of their larvae. Some produce larvae which are at first strongly attracted by light, and presumably rise close to the sea surface during the day. Wind direction is often landward during the day-time, driving the surface water towards the coast, and so it is likely that positive phototaxis improves the chances of pelagic larvae returning to the shore. It has been mentioned earlier that the larvae of many benthic species can discriminate between substrates and can for a time delay settlement until favourable conditions are encountered (pages 160–1). Some tend to settle gregariously, often in response to the presence of successfully metamorphosed members of the species, for example barnacles, *Sabellaria*, serpulid worms, mussels and some polyzoa. Settlement in shallow water may also be favoured by larval response to wave action; for example, cyprids of *Balanus balanoides* are reported to settle more readily under fluctuating water-pressure. When settlement occurs sublittorally, or at lower levels of the shore than are occupied by the adults, for example *Littorina* (*Melaraphe*) *neritoides* (page 192), the responses of the juveniles to various environmental stimuli may cause migration upshore towards the appropriate zone.

In many shore animals the planktonic phase is abbreviated or omitted, and this simplifies the problem of finding the correct shore-level. For instance, *Arenicola marina*, one of the most successful littoral worms burrowing in muddy sand, has only a brief pelagic period[52]. In late autumn or early winter the gametes are shed from the burrow onto the surface of the sand during low spring tides, and here fertilization occurs. The fertilized eggs may be dispersed to a limited extent by movements of the water, but they are heavier than water and slightly sticky, tending to adhere to the surface of the sand. They hatch after about 4–5 days and the larvae, although capable of swimming, seem from the outset to burrow into the deposit wherever the substrate is suitable.

Others completely eliminate pelagic stages by developing directly from egg to miniature adult form. These eggs are well-charged with yolk to enable the young to hatch in an advanced state. *Littorina obtusata* deposits its eggs in gelatinous masses on the surface of seaweed. Its young occasionally emerge as advanced veliger larvae, but probably more often do not hatch until after the velum has been resorbed and then appear as tiny crawling winkles. The dogwhelk, *Thais (Nucella) lapillus*, forms vase-shaped egg capsules which often occur in large numbers stuck to the underside of stones or sheltered rock surfaces. Each capsule contains several hundred eggs, but eventually only a dozen or so whelklets crawl out of each capsule, the rest of the eggs serving as food for the first few to hatch.

Eggs laid on the sea-shore are exposed to all the vicissitudes of this environment, and certain shore creatures contrive in various ways to give their eggs some protection. For example, several species of inshore fish guard their eggs, such as *Centronotus gunnellus* and *Cottus bubalis*. They lay sticky masses of large eggs from which advanced young are born, and one of the parent fish, usually the male, remains close to the eggs until they hatch, courageously protecting them against marauders. Others enclose the eggs upon or within their bodies. Amphipods and isopods, often very numerous on the shore, retain their eggs within a brood-pouch from which fully-formed young emerge. In *Littorina saxatilis* the eggs remain in the mantle cavity until they hatch as minute winkles. The tiny bivalve, *Lasaea rubra*, often abundant in rock crevices and empty barnacle shells, incubates its eggs and young within the gills until they are sufficiently developed to crawl out and maintain themselves near the parent. The viviparous blenny, *Zoarces viviparus*, gives birth to well-formed fish about 4 cm in length.

Physiological and behavioural adaptations are necessary to withstand the fluctuating nature of the shore environment[35]. The wide and rapid changes of temperature and salinity that occur on the shore surface during low tide require wide eurythermy[61] and euryhalinity in the exposed population. They must also be capable of making appropriate adjustments of behaviour in response to changes in their surroundings. *Patella*[18,19], if wetted with fresh water, pulls its shell hard down and remains still; but if repeatedly splashed with sea-water it begins to wander about. *Arenicola marina*[67], before commencing the intermittent irrigation cycles which replace the water in its burrow, moves backwards up the rear end of the burrow and appears to test the quality of the surface water with its tail, modifying the subsequent sequence of irrigation activity accordingly. Animals that live among algal fronds are usually coloured to match their surroundings, and some have considerable powers of colour change to match different algal species.

Appropriate changes of activity are required to meet the profoundly different conditions of submergence and exposure to air. Movement, feeding or reproduction are only possible for many littoral species during the periods when they are covered by water; and when uncovered, they become more or less inactive. In this quiescent state their respiratory needs are reduced, water is conserved and there is less danger of attracting the attention of sea birds and terrestrial predators. Restriction of movement as the shore dries during low tide may also help to confine free-living forms to their appropriate zones. Some shore animals, for example *Carcinus maenas*[51], *Eurydice pulchra*[37] and *Blennius pholis*[35], even when removed from the shore, display cyclical changes of activity having a tidal frequency, apparently controlled by endogenous rhythms.

A 12-h cycle of activities related to the tidal cycle has been observed in the tiny estuarine gastropod, *Hydrobia (Peringia) ulvae*[56], on mud flats in Europe. When these animals are first uncovered by the receding tide, the majority are found crawling on the surface of the mud. Later, most of them begin to burrow just below the surface, and eventually come to rest with proboscis extruding to the surface for feeding upon superficial deposits. Before the rising tide reaches them, they return to the surface and begin to float in the water left standing between ripple marks in the mud, hanging upside down from the surface film attached to a raft of mucus which serves the dual function of providing buoyancy and trapping food particles. As the tide rises and falls, many of the snails remain floating and are carried at first shorewards, and later back to their original level. They then sink and begin to crawl on the mud surface, and the cycle is repeated.

Free-living shore creatures are in some danger that their own movements may carry them out of their proper zones into levels too high or too low on the shore, or into positions too exposed to wave, wind or sun. Studies of their behaviour reveal some of the mechanisms whereby they find and keep within suitable parts of the shore. For example, a notable feature of the movements of the adult *Patella vulgata* is its 'homing' behaviour, usually returning to a particular site or 'home' on the rock after foraging for food, often over a distance of several feet. *Patella* feeds chiefly by scraping the surface with its long, toothed radula, rasping off the microscopic film of algae which forms a slimy coating on the rocks. During the day it seldom moves about unless submerged, but at night *Patella* can often be found crawling on moist surfaces while uncovered. Homing behaviour is most strongly developed in individuals living high on the shore. On returning to its home, *Patella* always settles in the same position and the margin of the shell grows to fit the rock surface very accurately. On soft rocks the home is often marked by a ring-shaped groove, the 'limpet scar', conforming to the shell margin and presumably worn into the rock by slight movements of the shell. Homing obviously ensures that the animal maintains its zonational position, and the exact fit between shell and rock reduces water-loss during exposure, and also lessens the danger of the shell being prised off the rock by predators.

It is uncertain how *Patella* finds its way back to its home, but in some animals it is evident that the direction of their movements is related to factors such as light (phototaxes), gravity (geotaxes), lateral contact (thigmotaxes), humidity (hydrotaxes) or direction of flow of water (rheotaxes)[6]. This is a complex field of study because animal behaviour is seldom altogether consistent, varying from one individual to another, sometimes changing at different stages of the life-history or reproductive cycle, and often being modified or even reversed by alterations in the condition of the animal or the environment; for example, by changes of temperature or salinity, the animal's need for food, its state of desiccation or its previous experiences. None the less, it has been demonstrated that some shore creatures display patterns of movement which carry them into situations for which they are well suited, and enable them to remain in appropriate zones despite their need to move about over the shore in search of food or for mating.

Lasaea rubra[48,49], a European species, for instance, is influenced by light, gravity, and contact. This tiny bivalve is widely distributed throughout the littoral zone, extending to a high level and occurring mainly in the protection of crevices, empty barnacle shells and tufts of Lichina. It makes temporary attachment by means of byssus but is capable of moving freely over the surface by using its extensible foot to crawl on a mucus film. On a level surface Lasaea moves away from light, but on a sloping surface it climbs even against the light. Its response to lateral contact overrides the effects of both light and gravity, causing the animal to move into crevices and small holes. Laboratory experiments have shown that Lasaea will crawl into a narrow hole even downwards towards bright light.

The periwinkle Littorina obtusata is most numerous under cover of middle-shore algae. On level surfaces it usually moves away from light. L. saxatilis extends high in the littoral fringe, showing a strong tendency to move towards light and to climb. L. littorea is widely distributed across the middle shore in both sheltered and moderately exposed situations. Its movements have been studied on U.S. and English mud flats[53,54], and found there to be related to the direction of the sun. In this locality the feeding excursions occur mainly during the periods shortly after the winkles are uncovered by the receding tide, or submerged when the tide returns. The majority of winkles on the mud move at first towards the general direction of the sun, but later they reverse their direction. They therefore tend to retrace their course, their overall movement following a roughly U-shaped path which brings them back approximately to their starting point. These animals accustomed to a horizontal surface show no geotaxic responses, but experiments with other specimens collected from the vertical faces of groins demonstrate responses to both light and gravity, these too moving over a U-shaped track, at first downwards and later upwards. Looped tracks have been reported for a number of other shore creatures, in some cases orientated to light, in others to gravity, and such behaviour has obvious advantages in enabling free-living animals to range about over the shore without moving far out of the levels in which they find favourable conditions.

The adults of *Littorina neritoides* (in Europe) occur high in the littoral fringe, but their eggs and larvae are planktonic, and settlement is mainly on the lower shore. The attainment of adult zonation is brought about by a combination of responses to light and gravity, modified by immersion. This winkle is negatively geotactic, and climbs upwards on rock surfaces. It is also negatively phototactic, and so will move into dark crevices. But the reaction to light reverses if the animal is immersed in water while it is upside down. If its crevice becomes submerged, *Littorina neritoides* therefore tends to crawl out along the ceiling towards the light, and then climb higher on the shore.

Some shore creatures have sufficiently well-developed vision to be able to see nearby objects and direct their movements by sight. An example is the sand-hopper *Talitrus saltator*[69], which burrows in upper shore sand in the day-time and emerges at night during low water to feed on the surface. These feeding explorations carry it well down the shore, but it eventually finds its way back to high-water level. If removed from its burrow during day-time and released lower on the shore on a firm, unbroken sand surface, irrespective of slope of the beach or direction of sun or wind it tends to move over the surface towards the back of the beach, where it burrows on reaching the dryer, looser sand. If both eyes are covered, the movements of *Talitrus* released low on the shore are haphazard and show no tendency to carry it back upshore. Experiments both on the beach and in the laboratory suggested that *Talitrus* is capable of seeing shapes (form vision), that certain shapes attract it, and that its movements towards the top of the beach are probably associated with its ability to see the line of the backshore, even in dim, night-time illumination.

Some experiments with periwinkles on a rocky shore have demonstrated that they too direct their movements in relation to what they see of their surroundings[33]. From an examination of the structure of the eye of *Littorina littorea*[55] it appears that this can form sharp images of distant objects in air, and can probably accommodate to bring near objects into focus.

Pardi[58] and others have studied the movements of *Talitrus* from various parts of the coastline when placed above high-water level. They have demonstrated that the animals generally move in a direction which would carry them towards the sea in the localities from which they were taken, even when removed to areas remote from the sea. Their orientation is based on their sight of the sun or, in shade, on their perception of the polarization of light from the sky. Once they are exposed to sunlight, the orientation shows even in animals bred in captivity in uniform light, which have never had previous sight of the sun or contact with the shore.

Environments with fluctuating conditions favour the evolution of species that exhibit wide variations, both physiological and anatomical. Where an unstable habitat also includes environmental gradients, different forms of a species are likely to occupy different zones. On the sea-shore it is obvious that many of the dominant species are highly variable, with differences

192

which can to some extent be correlated with differences of zonation. A notable example is *Littorina saxatilis* which shows great variety of shell size, colour, thickness, shape and surface texture. For this remarkably polymorphic species, fifteen varieties comprising six subspecies have been described with differences of distribution (*see* page 195). Other common polymorphic species on the shore include *Littorina obtusata*, *Thais lapillus* and *Mytilus edulis*, with some evidence of different forms in different localities.

ROCKY SHORES

Rocky shores exist where the effect of waves on the coastline is mainly erosive, wearing down the softer materials and carrying them away, leaving the hardest rocks exposed. Most of the substrate is therefore stable and permanent, forming a secure surface upon which can grow a variety of organisms requiring attachment; for example large algae, barnacles, mussels and limpets. The appearance of the shore depends largely upon the type of rock exposed. Horizontal strata often erode to a stepped series of fairly uniform level platforms which provide little shelter from the waves. Tilted strata running across the shore usually produce a very varied shore with numerous protruding rock ledges and overhangs, and deep pools in the gullies between them. Certain types of rock erode to a smooth surface, while some laminated rocks readily gape to form deep narrow fissures.

Many rocky shores are heavily populated. The agitation of the water keeps it well oxygenated, and favours plant growth by continually replenishing the supply of nutrients. The plants provide a primary food supply for animals, and copious additional food is available from the plankton.

Rocks present a variety of habitable environments—exposed rock faces, sheltered overhangs, crevices, deep or shallow pools, silt within fissures or under boulders, in the shelter of algae or in their ramifying holdfasts—each offers a domain which some species can occupy.

The size and composition of rocky shore communities is profoundly influenced by the intensity of wave action because this is one of the major factors determining the amount and type of algal growth on the rocks. Where wave-intensity is moderate, large algae cover the shore and give shelter to many small animals which cannot tolerate complete exposure to air and sun, for example coelenterates, sponges, polyzoa and small crustacea. Stronger waves prevent the growth of plants, and the rock surface then becomes covered mainly with barnacles and limpets, or sometimes at the lower levels by mussels. In extreme conditions of wave exposure, rock faces are swept virtually bare and the population is restricted to fissures and crevices. Because of the wetting effects of splash, heavy wave action tends to raise the levels to which sublittoral and littoral populations extend up the shore, and greatly increases the width and height of the littoral fringe (*Figure 7.11*). Where sand is deposited between rocks, their lower parts may be kept bare by the scouring effects of wave-tossed sand.

The intensity of wave action is a difficult parameter to evaluate. Attempts have been made to measure it with dynamometers[38], but an alternative approach has been to define degrees of wave exposure in terms of their biological effects. By studying the changes of population between exposed headlands and sheltered inlets, numerical values for wave intensity can be assigned to particular patterns of population on the assumption that the differences are due mainly to wave effects, i.e. a biologically defined scale of exposure to wave action[20].

The zonation of plants and animals is often clearly visible on rocky coastlines all over the world[20,50,60,64]. Where waves are not too violent to allow the growth of seaweeds, the rocks across much of the shore have a brownish-green covering of large fucoids, with bands of different colour at different levels indicating the dominance of particular species. Above the level of these large algae there is often a strip of relatively bare rock on which there are few plants except the black tufts of the lichen *Lichina pygmaea*. On south-west shores the barnacle *Chthamalus stellatus* may extend up to this level. Higher still, in the upper part of the littoral fringe, there is a black, tar-like streak of blue-green algae and the encrusting lichen *Verrucaria maura*. Above this the terrestrial vegetation begins, sometimes as a light-coloured lichen zone tinged with yellow or orange by encrustations of vivid species such as *Xanthoria* and *Caloplaca*.

Examining the fucoids more closely, we find certain species restricted within fairly close levels. Usually at the highest level there is the Channelled Wrack, *Pelvetia canaliculata*, forming a narrow band in the lower part of the littoral fringe where it is wetted by salt spray but only occasionally submerged. Just below *Pelvetia*, in the lower limit of the littoral fringe, there is often a narrow zone of *Fucus spiralis*. The greater part of the eulittoral zone is more or less covered with a mixed growth of *Ascophyllum nodosum* and *F. vesiculosus*, the former predominating in sheltered areas and the latter in regions more open to waves. Overlapping with *Ascophyllum* and *F. vesiculosus*, but extending below them into the lower part of the eulittoral zone, is a belt of Toothed Wrack, *F. serratus*. On many shores the lower limit of the eulittoral is indicated by a narrow band of *Himanthalia elongata*. Conspicuous plants in the sublittoral fringe, usually uncovered only at spring low-tides, are species of *Laminaria*, or, on the most wave-beaten rocks, *Alaria esculenta*.

The zonation of littoral animals is well illustrated by some of the species that dwell on the surface; for example, the barnacles that cover the rocks where wave action is too strong for plants. The most diverse populations of shore barnacles in the British Isles occur on the south-west peninsulas of England and Wales where at least six species are to be found, namely, *Chthamalus stellatus*, *Balanus balanoides*, *B. perforatus*, *B. crenatus*, *Verruca stroemia* and *Elminius modestus*. Their distribution can be summarized as follows:

Chthamalus spp. reach the highest level, sometimes extending above *Pelvetia* into the littoral fringe. These are a Lusitanian species extending northwards from the Mediterranean along Atlantic shores. The British Isles and New England are their northernmost limits. They are the dominant barnacle of the upper eulittoral zone in the south-west (page 69) of England, and occur in reduced numbers further north on western coasts and eastwards along the English Channel. In the U.S. they occur as far north as New England. *Balanus balanoides* is a boreal form flourishing in colder water than that of *Chthamalus*, and is dominant on the east coasts of Britain and the U.S. It does not reach as high a level as *Chthamalus*, and scarcely enters the littoral fringe. In both America and Europe the distribution of these species overlaps, and they compete for space on the shore. Where conditions of moderate shelter or lower temperature favour *Balanus balanoides*, *Chthamalus*, if present, tends to be restricted to a narrow fringe zone above the level of *B. balanoides*. Stronger wave-action broadens the *Chthamalus* zone at the expense of *Balanus*. On Devon, Cornwall and Pembrokeshire coasts, both species are less numerous in the lower eulittoral, where they compete with the larger *B. perforatus*. This southern form extends from the lowest levels of the shore up to about the middle of the eulittoral zone.

B. crenatus is a sublittoral species, intolerant of exposure, which may be found in rock pools or under algal cover on the lower shore all round the North Atlantic. *Verruca stroemia* is a eurybathic form with an extensive range from deep levels well down the continental slope up to the lower shore, where it is restricted mainly to pools or the undersides of stones.

Elminius modestus[26], a recent introduction to Europe which appeared in Chichester harbour and the Thames estuary during World War II, is an Australasian species which is thought to have been brought to Britain on ships' hulls. It has now spread around much of the British coastline, and has crossed the seas to Ireland and the French coast, and extends northwards along the shores of Belgium, Holland and the North Sea coasts of Germany and Denmark. In sheltered areas it is ousting the native population of *B. balanoides*, being able to survive at levels at least as high and as low as *Balanus* and having advantages in its rapid growth rate and extended breeding season through the summer months, and its wide tolerance of desiccation, silting, and variations of temperature and salinity.

The periwinkles, *Littorina spp.*, are very numerous on rocky shores. Four species occur in the British Isles, *L. (Melaraphe) neritoides* (and the following three species are also found in the United States), *L. saxatilis*, *L. obtusata* (*littoralis*) and *L. littorea*. *L. saxatilis* is a very variable species with distinct varieties at different levels of the shore and in different localities. Four sub-species have been described from the British coast[36]. *L. saxatilis tenebrosa* lives mainly in the littoral fringe, together with the adults of *L. (Melaraphe)* *neritoides*. Both of these are most numerous where violent wave action produces a wide fringe zone. *L. saxatilis neglecta* is a very small eulittoral form dwelling in crevices, empty barnacle shells and amongst mussel byssus, and

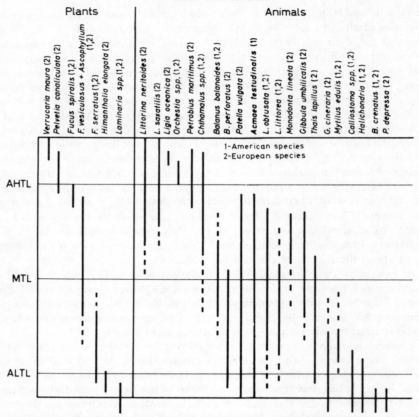

Figure 7.10. *Zonation of some common plants and animals of rocky shores exposed to moderate wave intensities along south-west peninsulas of England and Wales and the New England coast of U.S.*

is also most common on shores of strong wave action. *L. saxatilis jugosa* occurs mainly at the *Pelvetia* level on relatively sheltered shores. *L. saxatilis rudis* is a eulittoral form extending over the middle shore in sheltered localities. *L. obtusata* and *L. littorea* are also eulittoral species abundant in shelter. *L. obtusata* lives mainly under cover of the fucoids, and usually extends a little higher than *L. littorea*. The latter is often found on bare rocks, and is especially numerous in gulleys or on the sheltered faces of boulders.

The common limpet, *Acmaea testudinalis*, extends throughout the eulittoral zone on virtually all rocky shores around New England. In Britain, *Patella vulgata* is the dominant eulittoral limpet. In the south-west of England its lower limits overlap the uppermost distribution of *P. intermedia* and *P. aspera*, especially on strongly wave-washed shores.

Figure 7.10 indicates in a general way the zonation of the more conspicuous

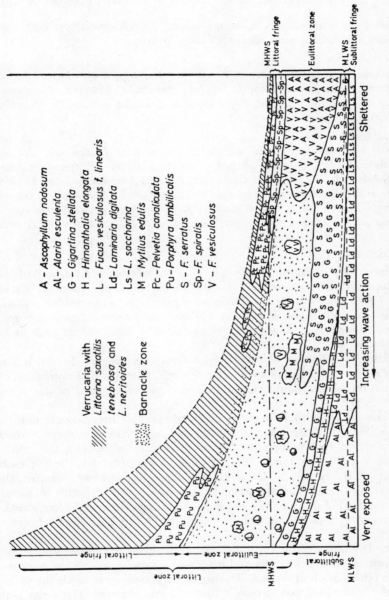

Figure 7.11. Diagram to illustrate how the distribution of some dominant plants and animals of rocky shores varies with wave intensity

A – Ascophyllum nodosum
Al – Alaria esculenta
G – Gigartina stellata
H – Himanthalia elongata
L – Fucus vesiculosus f. linearis
Ld – Laminaria digitata
Ls – L. saccharina
M – Mytilus edulis
Pc – Pelvetia canaliculata
Pu – Porphyra umbilicalis
S – F. serratus
Sp – F. spiralis
V – F. vesiculosus

Verrucaria with
Littorina saxatilis
tenebrosa and
L. neritoides

Barnacle zone

fauna and flora on an algal-covered rocky shore in the British Isles. *Figure 7.11* indicates the distribution of certain species in various intensities of wave action.

Some of the common species compromising the communities of combined American and British rocky shores are listed below.

The Littoral Fringe

Verrucaria maura, Lichina confinis, Xanthoria spp., L. pygmaea, Myxophyceae, *Pelvetia canaliculata, Fucus spiralis, Porphyra umbilicalis, Littorina (Melaraphe) neritoides, L. saxatilis tenebrosa, Ligia oceanica, Orchestia gammarella, Petrobius (Machilis) maritimum.*

Eulittoral Zone

Ascophyllum nodosum, Fucus vesiculosus, F. serratus, Himanthalia elongata, Laurencia pinnatifida, Corallina officinalis, Gigartina stellata, Rhodymenia palmata, Balanus balanoides, B. perforatus, Chthamalus stellatus, Elminius modestus, Littorina saxatilis rudis, L. obtusata, L. littorea. Patella vulgata, Thais lapillus, Gibbula cineraria, G. umbilicalis, Acmaea testudinalis, Monodonta lineata, Actinia equina, Anemonia silcata, Mytilus edulis, Porcellana platycheles, Carcinus maenas, Blennius pholis.

Upper Sublittoral Zone

Laminaria spp., Alaria esculenta, Corallina officinalis, Chondrus crispus, Lithothamnia, *Balanus crenatus, Verruca stroemia, Gibbula cineraria, Patella aspera, Patina pellucida, Anemonia sulcata, Asterias.spp., Psammechinus miliaris, Galathea spp., Porcellana longicornis, Halichondria panicea, Alcyonidium spp., Archidoris pseudoargus, Acanthodoris pilosa, Aeolidia papillosa, Mytilus edulis, Metridium senile.*

Zonation is a feature of shore populations everywhere. In the bibliography at the end of this chapter are a few references giving information on the zonation of intertidal communities in some other parts of the world [2,10,14,40–42,63,64,72].

Inaccessible cliffs, rocky stacks and islands provide breeding grounds for innumerable sea birds[5]. The following are some common species which nest mainly on rock ledges: the herring full (*Larus argentatus*), the greater black-back gull (*L. marinus*), the kittiwake (*Rissa tridactyla*), the guillemot (*Uria aalge*), the razorbill (*Alca torda*), the shag (*Phalocrocorax aristotelis*), the cormorant (*P. carbo*) and the gannet (*Sula bassana*). The puffin (*Fratercula arctica*) is a bird of the open sea and forms breeding colonies on certain rocky islands around the far North Atlantic coasts, where it makes its nest in burrows in the turf.

At the beginning of the breeding season, the grey seal (*Halichoerus grypus*) comes ashore onto rock ledges and inaccessible beaches where the pups can be born out of reach of the sea. During the first two weeks of life the young avoid the water and may drown if they fall into the sea. The adult seals usually stay ashore for several weeks at this time.

SANDY SHORES

The size of particles to which the name 'sand' is applied has been given (page 152) as coarse sand—2·0–0·2 mm; fine sand—0·2–0·02 mm. Seashore sands usually include a variety of types and sizes of particles. The main constituent of sand on many coasts is silica fragments. Our yellow beaches often consist almost entirely of coarse siliceous sand. Grey, muddy beaches contain silica particles mixed with silt, clay and organic debris. Various other substances also contribute to sandy deposits; for example, fragments of shell, diatoms, calcareous algae, foraminifera, and, in low latitudes, coral.

The surface of a sandy beach is liable to disturbance by wind and wave, and provides no firm anchorage for superficially-attached plants and animals. Where the sand contains embedded stones, certain algae can grow attached to these, for example, *Chorda filum*. In sheltered areas the deposit may be sufficiently stable to allow rooting plants such as *Zostera* or *Spartina* to become established. A microscopic vegetation of diatoms and coloured flagellates sometimes exists in the interstices between the superficial sand grains. At low tide these form green or brown patches on the surface of the shore, and in some cases appear capable of rhythmical movement, retreating below the surface each time the tide advances. Along the backshore, accumulations of wind-blown sand may form sand dunes, and when these are sufficiently stable they become consolidated by a characteristic dune vegetation, for example American beach grass (*Ammophila breviligulata*: Fernald).

The animal population of sandy shores mainly dwells below the surface, but includes some species which at times emerge to crawl or swim. Along the upper parts of some sandy beaches is a zone where the sand dries out and air penetrates during intervals between periods of submergence. This is the littoral fringe of the sandy shore, which in the British Isles is often inhabited by the burrowing amphipod *Talitrus saltator* (*Talorchestia spp.*, in the U.S.), where there is much organic debris deposited along the strand line. It is an air-breather, usually restricted within a narrow strip of upper shore except when it moves over the surface for feeding at night (page 192). The strand line and adjacent sand also houses a diverse population of insects, for example *Bembidion laterale*, *Chersodromia arenaria*, larvae of *Coelopa frigida*.

Lower on the shore the superficial layers of sand may dry briefly during low tide, but capillary forces hold a water table some height above sea-level depending on the size of spaces between the sand grains. In sand containing much fine material the deposit remains waterlogged throughout the tidal cycle, but in coarse sands the water table may drop considerably as the tide recedes. Even where there is air in the sand its humidity is high, and burrowing creatures are not in appreciable danger of desiccation. The deeper-dwelling forms are also well insulated from surface fluctuations of temperature and salinity, which seldom produce much effect below a depth of a few inches[24,59].

Wave action exerts a major control over the distribution of sand populations because it influences, directly or indirectly, many important features of the substrate, including its stability, particle size, gradient, drainage, oxygenation and organic content. Beyond a limit, strong wave action produces a virtually barren shore because no sizable organisms can withstand the crushing effects of deep churning in a coarse, shifting beach. In less wave-beaten conditions the sand is more stable, and survival becomes possible first for creatures that can burrow in fairly coarse material and do not require any permanent tube or burrow. These forms must be sufficiently robustly built to withstand pressures from sand movement, or must be able to avoid them by burrowing at a safe depth, or escaping by swimming. Coarse sand is also found in calmer conditions where the water is fairly fast-moving, for example the shores of tidal channels. Among the inhabitants of the coarser sands of the shores are *Donax spp.*, *Tellina spp.*, *Ensis spp.*, *Natica spp.*, *Portumnus spp.*, *Mercenaria mercenaria*, *Astropecten spp.*, *Hippa talpoida*.

With increasing shelter, movements of sand by wave action extend less deeply until only a superficial layer is disturbed. Below this the substrate is quite stable. Sheltered conditions permit the settlement of smaller particles and also favour the deposition of organic material, the latter providing food for a more numerous and diverse population than occurs in coarser, less stable sand. Sulphide-darkening of the subsurface layers indicates the activity of anaerobic bacteria in a deoxygenated medium. The burrowing population is then dependent upon oxygen obtained from the overlying water by drawing a current through the burrow, or extruding the respiratory organs above the surface of the sand. Where the surface dries at low tide, there must be some means of storing sufficient oxygen to survive this period.

The macrofauna of muddy sands of the shore includes *Arenicola marina*, *Amphitrite johnstoni*, *A. edwardsi*, *Nerine spp.*, *Notomastus latericeus*, *Scolecolepis spp.*, *Glycera spp.*, *Nephtys spp.*, *Venerupis spp.*, *Leptosynapta inhaerens*, *Mercenaria mercenaria*, *Palaemonetes vulgaris*, *Carcinus maenas*, *Crangon vulgaris*. There is also a diverse microfauna of flattened and threadlike forms living in the interstices of the sand[65], including the smallest known species from most of the invertebrate phyla.

Factors which vary with shore level, and influence the zonation of burrowing populations, include the duration of submergence which for many species determines the time available for feeding and respiratory exchange: also the particle-size of the substrate (page 179) and the range of hydrostatic pressure. A general indication of the zonation of some sand-dwelling species is given in *Figure 7.12*.

Muddy Shores[21]

The term 'mud' is loosely applied to deposits containing a high proportion of silt or clay particles. The finest particles settle only from still water, and

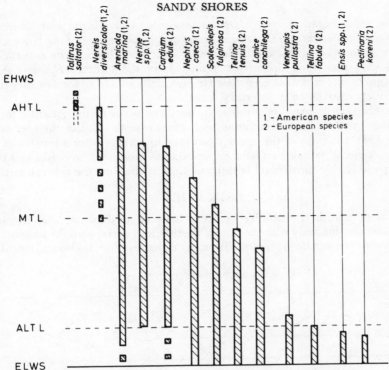

Figure 7.12. Zonation of some common inhabitants of intertidal sands

shores of mud are therefore found where conditions are normally calm and
without strong currents, for example at the landward end of sheltered inlets.
A muddy shore often merges into salt marsh where *Spartina* and other
halophytes become established. Except for drainage channels, these shores
have little slope, and extensive areas of 'mud flat' may be exposed as the tide
recedes. The mud surface seldom dries appreciably, and a layer of standing
water may be left at low tide. In these conditions organic debris readily
settles, and the organic content of the mud is correspondingly high, often
about 5 per cent of the total dry weight. The deposit usually compacts into a
soft, stable medium, easy for burrowing, in which permanent burrows and
tubes remain undisturbed. Because there is little exchange of interstitial
water the mud beneath the extreme surface layer becomes completely
deoxygenated and its sulphide content is high.

The population of sea-shore mud is made up of forms which can readily
tolerate silt, among which are *Nereis virens*, *Sabella spp.*, *Mya truncata*, *Capitella
capitata*, *Cereus pedunculatus*, *Upogebia deltaura*, *Nassa absoleba* and *Modidus
demissus*. The populations of muddy sand merge gradually into those of finer
mud, and the muddy shore population often includes *Arenicola marina*,

Cardium spp., *Amphitrite spp.*, *Marphysa sanguinea*, *Mya arenaria* and *Venerupis spp. Nassarius obsoleta* and *Ocenebra erinacea* (in Europe), may be found on rocky shores, occur also in soft mud, ploughing their way through the superficial layers with their long siphons extruding above the surface. *Hydrobia spp.*, *Nereis diversicolor* and *Macoma baltica* are essentially estuarine forms which are also widely distributed in intertidal mud.

Certain remote areas of mudflat in northern areas of the Atlantic form breeding grounds for the common seal, *Phoca vitulina*. Unlike the grey seal (page 198), the pups of this species can take to water within a few hours of birth. A gravid female can land on an exposed mud bank to calve, and the young seal is able to swim off with its mother by the time the tide returns.

ESTUARIES

At a river mouth the salinity gradient between fresh water and the sea fluctuates continuously with the state of the tide, and varies with the amount of fresh water coming downstream. In many estuaries there is also an irregular

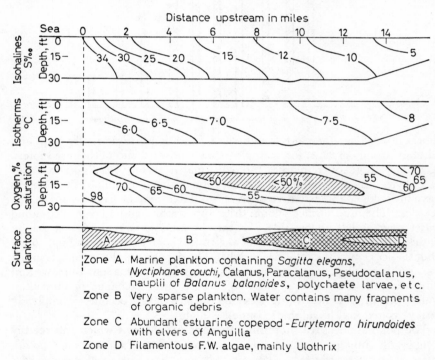

Zone A. Marine plankton containing *Sagitta elegans*, *Nyctiphanes couchi*, Calanus, Paracalanus, Pseudocalanus, nauplii of *Balanus balanoides*, polychaete larvae, etc.

Zone B Very sparse plankton. Water contains many fragments of organic debris

Zone C Abundant estuarine copepod - *Eurytemora hirundoides* with elvers of Anguilla

Zone D Filamentous F.W. algae, mainly Ulothrix

Figure 7.13. Section of Tyne estuary showing isohalines, isotherms, percentage oxygen saturation and distribution of dominant planktonts at 2 m depth

(Data obtained during a one-day students' field trip *18th March 1965*)

vertical salinity gradient between surface and bottom, with fresh water forming a layer above the denser salt water. Usually the estuary bottom and lower shore experience wider variations of salinity than the higher shore levels.

River water often carries large amounts of silt which become deposited at the river mouth. The inflow of sea-water along the floor of an estuary also tends to carry sediments into the estuary mouth, adding to the material deposited by the river. In this way areas of mud flat are formed beneath expanses of calm shallow water where there is little movement beyond the rise and fall of the tide. In these shallows, cooling and warming of the water produce wider changes of temperature than occur in river or sea. The lowest temperatures frequently coincide with the lowest salinities because in winter the outflow of fresh water may be greatly increased by rainfall or the rapid melting of snow. In summer, high temperatures and high salinities exist together when drought periods reduce the amount of fresh water at the same time as sunshine is heating and evaporating the water. Where industrial plant is sited near a river mouth, cooling installations may pour out quantities of warm water, so that the temperature of the estuary becomes continuously raised.

In narrow parts of an estuary, tidal ebb and flow of great volumes of water can produce powerful currents and heavy scouring of the bottom, particularly when the water carries much silt. Pollution by sewage or industrial effluents produces various effects. Addition of organic matter to the water may promote rapid bacterial growth, leading to a reduction in oxygen (*Figure 7.13*) especially where the water is stratified. The end-products of organic decomposition enrich the water with plant nutrients which may sustain a high level of production where conditions are favourable. Effluents containing solid matter reduce illumination and increase silting. The effects of chemical wastes depend upon their nature, but toxic materials can have disastrous effects upon the estuarine fauna and flora. The prevalence of fine silts in estuarine deposits leads to severe deoxygenation in the mud.

The population of estuaries therefore face peculiar and difficult conditions. At the seaward end there is always some penetration by marine species, and estuarine shores near the mouth are commonly inhabited by ordinary littoral forms, for example, *Balanus balanoides*, *Chthamalus stellatus*, *Acmaea testudinalis*, *Littorina littorea*, *L. saxatilis*, *Mytilus edulis*, *Thais lapillus*, *Crangon vulgaris*, *Arenicola marina*, *Palaemonetes vulgaris* and *Carcinus maenas*. The distance these extend upstream depends partly upon their powers of osmotic adjustment or osmo-regulation, and partly upon the protection afforded by shells, tubes or deep burrows into which some species retire during periods when the salinity falls below a safe level. In mid-estuary, where the widest and most rapid fluctuations of salinity occur, only extremely euryhaline forms can survive. This typically estuarine community (*Figure 7.14*) often includes *Enteromorpha intestinalis*, *Fucus ceranoides*, *Corophium volutator*, *Hydrobia spp.*, *Nereis diversicolor*, *Scrobicularia plana*, *Macoma baltica*, *Carcinus maenas*, *Sphaeroma rugicauda*,

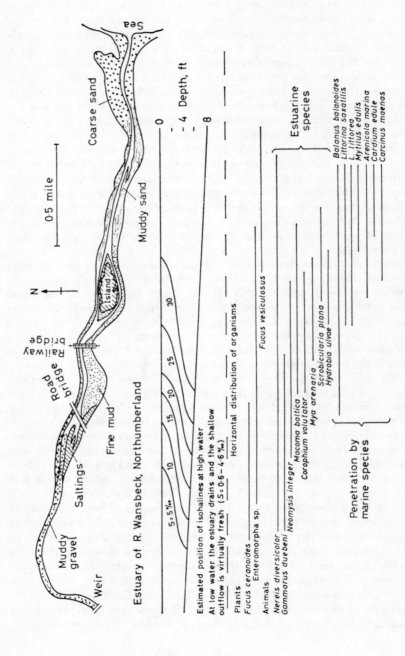

Figure 7.14. *Distribution of the benthos of an estuary in England*
(Data from a one-day students' field trip *1st March 1967*)

Gammarus spp., *G. duebeni* and *Balanus improvisus*. Certain mysids are numerous in brackish water, for example *Neomysis integer*, and in some estuaries there are often shoals of whitebait, i.e. small herrings and sprats (page 241). The plankton sometimes contains dense patches of the copepod (*Figure 7.13*).

In the upper parts of estuaries the gymnoblast hydroid *Cordylophora caspia* occurs on stones or wooden piers. The prawn *Palaemonetes vulgaris* is some-times very numerous in salt-marsh pools. At these levels, part of the popula-tion is of freshwater or terrestrial origin. There are often many larvae of midges (*Chironomus*) and mosquitoes (*Aedes*), also oligochaetes (*Tubifex*) and a variety of beetles (*Colymbetes, Ochthebius*) and bugs (*Notonecta, Sigaria*).

The distribution of many estuarine species changes seasonally. The flounder *Pseudopleuronectes americanus*, a common estuarine fish, migrates up-stream in summer but returns so the sea during the colder months for spawning. *Carcinus maenas* and *Crangon vulgaris* also move up estuaries in summer and seawards in winter, their movements being influenced by changes of both temperature and salinity because these species osmo-regulate less effectively as the temperature falls. Others make the reverse movements. *Pandulus montagui* goes out to sea in summer and into estuaries in winter.

River mouths where there are wide expanses of sand and mud are im-portant feeding grounds for many species of birds, especially in winter. Among these are the heron (*Ardea cinerea*), the oystercatcher (*Haematopus ostralegus*), the curlew (*Nemenius arquata*), the dunlin (*Calidris alpina*) and a variety of plovers, sandpipers, ducks and geese.

Mangrove Swamps

In tropical and subtropical areas a characteristic feature of the upper levels of estuaries and sheltered parts of the sea-shore is the mangrove swamp forest. This develops on mud flats which are exposed at low tide. The genus *Rhizophora* is a common mangrove tree which grows to large size, supported above the surface of the mud on a number of downcurving prop-roots resembling flying buttresses. These prop-roots contain air-spaces which provide oxygen for the underground root system embedded in the water-logged, oxygen-deficient mud. In some genera there are also aerial roots growing vertically out of the mud as slender, erect structures. There may also be 'pillar-roots' which support the branches. This elaborate rooting system reduces water movement and entraps and stabilizes the mud so that the mangrove forest tends to increase in extent, forming broad flat areas of swamp cut by drainage channels through which the sea flows with rise and fall of the tide. The spread and consolidation of the forest is assisted by a reproductive peculiarity of some mangrove species. These have seeds which germinate while still within the fruit borne on the tree, producing a long slender radicle growing downwards from the branch sometimes as much as 60 cm before dropping from the tree to stick into the mud. As the forest advances, the vegetation becomes zoned between land and sea with different species of mangrove at each level and different communities of smaller salt-marsh plants between the trees.

The mangrove swamp harbours a complicated community of animals. The roots of the trees provide a secure substrate for a variety of attached animals, especially barnacles, bivalves, serpulid worms and tunicates. Fish and free-living molluscs and crustaceans find shelter in the crannies between the roots. In the mud are large numbers of burrowing crabs, molluscs and fish, and the branches of the trees contain insects, lizards, snakes and birds. Details of some shores of this type, common at low latitudes, can be found in references at the end of this chapter[44,66].

Books   REFERENCES AND FURTHER READING

[1] Barrett, J. H. and Yonge, C. M. (1964). *Pocket Guide to the Sea-Shore.* London; Collins

[2] Dakin, W. J. (1953). *Australasian Seashores.* London; Angus and Robertson

[3] Clancy, Edward P. (1968). *The Tides—pulse of the Earth.* New York; Doubleday & Co.

[4] Eales, N. B. (1967). 4th Edn. *The Littoral Fauna of the British Isles.* Cambridge University Press

[5] Fisher, J. and Lockley, R. M. (1954). *Sea Birds: an Introduction to the Natural History of the Sea Birds of the North Atlantic.* London; Collins

[6] Fraenkel, G. S. and Gunn, D. L. (1961). *The Orientation of Animals.* Dover edition

[7] Green, J. (1968). *The Biology of Estuarine Animals.* London; Sidgwick and Jackson

[8] Bascum, W. (1964). *Waves and Beaches.* New York; Doubleday & Co.

[9] Lewis, J. R. (1964). *The Ecology of Rocky Shores.* London; English University Press

[10] Morton, J. and Miller, M. (1968). *The New Zealand Sea Shore.* London and Auckland; Collins

[11] Newell, R. C. (1970). *The Biology of Intertidal Animals.* London; Logos Press

[12] Nicol, J. A. C. (1960). *The Biology of Marine Animals.* London; Pitman

[13] Pilkington, R. (1954). *The Ways of the Sea.* (A simple account of waves, tides, etc.) London; Routledge and Kegan Paul

[14] Ricketts, E. F. and Calvin, J. (1962). Revised Hedgpeth, J. W. *Between Pacific Tides.* Stanford, California; Stanford U.P.

[15] Southward, A. J. (1965). *Life on the Sea-Shore.* London; Heinemann

[16] Steers, J. A. (1953). *The Sea Coast.* London; Collins

[17] Yonge, C. M. (1949). *The Sea Shore.* London; Collins

Papers

[18] Arnold, D. C. (1957). 'The Responses of the Limpet, *Patella vulgata*, to Waters of Different Salinities.' *J. mar. biol. Ass. U.K.* **36**, 121

[19] Arnold, D. C. (1959). 'The Reactions of the Limpet, *Patella vulgata*, to certain of the Ionic Constituents of Seawater.' *J. mar. biol. Ass. U.K.* **38**, 569

[20] Ballantine, W. J. (1961). 'A Biologically Defined Exposure Scale, for the Comparative Description of Rocky Shores.' *Fld Stud.* **1** (3), 1

[21] Bassindale, R. and Clark, R. B. (1960). 'The Gann Flat, Dale: Studies on the Ecology of a Muddy Beach.' *Fld Stud.* **1** (2), 1

[22] Boney, A. D. (1965). 'Aspects of the Biology of Seaweeds of Economic Importance.' *Adv. Mar. Biol.* **3**, 105

REFERENCES AND FURTHER READING

[23] Brafield, A. E. and Newell, G. E. (1961). 'The Behaviour of *Macoma balthica*.' *J. mar. biol. Ass. U.K.* **41,** 81

[24] Bruce, J. R. (1928). 'Physical Factors on the Sandy Beach. Pt. I. Tidal, Climatic and Edephic. Pt. II. Chemical Changes. Carbon Dioxide Concentration and Sulphides.' *J. mar. biol. Ass. U.K.* **15,** 535 and 553

[25] Connell, J. H. (1961). 'Effects of Competition, Predation by *Thais lapillus*, and other Factors on Natural Populations of the Barnacle, *Balanus balanoides*.' *Ecol. Monogr.* **31,** 61

[26] Crisp, D. J. (1958). 'The Spread of *Eliminius modestus* in North-West Europe.' *J. mar. biol. Ass. U.K.* **37,** 483

[27] Crisp, D. J. (1961). 'Territorial Behaviour in Barnacle Settlement.' *J. exp. Biol.* **38,** 429

[28] Crisp, D. J. and Southward, A. J. (1958). 'The Distribution of Intertidal Organisms along the Coasts of the English Channel.' *J. mar. biol. Ass. U.K.* **37,** 157

[29] Crisp, D. J. and Stubbings, H. G. (1957). 'Orientation of Barnacles to Water Currents.' *J. Animal Ecol.* **26,** 179

[30] Davidson, P. E. (1968). 'The Oystercatcher, a Pest of Shellfish.' In '*The Problems of Birds as Pests*'; Editors Murton, R. K. and Wright, E. N. London and New York; Academic Press

[31] Davies, P. S. (1966). 'Physiological Ecology of *Patella*. I. Effect of Body Size and Temperature on Metabolic Rate.' *J. mar. biol. Ass. U.K.* **46,** 647.

        (1967). 'II. Effect of Environmental Acclimation on Metabolic Rate.' *J. mar. biol. Ass. U.K.* **47,** 61

[32] Evans, F. G. C. (1951). 'Analysis of the Behaviour of *Lepidochitona cinereus*.' *J. Animal Ecol.* **20,** 1

[33] Evans, F. (1961). 'Responses to Disturbance of the Periwinkle *Littorina punctata* on a Shore in Ghana.' *Proc. zool. Soc. Lond.* **137,** 393–402

[34] Fairbridge, R. W. (1960). 'The Changing Level of the Sea.' *Scient. Am.* May

[35] Gibson, R. N. (1969). 'The Biology and Behaviour of Littoral Fish.' *Oceanogr. Mar. Biol. Ann. Rev.* **7,** 367

[36] James, B. L. (1968). 'The Distribution and Keys of Species in the Family Littorinidae and their Digenean Parasites in the Region of Dale, Pembrokeshire.' *Fld Stud.*, **2** (5), 615

[37] Jones, D. A. and Naylor, E. (1970). 'The Swimming Rhythm of the Sand Beach Isopod *Eurydice pulchra*.' *J. Exp. Mar. Biol. Ecol.* **4,** 188

[38] Jones, W. E. and Demetropoulos, A. (1968). 'Exposure to Wave Action: Measurements of an Important Ecological Parameter on Rocky Shores in Anglesey.' *J. Exp. Mar. Biol. Ecol.* **2,** 46

[39] Knight-Jones, E. W. and Stevenson, J. P. (1950). 'Gregariousness During Settlement in the Barnacle *Elminius modestus*.' *J. mar. biol. Ass. U.K.* **29,** 281

[40] Knox, G. A. (1960). 'Littoral Ecology and Biogeography of Southern Oceans.' *Proc. Roy. Soc. B.*, **152,** 577

[41] Knox, G. A. (1963). 'Biogeography and Intertidal Ecology of the Australasian Coasts.' *Oceanogr. Mar. Biol. Ann. Rev.* **1,** 341

[42] Lawson, G. W. (1966). 'Littoral Ecology of West Africa.' *Oceanogr. Mar. Biol. Ann. Rev.* **4,** 405

[43] Longbottom, M. R. (1970). 'The Distribution of *Arenicola marina* with Reference to Effects of Particle Size and Organic Matter in Sediments.' *J. Exp. Mar. Biol. Ecol.* **5,** 138

[44] Macnae, W. (1968). 'A General Account of the Fauna and Flora of Mangrove Swamps and Forests in the Indo-West Pacific Region.' *Adv. Mar. Biol.* **6,** 73

[45] McIntyre, A. D. (1970). 'The Range of Biomass in Intertidal Sand, with Special Reference to the Bivalve, *Tellina tenuis*.' *J. mar. biol. Ass. U.K.* **50,** 561

[46] McLusky, D. S. (1967). 'Some Effects of Salinity on the Survival, Moulting and Growth of *Corophium volutator*.' *J. mar. biol. Ass. U.K.* **47**, 607

[47] Meadows, P. S. and Reid, A. (1966). 'The Behaviour of *Corophium volutator*.' *J. Zool. Lond.* **150**, 387

[48] Morton, J. E. (1960). 'Responses and Orientation of the Bivalve, *Lasaea rubra*.' *J. mar. biol. Ass. U.K.* **39**, 5

[49] Morton, J. E., Boney, A. D. and Corner, E. D. S. (1957). 'The Adaptations of *Lasaea rubra*, a Small Intertidal Lamellibranch.' *J. mar. biol. Ass. U.K.* **36**, 383

[50] Moyse, J. and Nelson-Smith, A. (1963). 'Zonation of Animals and Plants on Rocky Shores around Dale, Pembrokeshire.' *Fld. Stud.* **1** (5), 1

[51] Naylor, E. (1958). 'Tidal and Diurnal Rhythms of Locomotory Activity in *Carcinus maenas*.' *J. exp. Biol.* **35**, 602

[52] Newell, G. E. (1948). 'A Contribution to our Knowledge of the Life History of *Arenicola marina*.' *J. mar. biol. Ass. U.K.* **27**, 554

[53] Newell, G. E. (1958). 'The Behaviour of *Littorina littorea* under Natural Conditions and its Relation to Position on the Shore.' *J. mar. biol. Ass. U.K.* **37**, 229

[54] Newell, G. E. (1958). 'An Experimental Analysis of the Behaviour of *Littorina littorea* under Natural Conditions and in the Laboratory.' *J. mar. biol. Ass. U.K.* **37**, 241

[55] Newell, G. E. (1965). 'The eye of *Littorina littorea*.' *Proc. zool. Soc. Lond.* **144**, 75

[56] Newell, R. (1962). 'Behavioural Aspects of the Ecology of *Peringia ulvae*.' *Proc. zool. Soc. Lond.* **138**, 49

[57] Newell, R. (1965). 'The Role of Detritus in the Nutrition of Two Marine Deposit Feeders, the Prosobranch *Hydrobia ulvae* and the bivalve *Macoma balthica*.' *Proc. zool. Soc. Lond.* **144**, 25

[58] Pardi, L. (1960). 'Innate Components in the Solar Orientation of Littoral Amphipods.' *Cold Spring Harb. symp. quant. Biol.* **25**, 395

[59] Reid, D. M. (1930). 'Salinity Interchange between Sea Water in Sand, and Overflowing Fresh Water at Low Tide.' *J. mar. biol. Ass. U.K.* **16**, 609

[60] Southward, A. J. (1958). 'The Zonation of Plants and Animals on Rocky Sea Shores.' *Biol. Rev.* **33**, 137

[61] Southward, A. J. (1958). 'Temperature Tolerances of some Intertidal Animals in Relation to Environmental Temperatures and Geographical Distribution.' *J. mar. biol. Ass. U.K.* **37**, 49

[62] Stephen, A. C. (1953). 'Life on Some Sandy Shores.' p. 50, *Essays in Marine Biology*. Edinburgh and London; Oliver and Boyd

[63] Stephenson, T. A. (1942). 'Causes of Vertical and Horizontal Distribution of Organisms between Tidemarks in South Africa.' *Proc. Linn. Soc. Lond.* **154**, 219

[64] Stephenson, T. A. (1953). 'The World between Tidemarks.' p. 73, *Essays in Marine Biology*. Edinburgh and London; Oliver and Boyd

[65] Swedmark, B. (1964). 'The Interstitial Fauna of Marine Sand.' *Biol. Rev.* **39**, 1

[66] Thom, B. G. (1967). 'Mangrove Ecology and Deltaic Geomorphology.' *J. Ecol.* **55**, 301

[67] Wells, G. P. (1949). 'The Behaviour of *Arenicola marina* in Sand.' *J. mar. biol. Ass. U.K.* **28**, 465

[68] Wells, M. J. (1965). 'Learning by Marine Invertebrates.' *Adv. Mar. Biol.* **3**, 1

[69] Williamson, D. I. (1951). 'Studies in the Biology of Talitridae: Visual Orientation in *Talitrus saltator*.' *J. mar. biol. Ass. U.K.* **30**, 91

[70] Wilson, D. P. (1968). 'The Settlement Behaviour of the Larvae of *Sabellaria alveolata*.' *J. mar. biol. Ass. U.K.* **48**, 367

[71] Yonge, C. M. (1953). 'Aspects of Life on Muddy Shores.' p. 29, *Essays in Marine Biology*. Edinburgh and London; Oliver and Boyd

[72] Yonge, C. M. (1963). 'The Biology of Coral Reefs.' *Adv. Mar. Biol.* **1**, 209

CHAPTER 8

SEA FISHERIES

FISHING METHODS

Even with the refinements of modern science and technology, fishing remains an essentially primitive method of obtaining food. In our exploitation of the fish stocks of the sea, we still behave mainly as nomadic hunters or trappers of natural populations of animals living in the wild state. There are broadly three ways of capturing fish: they may be scooped out of the water by means of a bag of fishnet such as a trawl or seine; they may be enticed to bite upon a baited hook attached to a fishing line; or they may be snared or entangled in some form of trap such as a gill-net. In whaling and swordfishing the quarry is pursued and speared with a harpoon.

Each method is modified in innumerable ways to suit local conditions, and the following brief account refers to the fishing techniques used commercially on both sides of the Atlantic[1]. Here the fisheries comprise two major groups, the *demersal* and the *pelagic* fisheries. Demersal fishing takes place on the sea-floor for species which live mainly close to the bottom, for example cod, haddock, whiting, yellow tail, ocean perch and plaice and sole. A main method of demersal fishing is by trawl. Seines and long lines are also used. Pelagic fisheries seek shoals of fish which, while they may roam throughout a considerable depth of water, are caught near the surface, for example herring, mackerel, bluefin tuna and menhaden. The chief pelagic fisheries make use of gill-nets, ring-nets and purse-seines.

Demersal Fishing

The Beam Trawl

Beam trawls have been in use for several hundred years, and were important in the days when fishing vessels were sail-driven. Although less used in modern commercial fisheries, beam trawls are still sometimes operated from small boats by inshore fisheries where the modern otter trawl is too large. Being easy to handle from small craft, small beam trawls also have applications in biological work.

The beam trawl (*Figure 8.1*) is a tapering bag of netting which can be towed over the sea-bed. The mouth of the bag is held open by a long beam, the ends of which are supported about 2 ft above the sea-bed by a pair of strong metal runners. The upper leading edge of the bag is attached to a strong headrope lashed to the beam. The under part of the bag, which drags over the bottom, is attached to a considerably longer groundrope.

As the trawl is towed, the groundrope trails behind the headrope so that fish disturbed on the bottom by the groundrope are already enclosed under the upper part of the net.

The hind part of the trawl-net tapers to a narrow sleeve of stronger, finer-mesh netting known as the cod end, within which the captured fish accumulate. The rear of the cod end is tightly closed by a rope, the cod line. When the trawl is hauled up to the vessel the catch can be released by untying the cod line, letting the fish out of the back of the bag.

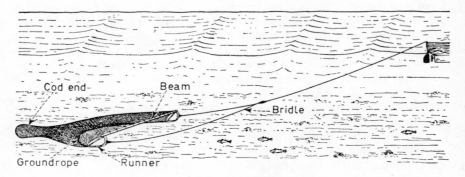

Figure 8.1. The beam trawl

The size of a beam trawl is limited by the length of beam which it is practicable to handle. The largest nets were sometimes as much as 50 ft in width. During the latter years of the nineteenth century, the introduction of steam trawlers capable of towing much heavier nets caused the beam trawl to become almost entirely superseded by the larger, more easily handled and more efficient otter trawl except for inshore use.

The Otter Trawl

This is now the chief method of demersal fishing. The otter trawl (*Figure 8.2*) has a bag of netting resembling that of the beam trawl in general shape, but considerably larger. The sides of the bag are extended outwards by the addition of wings of netting attached to large, rectangular, wooden 'otterboards'. These otterboards are towed by a pair of very strong steel cables, the warps, which are attached to the otterboards in such a way that the pressure of water causes the otterboards to diverge as they move, pulling the mouth of the net wide open horizontally. The under-edges of the otterboards slide over the sea-bed, and are shod with steel for protection.

The headrope, to which the upper lip of the trawl-net is laced, is usually some 100–130 ft in length, and bears numerous hollow metal floats which keep it a few feet above the bottom. Sometimes, elevator boards known as 'kites' are fitted to the headrope to increase the gape of the net. The lower lip and groundrope are considerably longer, about 140–180ft, and trail several feet behind the headrope during trawling. The groundrope is a heavy, steel-wire rope, carrying on its central part a number of large steel bobbins, about 2 ft in diameter, and on its lateral parts several large rubber discs. These help the trawl to ride over obstructions on the sea-bed. A bottom trawl cannot be used, however, on very rough ground or where wrecks may snag the net. Large nets cost over $1,000 and loss or damage are expensive.

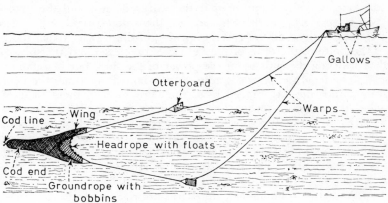

Figure 8.2. The otter trawl

At the junction of the body of the trawl and the cod end, there is a vertical flap of netting, the flopper, which hangs down and acts as a valve to prevent the escape of fish from the cod end if the trawl should stop moving. The under part of the cod end is covered with nylon to protect it from chafing on the bottom, and a cod line is tied to close the free end. The trawl wings may be joined directly to the otterboards, but nowadays trawls of the Vigneron Dahl type are commonly used, in which long cables about 180 ft in length are inserted between the wings and the otterboards. This simple modification gives the trawl a greater sweep, disturbing more fish and considerably increasing the catching-power of the gear.

For shooting and hauling the trawl, the sides of the trawler carry pairs of strong gallows supporting the blocks over which the trawl warps run. On most vessels, starboard is the traditional working side for the trawl; but many trawlers are now fitted with gallows on both sides, and carry two trawls, working each side alternately. On the deck is a powerful winch for winding the warps.

The deck of the trawler can be divided into several compartments, the fishpounds, by walls of movable partitioning about 2 ft high. These are put into position before fishing commences and the captured fish are sorted into them. On many ships the hull is divided into three fish-rooms for storing the catch and before leaving port the forward fish-room is filled with ice for preserving the fish. Nowadays an increasing number of vessels do not rely on ice but are fitted with refrigeration equipment.

When shooting the trawl, the vessel is manoeuvred with the working side to windward. The warps are paid out until the otterboards are well sub-merged, and the trawl net is slung overboard. As the warps are being run out, the vessel moves full-speed ahead on a course turning towards the working side, thereby ensuring that the warps do not foul the propeller. The two warps carry markers to indicate the amount paid out, and must be of equal length to ensure correct opening of the trawl. When the trawl strikes the bottom, the drag of the net slows the vessel and the subsequent speed of trawling depends upon the power of the ship, but is usually some 3 knots. The length of warps is adjusted to about three times the depth of the water, and they are braced in at the stern to keep them clear of the propeller, the strain of towing being taken on the winch brakes.

The first tow is usually a short one to determine the likely weight of catch. The duration of each tow is then judged by the size of the previous catch, and may be from as little as 15 min on exceptionally rich fishing grounds, to tows of several hours. The trawl must not be allowed to overfill because of the danger of splitting the net and losing the catch.

When the net is to be hauled, the warps are released from the stern and wound-in over the gallows by the winch. The vessel steams ahead gradually turning until the working side is to windward and the otterboards are drawn up to the gallows. The net is hauled to the side of the vessel and man-handled aboard. The cod end usually comes up to the surface with a rush, due to the distension of the swim-bladders of fish brought up quickly from the bottom. When the cod end comes alongside, it is brought aboard by a hoist from the mast-head, and suspended about 3 ft above the deck. The cod line is then untied and the fish discharged on to the deck.

Meantime, the deck has been divided into fishpounds and the fish-room floors have been covered with ice. Before storing, the fish are gutted and the heads removed from the larger species such as cod, pollock and haddock. The livers are often kept and boiled-down on board in large steam vats to extract the oil. The different species of fish are sorted into separate pounds, thoroughly washed by hosing, and are then passed down a chute into the fish-rooms where they are packed with ice shovelled around them.

The American North Atlantic Trawler Fleet

Introduced into New England water in 1905, Otter Trawling still remains a most efficient method of offshore fishing, and accounts for 46% of the fish landed in the region. The trawlers fish the Continental Shelf and range

212

from the Grand Bank south to Virginia. The largest vessels may bring in catches of 9,000 kg, and though each trip is made to catch one or two particular species, virtually every species of fish native to the shelf will be taken by the otter trawl at one time or another.

There about 900 North Atlantic vessels in the otter trawl fleet. The vessel sizes vary greatly and the total number of fishermen on all vessels is about 4,000. The New England States represent about 80% of the industry in which North Atlantic trawl-caught fish is landed. In fact, about 55% of all domestically produced fillets and steaks consumed in the United States, are supplied by this New England fishery. During 1967 the landings of trawl-caught fish in the North Atlantic amounted to 240 million kilograms and was sold for $41 million. These figures represent about 70% of the total quantity, and about 80% of the total value, of all U.S. trawl-caught fish.

Trawlers used to be powered by coal-fired, steam reciprocating engines. Apart from their proved reliability, these engines had the special advantage for fishing vessels that they could be operated over a wide range of engine speeds. They are, however, greedy for fuel and correspondingly expensive to run, and are being superseded in modern trawlers by more efficient diesel engines with lower running costs. The diesel engine runs best at a fixed speed; and the introduction of the variable-pitch propeller, enabling the engine speed to remain constant during manoeuvring, has been largely responsible for the acceptance of this type of engine on fishing vessels. Trials are also proceeding in the use of diesel-electric propulsion for trawlers. More efficient engines together with the development of advanced electronic equipment for navigation and fish detection have greatly increased the speed and catching power of modern trawlers.

The stern trawl has been an important development in fishing technique during recent years, requiring departures from the traditional design of trawlers. The stern trawler has a slipway built into the stern of the vessel, up which the net is hauled to the deck. This allows greater mechanization of the shooting and hauling of the trawl, making the trawlerman's life less arduous and speeding up the handling of the net between tows.

A major problem facing the British trawler fleet is the scarcity of fish on the nearer fishing grounds, due to their gradual exhaustion by intensive fishing. Alarmed at the depletion of this important natural resource, some nations with fishing grounds near their coasts have attempted to conserve them by extending their territorial fishing-limits, and British trawlers are now barred from some of the sources of fish that used to be most profitable to them. Consequently, they have to travel further afield to obtain their catches, and about half the fish landed in the British Isles is now brought in from the distant fishing grounds around Iceland, Greenland and in the Arctic.

Long voyages not only increase the cost of fishing but also raise severe problems connected with the preservation of the catch. Fish is a rapidly perishable commodity, and deteriorates fairly quickly even when stored in ice. It can be kept in fairly good condition in ice for about a week, but within a fortnight it is stale. This effectively limits the duration of fishing for distant-water trawlers if they rely solely on ice for preservation. Although modern fishing vessels have powerful engines and can make good speed, most of the distant-water fishing grounds lie several days voyage from port, by which time a considerable part of the catch has deteriorated to some extent and only the last-caught fish can reach the market in first-class condition. The average round trip of distant-water trawlers lasts from 1 to 5 weeks and these vessels are seldom able to fish long enough to fill their holds because, by that time, too much of the earlier-caught fish would be spoilt. The trawler skipper is therefore kept informed by radio of market conditions and the movements of other fishing vessels, and is advised when to cease fishing and dash for port to obtain the best price for the catch.

Clearly, something better than ice is now needed to preserve fish between the time they are caught and the time they reach the consumer. Some improvement in the period of effective preservation in ice is obtainable by the addition of antibiotics, but the most notable advance in fish-preservation has been the development of deep-freezing techniques. At temperatures around $-30°C$ fish can be kept without serious deterioration for long periods, but for good results they must be frozen as soon as possible after capture, certainly within 3 days. This presents an obvious problem where the fishing grounds are several days' voyage from port unless it is possible to deep-freeze the fish at sea. To do this, some trawlers now carry refrigeration equipment so that some or all of their catch can be frozen immediately, and the duration of their voyage can thus be extended. It is not essential to freeze the entire catch because the later-caught fish can be brought to market in good condition in ice.

When fish are frozen aboard, they are often preserved as whole fish which may be thawed later and dealt with through the normal wet-fish marketing channels. These fish are also suitable for smoke-curing. Alternatively, the fish may be filleted before freezing. This requires a staff of fish-processors in addition to the normal complement of fishermen, and a vessel larger than the normal trawler is needed. In consequence, various types of factory-ship have been developed. This may be an enlarged trawler, equipped both to catch and to process the fish, or it may be a larger ship equipped solely for processing and storing great quantities of fish, and operating in conjunction with a fleet of trawlers which bring their catches to it. On these vessels the fish are filleted and frozen, the oil extracted from the livers and the offal converted into fish-meal for cattle food or fertilizer. There are now many foreign-owned stern trawlers freezing their catch at sea.

At present about 25 per cent of the fish supply of the U.K. is deep-frozen. Preservation by freezing has advantages in addition to making possible longer fishing voyages in distant waters. Because frozen fish can be kept for long periods without deterioration, the market supply can be regulated independently of seasonal variations in the quantities of fish caught. Freezing has also opened the market to a wider range of species, because the 'fishstick' has proved a very popular commodity. This is a block of frozen filleted fish, coated in batter and requiring only to be cooked through before serving. Presented in this form, various nutritious fish can be marketed which are not ordinarily bought by the general populace.

The Spanish Trawl (Pareja)

The Spanish trawl (*Figure 8.3*) is a very large net, similar in general principle to other bottom trawls, but towed by two vessels working together. This trawl has no otterboards, the mouth of the net being held open horizontally by the lateral pull of the warps, one of which is attached to each vessel.

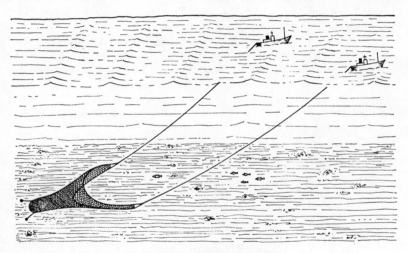

Figure 8.3. The Spanish trawl

The net has a wide sweep and great catching-power, the headrope sometimes being as much as 300 ft in length. The mesh is constructed of lighter material than the otter trawl, and generally produces a catch in rather better condition. In spite of its great size, this net can be towed by a pair of relatively low-powered vessels.

The Spanish trawl can be operated in deep water, and, although not widely used by North Atlantic fishermen, it has found some favour on British ships fishing for hake down to 600 m on the continental slope.

The Danish Seine

The Danish seine (*Figure 8.4*) is a light-weight net for taking fish from the bottom, mainly used in shallow water by small motor vessels. The net consists of a central bag with lateral wings extending 80–100 ft on either side. The groundrope is weighted, and the headrope buoyed by small floats.

There are several ways in which this net may be operated. Usually one end of a warp is attached to an anchored buoy, and the vessel steams down tide letting out a great length of warp, sometimes as much as 2 miles. The vessel now turns at right angles and shoots the net across the tide. A second warp is attached to the net, and the vessel turns and steams back to the buoy running out the second warp. Both warps are then winched in, dragging the net over the sea bottom. The converging warps sweep the fish between them into the path of the net, gradually drawing the wings of the net together to enclose the catch, and finally the net is hauled up to the vessel.

Alternatively, the warps and net may be laid in a circular course, and then towed behind the vessel for some distance before hauling. This is termed fly-dragging.

The Danish seine carries no devices for clearing obstacles on the sea-bed, and is therefore only used where the bottom is fairly smooth. The condition of the catch is usually better than that obtained by trawl, and because many of the boats make only short voyages the catch is fresh. Fish may be brought up alive and can be carried live to port in sea-water tanks. In the British fishing fleet there are some 650 boats of this type, supplying an important part of the catch.

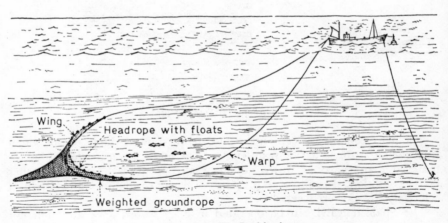

Figure 8.4. The Danish seine

Long Lines

Long-line fishing by North Atlantic vessels is confined mainly to areas where the sea bottom is too deep or too rough to be suitable for trawling. Line fishing vessels (i.e. line trawlers) operate around the banks of the North

Atlantic and down the continental slope to the north and west of the British Isles, and further afield off Rockall, Faroes, Iceland and Greenland. Halibut and cod comprise an important part of their catch.

Long-line fishing gear comprises a very long length of strong line bearing at intervals numerous short lengths of lighter line, the snells, which carry baited hooks (*Figure 8.5*). In laying the line, one end is anchored to the sea bottom and its position marked by a buoy. The liner steams along a straight course running out the full length of the line along the bottom, and the other end is marked by a second buoy. As the line is laid, the hooks are baited by hand with pieces of fish or squid. A long line may be nearly a mile in length, and a single vessel may sometimes shoot as many as 30 lines, carrying many thousands of hooks. Each line is left on the sea bottom for a few hours before hauling. As the line is brought aboard, the fish are removed from the hooks and the line carefully coiled for use again.

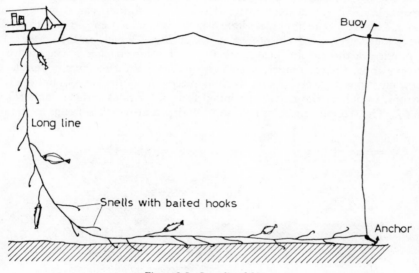

Figure 8.5. Long-line fishing

Various miniature versions of long-line fishing, for example small lines and haddock lines, are used in inshore waters around the British Isles, using shorter lines and small hooks suited to smaller fish.

Fishing by line is extremely laborious. Each hook has to be individually baited, and each captured fish removed from the hook. If the line becomes caught or entangled on the sea-bed, its recovery may take many hours; or the line and its catch may be lost. Line fishing is, however, widely practised throughout the world. The gear is simple and relatively inexpensive, and can be operated from small boats which need not be specially designed for this purpose. It is therefore well suited to the needs of communities where little capital is available for investment in fishing.

This primitive technique survives in the fishing industries of wealthier nations by avoiding direct competition with trawlers, exploiting fishing grounds where trawls cannot be operated and concentrating on valuable species such as halibut, which are not caught in quantity by trawl. To some extent, line fishing can be made selective by choice of bait and size of hook. Line-caught fish often fetch a high price because, being individually caught and handled, they can be brought to market in specially good condition. The capital costs of line fishing are kept low by the conversion to 'lining' of various obsolescent craft, such as old schooners.

Pelagic Fishing

Gill-Nets

Gill-nets are used to catch fish which form shoals near the surface, for example herring, mackerel, menhaden and sometimes tuna. A gill-net consists of a series of rectangular, light-weight nets joined end-to-end to form a very long vertical curtain of netting which hangs loosely in the water. The top edge of the curtain bears cork floats; the lower edge is weighted by a heavy rope, the messenger, by which the net is attached to the vessel. At the junctions between the individual pieces of netting are buoy-ropes, or strops, by which the net is suspended from a series of surface floats (*Figure 8.6*). The size and mesh of net, and the depth at which it hangs in the sea,

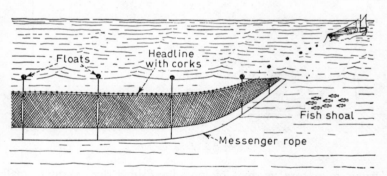

Figure 8.6. Gill-nets

are chosen to suit the particular fish sought. Herring gill-nets commonly have a mesh of 30–40 rows/yard, and the top edge of the net is usually suspended about 18 ft below the surface. The vertical dimension of the curtain is about 15 yards, and its total length may be more than a mile.

Gill-nets are shot shortly before darkness. The fishing vessel, known as a drifter, moves slowly down wind while the net is paid out by hand until the full length has been put into the water. The messenger rope from the foot of the net is brought to the bow of the drifter, which swings bow to wind and simply drifts attached to its long curtain of netting. During darkness, shoals of fish ascend into the surface layers, and become entangled as they attempt to swim through the net. The mesh size is selected so that the head of the fish passes easily through the net but the larger middle part of the body will not go through. When the fish try to wriggle out backwards, the net catches behind the gill-covers so that the fish are unable to escape. This method of capture is fairly selective, retaining only fish within a particular range of sizes. Smaller fish can swim right through, while larger fish may not pass sufficiently far into the mesh for the net to slip behind the gill-covers.

The net is hauled a few hours before dawn. The strain of hauling is taken on the strong messenger rope, which is wound in by a capstan or winch on the deck of the drifter. The net is carefully drawn in by hand over special rollers fitted on the side of the drifter. As the net comes aboard, the fish are shaken out and fall through hatches into the hold. Unless the catch is a poor one, herring drifters usually return to port after each night's fishing to sell their catch as early as possible.

Gill-nets are becoming superseded by ring nets, purse seines and pelagic trawls.

Ring Nets and Purse Seines

A ring net is a curtain of fine-mesh net hung vertically in the water to encircle surface shoals of fish. Around New England it is used mainly for capturing herring, although the technique is applicable to many other species which form shoals near the surface. It is specially suitable for inshore and enclosed waters such as sounds and estuaries where the enormous lengths of net used by drifters are unmanageable.

The size and mesh of ring nets are selected to suit local conditions and the species sought. Herring ring nets are up to 200 yards in length and 30–40 yards deep. The top of the net is buoyed by corks and floats at the surface. The lower edge of the net is attached to a weighted sole rope (*Figure 8.7*). A fine-mesh net is used because the fish are not entangled, as in a drift net, but are trapped by being surrounded by netting.

Ring nets are usually laid by a pair of vessels operating together. When a shoal of fish has been detected, nowadays often by echo-location, one end of the net is secured to one vessel while the other steers a semi-circular course paying out the net. The two vessels then steam on parallel courses, towing the net and finally turn to meet and enclose the shoal. A pair of messenger ropes attached to the lower edge of the net are winched in, closing the bottom of the net below the school so that the fish cannot escape by swimming underneath. The net is gradually hauled on board until the captured fish are densely crowded within the central part of the net. They are then scooped out of the net and brought aboard by a dip-net known as a brailer or by means of a large suction hose.

Purse seines are similar in principle to ring nets but much larger, sometimes as much as 400 yards in length and 90 yards in depth. Such huge nets

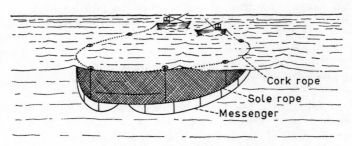

Figure 8.7. A herring ring net

require calm water for easy handling. They have proved very successful in capturing herring in the sheltered Norwegian Fjords, and are used by North American fishermen in the river-mouth salmon fishery of British Columbia, but there are few parts of the British coastline where they are suitable.

Some Recent Developments in Fishing

Plastic Nets

Synthetic resins have many applications in fishing as materials for nets, ropes and floats, and have several advantages over traditional materials. They are imperishable and can be made very strong. Plastic nets are therefore long-lasting, they can be made larger than twine nets and heavier weights of fish can be hauled up without tearing. Synthetic fibres are available in various densities, both lighter and heavier than water. By selecting materials of different densities for different parts of a net, its correct orientation in the water can be ensured. The use of transparent materials which are invisible in water may increase the efficiency of some fishing gear.

The virtual indestructibility of plastic materials may in time pose problems for fishermen. Nets lost at sea do not rot away, but remain as hazards to destroy fish or entangle fishing gear.

Echo-Location for Fish Detection

In the past, fishermen had to rely on experience and judgment for finding fish. Now a variety of sonic equipment is available to detect fish shoals. Both echo-sounding and echo-ranging instruments (*see* pages 43–5) have been developed for use on fishing vessels. An experienced operator can distinguish on the trace of the echo-recorder the signals reflected by fish, and may in some cases be able to identify the species by the characteristic pattern they produce.

Pelagic Trawling (mid-water trawls)

Floating trawls were developed mainly by Scandinavian and Icelandic fishermen and are now being increasingly used by British vessels, especially for catching herring. These mid-water trawls have applications not only in fishing for the usual pelagic species but also for the exploitation of new sources of fish in oceanic areas or for the capture of demersal species during periods when they leave the sea-bed.

Pelagic trawls are still in an experimental stage of development. They usually take the form of a conical net towed by a pair of vessels (*Figure 8.8*). The net is kept open by the divergent pull of the warps and by various otterboards, elevators and depressors attached to the mouth. The depth of the trawl can be regulated by the length of the warps and the speed of the vessel, but its control requires considerable skill. Accurate location of the fish shoals is obviously essential. Sonic techniques are being developed to enable the position of the net and its relation to the fish shoals to be accurately known.

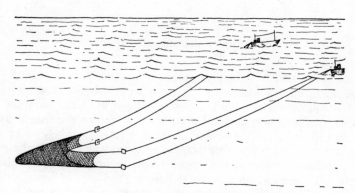

Figure 8.8. Pelagic trawling

The American Fishing Fleet

There are many different types of vessels in the American fishing fleet, ranging in size from inshore boats as little as 20 feet in length to factory ships of over 200 feet. The chief demersal and pelagic species comprising the recent landings of the New England fisheries, together with values and weights for each species, are summarized in Table 8.1.

SEA FISHERIES

TABLE 8.1

New England Fish Landings in 1968

Demersal Species	Value × $1,000	Weight × 1,000 lb
Flounders	12,169	106,231
Haddock	9,292	70,643
Cod	3,459	49,167
Whiting	2,843	85,091
Ocean Perch	2,377	61,323
Scup	2,110	12,135
Pollock	401	7,058
Butterfish	398	2,778
White hake	156	2,811
Cusk	84	1,350
Red Hake	52	1,304

Pelagic Species		
Herring	1,934	92,184
Menhaden	1,159	86,284
Mackerel	404	7,125

Experimental Fishing Methods

The Fish Pump

This novel method of fishing attempts to capture fish by sucking them out of the sea through a wide hosepipe in a powerful stream of water pumped aboard the fishing vessel. This is not as simple as it may appear. The disturbance of the water caused by pumping frightens the fish away, and only those quite close to the orifice of the hosepipe are likely to be drawn in. The fish must therefore be attracted in some way towards the hose so as to concentrate them near the opening before the pump is switched on. The method has been applied commercially in the Caspian Sea where there are species of Clupeonella which can be attracted close to the pump by light, but many fish will not come near to a powerful lamp. Another possibility is to attract fish to the pump electrically.

Electrical Fishing

When direct electric current (d.c.) is passed through water, a fish within the electrical field will turn and swim towards the anode. This is known as the anodic effect. The intensity of this effect depends upon the potential gradient to which the fish is exposed, and large fish are therefore influenced more than small ones. If the strength of the electric field is progressively increased, the fish eventually becomes paralysed and finally electrocuted.

These effects have been applied in various ways to the capture or enclosure of fish in fresh waters. In the sea, it is difficult to maintain electric fields of sufficient strength because of the high conductivity of the water. None the less, various experiments in marine electrical fishing have been conducted, and some commercial applications may ensue. For example, the anodic effect might be used to attract fish towards nets or fish pumps. As they approach the anode, the fish are likely to become paralysed by the increasing field strength and consequently unable to avoid capture.

Electrocution has been applied commercially to tuna fishing to cut short the struggles of fish caught by electrified hand lines, and electric harpoons have been used in whaling. It has recently been claimed that a system of electrodes attached to a shrimp trawl greatly increases the catch, pulses of current stimulating reflex muscle contractions which cause the shrimps to leap out of the sand or mud and become caught in the net.

Fish Landings in the North Atlantic

In 1969, the total value at first sale of fish landed in the United Kingdom was about $150 million while the value of landings in New England for 1968 was about $37 million. The weight and value of the main demersal and pelagic species of fish are summarized in Tables 8.1 and 8.2 and the chief fishing areas of the north-west Atlantic are depicted in *Figure 8.9*.

TABLE 8.2

U.K. Fish Landings in 1969

(Reproduced by courtesy of The White Fish Authority)

Species	Metric tons	Value (× £1000)	Value as % of total value of all wet fish	Percentage of weight taken by different gears							
				Bottom trawl	Pelagic trawl	Danish seine	Lines	Drift net	Ring net	Purse seine	Unspecified (a)
Cod	380,071	27,830	46·6	88·9	—	9·9	0·5	—	—	—	0·7
Haddock	143,040	11,412	19·1	57·6	—	40·5	0·3	—	—	—	1·6
Plaice	41,671	5,444	9·1	74·6	—	22·4	—	—	—	—	3·0
Whiting	33,601	1,968	3·3	40·9	—	54·0	0·1	—	—	—	5·0
Herring	123,698	3,287	5·5	—	43·6	—	—	14·3	31·1	11·0	—
Coalfish	47,314	1,525	2·6	97·2	—	2·3	0·3	—	—	—	0·2
Lemon soles	4,739	1,131	1·9	61·1	—	37·2	0·1	—	—	—	1·6
Skates and Rays	10,355	1,073	1·8	67·2	—	19·3	6·3	—	—	—	7·2
Hake	3,847	910	1·5	60·5	—	31·9	0·1	—	—	—	7·5
Dover sole	1,647	841	1·4	84·6	—	0·8	0·2	—	—	—	14·4
Halibut	1,520	616	1·0	69·0	—	3·7	26·8	—	—	—	0·5
Total 'top eleven'	791,503	56,037		66·4	6·8	16·4	0·5	2·2	4·9	1·7	1·2
Total all wet fish	884,337	59,678	93·9	63·0	11·4	15·4	0·9	2·1	4·4	1·5	1·3

a From landings by English vessels under 12·6 m (40 ft) in length and Scottish *Nephrops* trawl

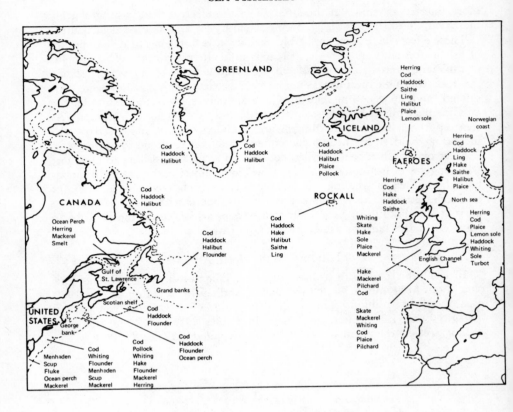

Figure 8.9. Important commercial species of fish caught in Atlantic areas

THE BIOLOGY OF SOME FOOD FISHES

The Cod—*Gadus morhua (callarias)*

Distribution

The cod (*Figure 8.10*) has an extensive range throughout the Arctic and the northern part of the North Atlantic. Although the isotherms do not set firm limits to its distribution, cod is most abundant in seas within the temperature range 0–10°C. It is found around Greenland (mainly on the west coast) and Iceland, in the Barents Sea and around Nova Zemlya. It occurs around the Faeroes, along the Norwegian coast, in the North Sea and Baltic, the Irish Sea, and the English Channel. On the eastern side of the Atlantic, the Bay of Biscay is as far south as cod extend in any numbers.

It is also found along the coasts of Labrador, Newfoundland and south along the North American coast as far as Virginia. In the northern part of the Pacific, a closely related form, *G. macrocephalus*, occurs over a wide area.

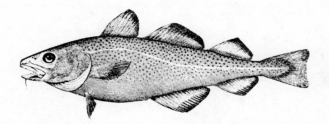

Figure 8.10. The Atlantic Cod, Gadus morhua

No other single species is of such importance as cod for human consumption. It is reckoned that between 300 million and 400 million are taken each year. They are mainly caught by trawl, and some are also captured by long lines. The fishmarket value of cod landed in the U.K. in 1969 was nearly $70 million, being nearly 47% of the value of all demersal landings. New England values were nearly $3·5 million and 1% respectively.

Life History

The following information relates chiefly to cod in the North Sea[5,6,43].

Spawning occurs between January and April, the peak spawning period in the North Sea usually being March to April. The fish collect in schools close to the sea-bed, mainly in water of between 60–100 m depth, and the female sheds between 3 and 7 million pelagic eggs, diameter 1·15—1·16 mm. The male cod sheds its milt (sperm) into the water and fertilization is external. The fertilized eggs are buoyant and gradually float to the surface.

Spawning areas are probably numerous and widespread. The detection of spawning grounds is made difficult by the close similarity between the eggs of cod and haddock, which also spawns at the same time in overlapping areas. It is only shortly before hatching that the eggs of the two species may be easily distinguished by differences of pigmentation. In the North Sea, cod eggs are mainly found to the north of the Dogger Bank, the more southerly waters being too shallow to be favoured by spawning shoals. From what is known of the distribution of cod larvae and the currents of the North Sea, it appears that the main spawning concentrations occur off Flamborough Head, in the Moray Firth and in the regions of the Ling Bank, the Great Fisher Bank, and Long Forties (*Figure 8.11*), but there are probably many other subsidiary spawning areas. Further afield, cod are known to spawn around Iceland, Faeroes and the Lofotens, also on the west coast of Greenland, on the Newfoundland Banks and along the Atlantic coast of North America.

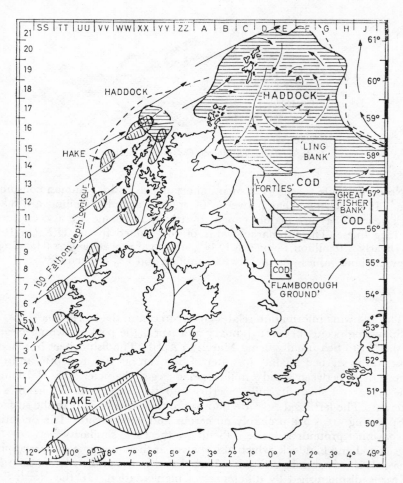

Figure 8.11. Main spawning areas of Cod, Haddock and Hake around the British Isles. The chart also forms a guide to the main nursery areas, towards which the pelagic fry are normally drifted by the currents, as indicated approximately by the arrows

(From Graham, M.[6] by courtesy of Edward Arnold)

After spawning the shoals disperse. The eggs take from 9 to over 20 days to hatch depending on the water temperature. In the central and southern North Sea, the eggs hatch after about 10–14 days. At the time of hatching the larva is about 4 mm long, the mouth has not yet formed and the animal is at first entirely dependent for food on the ventrally-attached yolk-sac beneath which it floats upside down. About a week later, the yolk-sac has become completely resorbed, and the mouth has perforated and the young fish begin to feed for themselves in the surface waters. At this early stage the nauplius larvae of copepods are a major part of their food.

The planktonic phase lasts for about 10 weeks, by the end of which time the young cod has grown to about 2 cm in length and increased in weight about forty times. Throughout this period copepods remain the chief food. In the North Sea, *Calanus*, *Paracalanus*, *Pseudocalanus* and *Temora* are important foods for the cod fry, which in their turn are preyed upon by carnivorous zooplankton, particularly ctenophores and chaetognaths.

At the end of the planktonic phase the cod fry disappear from the surface layers and go down to the sea-bed. In the North Sea there are nursery areas to the south-east of the Dogger Bank and around the Fisher Banks. The fish are not easy to find at this stage because they are quite small and occur mainly in areas where the sea bottom is rocky, making it difficult to operate nets. The change from planktonic to demersal life involves a change of diet. The young demersal cod feed at first on small benthic crustacea such as amphipods, isopods and small crabs. As the fish increase in size, they take larger and faster-moving prey, and the adult cod is an active hunter, chasing its prey and feeding principally on other fish such as herring, sand eels and haddock, and on squid. A variety of benthic annelids, crustacea and molluscs are also included in the diet.

The growth rate of cod varies in different areas. In the North Sea they have reached approximately 8 cm in length at the end of their first 6 months, 14–18 cm by the end of the first year, and 25–35 cm by the end of the second year. Further north the growth rate is slower. Off the Norwegian coast, cod only attain about 8 cm during their first year, reaching 30–35 cm by the end of the third year. The fish begin to be taken in trawl nets once they exceed about 25 cm in length.

In the North Sea, cod reach maturity when about 70 cm long at 4–5 years of age. They grow to considerable size, sometimes reaching about 1·5 m in length and weighing 30 kg or more.

Although tagging experiments (pages 235, 260) have revealed extensive migrations by individual cod throughout the North Atlantic and into the North Sea and Barents Sea, there does not appear to be any large-scale movement of populations between different areas apart from the tendency to congregate for spawning.

Age Determination

Markings on the scales of cod give an indication of age. The surface of the scale bears a large number of small calcareous plates or sclerites (*Figure*

8.12), the number of which increases as the scale grows. Sclerites formed during the summer are larger than those formed in winter, and the alternating bands of large and small sclerites indicate the number of seasons through which the fish has lived. It is sometimes difficult to distinguish the zones of summer and winter sclerites clearly, and this means of age determination is not completely reliable. The method is satisfactory up to 3 years of age, but as the fish become older the accuracy of the scale-age becomes less certain. It is therefore usual to apply the Petersen method (*see* page 264) to the analysis of cod populations, age being assigned to length. The age of each length-group is taken as the scale-age of the majority of fish in each length-group.

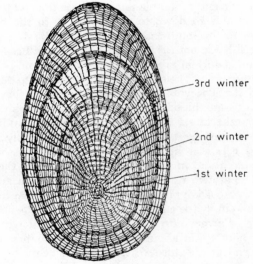

Figure 8.12. Growth zones in gadoid scale

3rd winter

2nd winter

1st winter

Growth rings are also present on the otoliths of cod, but are no more reliable than scales for determining age.

The age-groups of fish are commonly designated as follows:

Group O —Fish of less than 1 complete year of life.
Group I   —Fish between 1 and 2 years of age.
Group II —Fish between 2 and 3 years of age.
Group III—Fish between 3 and 4 years of age, and so on.

Haddock—*Melanogrammus aeglefinus (Figure 8.13)*

Distribution

The distribution of haddock is similar to that of cod, but with slightly narrower temperature limits, not extending quite so far as cod to the north or south. It is abundant in the northern part of the North Sea, and is also found around the Orkneys, Shetlands, Rockall, Faeroes and Iceland. It occurs on the west coast of Greenland, around Newfoundland and on the Atlantic coast of North America. Haddock comprise about 20 per cent of the auction value of fish landed in the U.K., approximately half of all Scottish landings, and nearly one third of the value of demersal fish landings in New England.

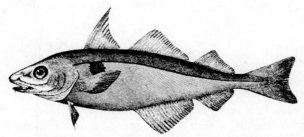

Figure 8.13. The Haddock, Melanogrammus aeglefinus

Life History[6]

The spawning period is January to June, the peak spawning in the North Sea being between March and April. The female haddock deposits up to 3 million eggs very like those of cod. The spawning schools usually congregate in rather deeper water than cod, mainly 80–120 m depth. This preference for deeper water restricts the haddock spawning areas of the North Sea to a more northerly distribution than those of cod *(Figure 8.11)*. When preparing to spawn, the North Sea haddock migrate northwards towards the deeper water to the north of the Fisher Banks.

After spawning the fish return to shallower water and form feeding schools in the central part of the North Sea. This movement of fish has its effect on the North Sea haddock fishery, which generally shows two peaks of landings (*a*) during February to May, when the best catches are made in the spawning areas in the northern part of the North Sea; (*b*) in September to October, when the main fishery occurs amongst the feeding shoals in the central part of the North Sea.

The buoyant eggs rise to the surface layers, and in the northern North Sea the eggs take some 14–20 days to hatch, a 4 mm larva with attached yolk-sac emerging. The miniature adult form is reached when about 2–2½ cm in length. Haddock tend to remain in the surface waters rather longer than cod, sometimes till 5 cm or more in length, but after that the fish become demersal and show a greater tendency than cod to congregate in shoals.

During their planktonic phase, haddock feed mainly on copepods. Once they have become demersal, haddock obtain much of their food by searching about in the deposit for crustacea, molluscs, annelids and echinoderms. Although they feed to some extent on fish, they do so far less than cod and there is thought to be no strong competition for food between these two closely related species.

Growth rates vary considerably in different areas, the southern part of the North Sea being a rapid growth area where haddock average about 45 cm in length at 5 years of age. In the northern part of the North Sea, haddock of the same age average only about 30 cm in length. They can grow to about 90 cm. Probably most haddock reach maturity during their third year.

Age Determination

As for cod.

Hake—*Merluccius merluccius* and Whiting— *M. bilinearis*.

Distribution

Whiting, or as it is more commonly called, Silver Hake, are found along the Continental Shelf of eastern North America and northward to the Grand Banks off Newfoundland, and southward to the region of South Carolina.

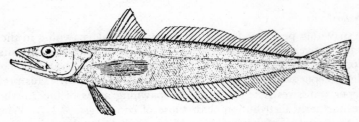

Figure 8.14. The Silver Hake, Merluccius bilinearis

However, it is most abundant between Cape Sable and New York. Whiting occurs further offshore and at greater depths as far south as the Bahama Islands. The Hake family is apparently represented by a number of sub-species which are not well understood. For example, the Whiting, or Silver Hake, is closely related to the European Hake (*Merluccius merluccius*) which has been studied in somewhat more detail and is described below.

Around the British Isles[8], the fish are mainly found off the west coast but may enter the northern part of the North Sea during the summer months when a migration occurs from beyond the Orkneys and Shetlands. The major part of the British fishery for hake takes place off the west coast of the British Isles by trawlers from Milford Haven and Fleetwood. Appreciable quantities of hake are also landed by lines. The commercial fishery mainly operates over shelf regions, but the fish can be trawled from depths well below the continental edge. In 1969, hake landed in the U.K. raised about $2·25 million at the fishmarkets.

Life History

Hake spawn in numerous areas off the west coast of the British Isles between April and October (*Figure 8.11*). During the early part of the season, spawning mainly occurs close to the continental edge, but as the season proceeds the spawning fish are found in progressively shallower water. The fish also spawn in the Mediterranean, off the Atlantic coast of Morocco and in various areas in the western part of the North Atlantic.

The female probably deposits between 500,000 and 2,000,000 eggs on or near the bottom. The eggs are small, diameter approximately 1·0 mm, and buoyant. They float up to the surface and hatch after about 7 days. From spawning areas to the west of the British Isles, the eggs and developing larvae are carried eastwards by the surface drift towards shallower coastal water.

Little is known of the biology of the early stages of hake . It seems that the fish remain pelagic for their first 2 years of life, and then descend to the sea-bed. They grow to about 10 cm by the end of their first year, 20 cm by the end of the second year and after that probably add about 7–8 cm/year, eventually reaching over 1 m in length.

There is a difference between the sexes in the age of onset of maturity. The majority of male hake mature in Groups III to VI when between 28–50 cm long, but females not until they are Group VI to VIII fish at about 65–75 cm.

Throughout life, hake feed almost entirely on pelagic prey. The young fish feed at first on copepods, later on larger planktonic crustacea such as euphausids, and also on small fish and small cephalopods. The diet of older fish consists almost entirely of fish and cephalopods. The hake is cannibalistic, and sometimes small hake comprise as much as 20 per cent of the food of the larger adults. In addition, pelagic fish such as *Gadus poutassou*, *Scomber*, *Caranx*, *Clupea spp.* etc. are taken. Bottom-feeding fish are seldom eaten and hake appear to feed mainly at night, making diurnal feeding migrations from the bottom to mid-depth water during darkness. Trawling for hake must therefore be performed during daylight while the fish are on the sea-bed.

Age Determination

There are annual growth rings in the otoliths of hake. The age analysis of the hake populations can be made by relating the different length groups of the population to the otolith rings characteristic of each length group (*see* pages 263–264).

Plaice—*Pleuronectes platessa*

Distribution

Plaice[17] (*Figure 8.15*) are bottom-living fish in shallow water of the north-east Atlantic area, extending from the Arctic to the Mediterranean, mainly found in depths of less than 80 m though occasionally occurring at depths of 120 m or more. They are plentiful throughout the North Sea, English Channel, Irish Sea and in the Bay of Biscay, and also in the Barents Sea and around Iceland and Faeroes.

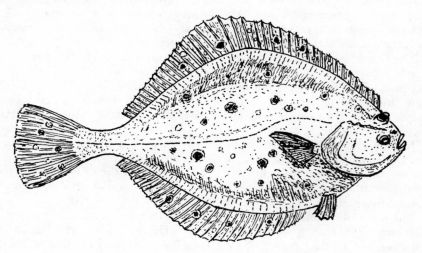

Figure 8.15. The Plaice, Pleuronectes platessa

Life History

For spawning[38],[39], plaice congregate in great numbers in particular spawning areas (*Figure 8.16*). In the southern North Sea, the chief spawning area is in the Flemish Bight between the Thames Estuary and the Dutch and Belgian coasts. A large part of the population of southern North Sea plaice make a southward migration during the early winter months towards the tongue of slightly warmer and more saline water which flows into the North Sea through the Strait of Dover. Here, spawning occurs between December and March. After spawning, the spent fish return northwards to their main feeding grounds in the central North Sea. Other spawning concentrations in the North Sea occur to the east of the Dogger Bank, off Flamborough Head and further north in the Moray Firth and around the Shetland Isles.

Spawning also occurs in the Irish Sea, the English Channel and the Baltic, mainly during February and March. In more northerly latitudes, plaice spawn a little later in the year, for example, in the Barents Sea in April, and around Iceland between March and June. There is no special uniformity of depth, temperature or salinity in the spawning areas, but they are never far from the coast.

A female plaice spawns between 10,000 and 600,000 eggs in a season (*Figure 8.35*). The eggs average about 1·9 mm in diameter. They are spawned and fertilized close to the bottom but are buoyant and float up to the surface. The eggs then increase slightly in density as the embryos develop, sinking to deeper levels by the time the larvae hatch. Those spawned in the Flemish Bight hatch after some 15–17 days. The larva is

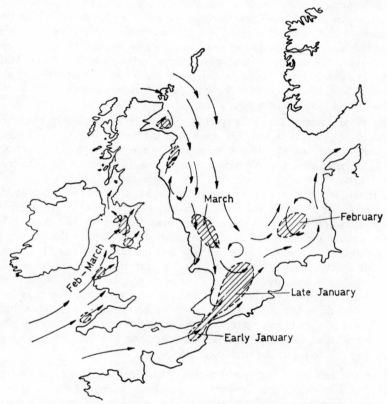

Figure 8.16. Main spawning areas of plaice around the British Isles. Arrows indicate directions of drift

symmetrical, 6·0–7·5 mm in length with a ventral yolk-sac which is resorbed within about 8 days. When active feeding commences, the larvae take chiefly flagellates and small diatoms. As the larvae grow, larger diatoms, molluscan larvae, early stages of copepods and the appendicularians *Fritillaria* and *Oikopleura* are taken, the appendicularians being a particularly important component of the diet of plaice larvae, often forming virtually the entire food.

Normal symmetry and a planktonic mode of life continue until about 30 days after hatching, by which time the food is usually predominantly copepods, but also includes a variety of invertebrate larvae. At this stage, when the plaice larva is about 10 mm long and 2 mm in height, its metamorphosis begins, and within the next 17 days the larva becomes gradually converted into a 'flatfish'. It becomes laterally flattened and acquires a new swimming position with its left side downwards. The skull is progressively transformed by the movement of the left eye to a new position dorsal and slightly anterior to the right eye on what now becomes the uppermost side of the body. Colour disappears from the new underside, and the upper parts develop the characteristic pigmentation and spots of the plaice. During this period of metamorphosis, the fish becomes demersal and grows to about 14 mm in length and 7 mm in 'height', i.e. the maximum distance between the bases of the dorsal and ventral fins.

The chief nursery areas for young plaice are along sandy stretches of coastline. Eggs and larvae spawned in the Flemish Bight are carried by the drift of the surface waters in a north-easterly direction towards the Dutch, German and Danish coasts, usually travelling some $1\frac{1}{2}$–3 mile/day. By the time their metamorphosis is complete, the young fish have drifted close to the coast where they settle upon the sea-bed. The extensive sandy expanses along these shores form a major nursery area for young plaice. Larvae spawned in various other places near the British Isles also drift towards regions of shallow sandy bottom, and the Wash and the greater part of the Lancashire and Cheshire coasts are other plaice nursery areas. Considerable fluctuations occur from year to year in the numbers of young plaice which successfully complete their metamorphosis. Of several factors which may influence survival at this stage, an important one may be the strength and direction of the wind, upon which depends the speed and direction of drift of the surface water. Poor survival may occur in years when a large proportion of the larvae fail to reach, or are carried beyond, the sandy regions required by the young fish.

Once the fish have become demersal, they feed at first on a variety of small benthic organisms including annelids, harpacticids, amphipods, small decapods, small molluscs etc. As they grow larger, molluscs form an increasingly important part of the food. The diet varies from place to place according to the nature of the food available but molluscs often account for about 25 per cent of the total. *Venus, Cardium, Spisula, Solen, Mactra* and *Tellina* are important plaice foods, and they also take the tectibranch *Philine* and numerous annelids, crustacea, echinoderms, actinians and sand-eels.

During their first year of life, the majority of plaice are to be found close

234

inshore, mainly in sandy areas in water of less than 5 m depth. With increasing size they migrate into deeper water. This movement usually occurs during the latter part of each summer. At the end of their first summer, the fish move into the 5–10 m zone and are thought to spend much of their first winter buried in the substrate. They are now 6–8 cm in length. In the following spring, they return to shallower water and, during the summer, range to and fro along the coast in search of food. As autumn approaches, they again migrate into deeper water, this time to about the 20 m line. This pattern of behaviour is repeated in each age group. Until their third summer, by which time they are about 20 cm in length, most plaice are found near the coast in water of less than 20 m depth. Thereafter, as they increase in age and size, they inhabit increasingly deep water further from the shore. Heinke has expressed some general observations of the distribution of North Sea plaice in the form of a 'law', as follows: 'In any part of the North Sea the size and age of plaice are inversely proportional to the density of plaice population but directly proportional to the depth of the water and to the distance from the coast.'

North Sea plaice reach a size of between 35 and 45 cm by about their sixth year of life. In most areas the females are slightly larger than males of the same age. Sexual maturity depends on size rather than age. In the North Sea most females mature when between 30 and 40 cm in length and 4–5 years of age, and males mature at 20–30 cm when most are 4 years old.

Age Determination

The age of plaice may be determined by examining the markings on scales or otoliths, the latter being the more satisfactory. There are three otoliths contained in the membranous labyrinth. The largest of these, the 'Sagitta', is a flattened, oval-shaped disc showing a white, opaque central nucleus around which lie rings of opaque material separated by translucent zones. Fish of less than one complete year of age, i.e. Group O fish, show only the central nucleus. Each completed year of life is represented by one opaque ring.

Tagging Experiments

Numerous tagging experiments have been performed on plaice. The fish are captured, identifiable tags attached, and records kept of the times and places of release. Several types of tag are used. The commonest consists of two plastic discs engraved with numbers for identification. One disc is threaded on a short length of wire which is pushed through the muscles at the base of the dorsal fin at its midpoint and attached to the other disc. This can be done without drawing blood and the fish are kept for a time in a sea-water tank to make sure that they seem unharmed. A small reward is offered for the return of a disc with details of the circumstances of recapture of the fish.

Tagging experiments have demonstrated that in some areas fishery for plaice is very intensive. In experiments in the North Sea, over 30 per cent of tagged plaice have been recaptured within 12 months of their release. By relating the percentage recapture of tagged plaice to the total landings of the commercial fishery, an assessment of the total plaice population of the North Sea has been made (*see* page 265).

Tagging experiments also show the extent of plaice migrations, and have revealed that these comprise two chief types of movement; (*a*) feeding migrations of individual plaice which follow no ascertained pattern except that they are usually restricted between certain depth-limits dependent upon the size and age of the fish, and (*b*) seasonal migrations of entire populations of mature plaice from wide areas towards the spawning grounds.

Fishery

Plaice are caught mainly by trawl or Danish seine, and make up about 4 per cent of the landings of British fishing vessels. They may be taken in shallow water all around the British coasts. A large part of the commercial catch comes from the southern North Sea.

The Lemon Sole—*Microstomus kitt*

Distribution

Like plaice, the lemon sole (*Figure 8.17*) is a flatfish of shelf areas of the north-east Atlantic ranging from the Arctic to the Bay of Biscay, living mainly within depths of 100 m but found in reduced numbers to nearly 200 m. It does not extend as far south as plaice and generally favours a rougher sea bottom, but the two species often occur together. Lemon soles are specially abundant in the north-west part of the North Sea off the east coast of Scotland, also around the Faroes and along the south-west coast of Iceland (*Figure 8.18*).

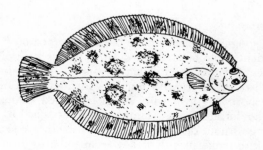

Figure 8.17.   The Lemon Sole, Microstomus kitt

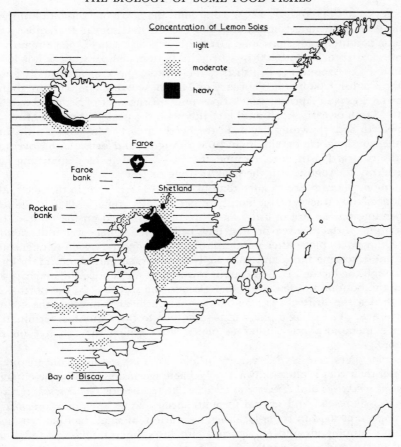

Figure 8.18.   Distribution and main concentrations of Lemon Soles
(From Rae, B. B., *The Lemon Sole* (1965), by courtesy of Fishing News (Books) Ltd)

Life History[14]

Lemon soles spawn widely throughout their range, gathering in small local concentrations wherever they are normally found. There are no huge assemblies of spawning lemon soles drawn from wide areas, as occur with plaice. Tagging experiments have indicated a tendency for the fish to swim against the current during the period preceding spawning. In the north-west North Sea this results in a limited northerly movement towards the

237

Orkneys and Shetlands or westward around the north of Scotland, but there is no great spawning migration. The fish do not appear to require very precise conditions for spawning. In the North Sea it takes place mainly at depths between 50 and 100 m when the bottom temperature is not lower than 6·5°C. Around the British Isles the earliest spawners are usually found in the English Channel in February or March, where the maximum abundance of eggs is in April to June. Spawning off the west of Scotland extends from March or early April until late July with the peak in April to May. In the North Sea, spawning begins in the north in early May and a little later further south, with a maximum in June to August, and lasting into November in the central North Sea. Around the Faroes and Iceland spawning lasts from May to August with the peak in June to July.

Lemon-sole eggs are smaller than plaice, 1·13–1·45 mm in diameter, and produced in rather greater numbers per female, some 80,000 to 700,000 depending on the size of fish. They are probably shed on the bottom, but float to the surface for two or three days before sinking back to middle depths. Hatching takes place after some six to ten days depending on temperature, and the emerging larva and yolk-sac is symmetrical and about 5 mm long. Although the larvae can be taken in tow-nets at all depths, the majority are near the bottom and few are found close to the surface. In the north-west North Sea the drift of eggs and larvae is southwards towards the central North Sea. The pelagic phase lasts some two to three months, terminating with a metamorphosis similar to plaice, after which the fish live on the bottom.

Lemon soles take a wide variety of food from the sea-floor, but the major part of their diet is almost always polychaete worms, especially the eunicids *Onuphis conchylega* and *Hyalinoecia tubicola*, the terebellids *Lanice conchilega* and *Thelepus cincinnatus* and several serpulid species. The anemone *Cerianthus* is an important food in some localities. A variety of small benthic crustacea (mainly amphipods and eupagurids), molluscs (mainly chitons and small gastropods) and some small ophiuroids are also taken. Small bivalves are eaten and the fish bite the siphons off larger lamellibranchs; but, compared to plaice, bivalves form a much smaller part of the diet of lemon soles and it is unlikely that the two species compete much for food.

Lemon soles grow rather slowly. The western North Sea off the British coast between Yorkshire and Aberdeen is a region where they grow relatively quickly, attaining about 25 cm in length at the end of their third year and 30–35 cm by the end of the sixth year. In most other areas of their distribution the lengths at these ages are nearer 15 and 25 cm. The fish can grow to nearly 50 cm and rare specimens have been caught between 55 and 67 cm, probably at ages between 15 and 20 years. The majority age of first spawning is 4 years for males and 5 years for females. Like several other species, after maturity the males have a slightly greater mortality rate than females so that in the older age groups the females progressively predominate.

Age Determination

The age of lemon soles is best determined by the annual growth rings in their scales. The otoliths are small, and in older fish the growth zones are more difficult to distinguish than in plaice.

Fishery

Lemon soles are caught by trawls and bottom seines. In the North Sea the catch is of special value to Scottish boats, and is taken mainly by seining.

Distribution   Herring—*Clupea harengus*

Herring (*Figure 8.19*) are widely distributed across the northern North Atlantic between Newfoundland and the British Isles, and also in the Arctic. Bigelow and Schroeder[43] indicate that the herring ranges off the European coast north to Norway, Iceland, Spitzbergen, and the White Sea, and south to the Straits of Gibraltar. The species is known on the American coast as far north as norther Labrador and the west coast of Greenland. It is commonly found as far south as Cape Cod and Block Island and is occasionally seen in the winter in small numbers as far south as Cape Hatteras.

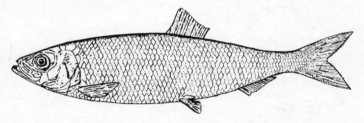

Figure 8.19. The Herring, Clupea harengus

A very similar form, *Clupea pallasii*, extends in the North Pacific from Japan to the coast of British Columbia, and also occurs in the North Atlantic. Herring-like fish are found in many other areas, including fresh water.

Life History[10,11,31,32]

When herring are preparing to spawn, they congregate in huge schools as they approach their spawning grounds, sometimes as many as 500,000,000 fish per square mile. They are the only commercially-important marine teleosts of British waters to lay demersal eggs. A spawning female deposits 10,000–60,000 eggs on shallow, gravelly parts of the sea-bed. The eggs are heavier than water, and form sticky masses which are usually laid among stones and weed to which they readily adhere.

Where they occur, spawning herring can be found at almost any time of year. The major spawning schools are of two principal types, Oceanic

239

and Shelf herring (*Figure 8.20*). Oceanic herring have an extensive range in deep water in the North Atlantic, Arctic and Norwegian sea. Those which approach the British coasts form winter-spring spawning schools between February and April around the Hebrides, Orkney and Shetland Isles and

Figure 8.20. Main spawning areas of herring around the British Isles

along the Irish and Scottish coasts. These schools occur mainly in water of oceanic-neritic type (*see* page 113) where the temperature lies between 5–8°C. The Shelf herring of the North Sea, English Channel, Minch and Irish Sea form mainly summer-autumn spawning shoals. The North Sea shoals spawn from the Shetlands to the east coast of Scotland from July to September, off Northumberland, Yorkshire and Lincolnshire between August and October, off the East Anglian coast between October and December, and in the eastern English Channel near the north French coast from December to January. These shoals are found in neritic water at 8–12°C. Slight morphological differences have been noted between the two groups. Both show variation in the number of vertebrae between 54 and 59, with a mean vertebral count slightly above 57 for the oceanic herring and below 57 for the shelf herring. There are also slight differences in the number of gill-rakers, number of keeled scales on the ventral surface and in the structure of the otolith. The oceanic herring are slower growing, later maturing, longer lived and reach a larger size than the shelf herring. It is now generally considered that the two groups, both of which can be further subdivided into several more or less distinct stocks, are biologically separate units with no appreciable interbreeding.

The precise location and limits of herring-spawning grounds have been difficult to determine because herring eggs are not easily found by dredging. It seems that the selected sites are quite patchily spread with eggs, each patch containing enormous numbers of eggs densely-packed in several layers on ground of small stones or gravel, usually at about 40 m depth. Apart from dredging, the situation of spawning grounds is approximately known from observing the position of ripe and newly-spent herring, the occurrence of herring larvae in tow-net hauls, and by noting the places from which trawlers bring up 'spawny' haddock, i.e. haddock engorged with herring eggs, on which they prey.

In water of 5–6°C, herring eggs hatch in about 22 days, at 11–12°C in 8–10 days. The newly hatched larva is about 6–8 mm long and at first depends on the food reserves of the yolk-sac. It must be able to feed itself within some 3–4 weeks. After hatching, the larva swims to the surface. During the period of resorption of the yolk-sac the larva develops a mouth and begins to feed at first mainly on diatoms, copepod eggs and early copepod larvae. As the larva becomes larger, it takes the later larval stages of copepods and the adults of small copepods such as *Paracalanus* and *Pseudocalanus*.

During the first weeks of life the fish are carried by the drift of the water and there is some mixing of broods from different areas. Within a few months, when the young fish become able to swim more strongly, they make a shoreward migration. They are often particularly numerous in large estuaries such as the Thames Estuary and the Wash. Here they form large schools known as 'whitebait' which often contain a mixture of young herring

241

and young sprats. At this stage, the food consists largely of small crustacea; for example, the larvae of shrimps and prawns, and estuarine copepods such as *Eurytemora hirundoides*. The young herring remain inshore until near the end of their first year of life, and then move out to deeper water. By now the fish are mostly some 4–8 cm in length.

Southern North Sea herring have an extensive nursery area around the Dogger Bank, and here the 2-year-old fish occur in great numbers. During their second year these fish grow to some 13–18 cm in length but are still extremely thin. They probably remain on the nursery ground until they become mature and join the adult spawning shoals.

In their third year these herring fatten, and feeding shoals of the immature fish often approach the English coast during the midsummer. Throughout life the food is predominantly planktonic, mainly copepods, but also includes chaetognaths, pteropods, hyperiid amphipods, appendicularians, decapod larvae and fish eggs. They sometimes feed on other fish, for example *Ammodytes*. The type of food varies seasonally to some extent and also varies from place to place according to the nature of the plankton.

By the end of their third year they have reached the 'fat herring' stage, and the flesh has become rich in oil for which they are particularly prized by man. During their fourth year, i.e. as Group III fish, some southern North Sea herring become sexually mature but the majority of shelf herring mature as Group IV or Group V fish. The oceanic fish of the northern part of the North Sea and the Norwegian coast mainly reach maturity between their fifth and eighth years. The maturing virgin fish leave the young fish shoals and join the adult spawning shoals.

Certain patterns of movement can be discerned in herring. For example, there are diurnal changes of vertical distribution associated with changes of illumination, the fish usually forming compact schools on or near the sea-bed during daylight, and approaching the surface during darkness to feed on the plankton. There is also the annual cycle of migration of the adult fish, which congregate in huge shoals as the breeding season approaches and move into shallow water to spawn, followed by a dispersal to deeper water and the formation of feeding schools.

Several factors influence the movement of the schools; in particular, the condition of the plankton. Where there is an abundance of suitable zooplankton food, the feeding shoals tend to amass, and the numbers of herring can sometimes be correlated with the numbers of *Calanus*. On the other hand, herring are seldom found in water heavily loaded with phytoplankton. It has been suggested that this is due to an 'exclusion' effect (*see* page 134) but because such water usually contains very little zooplankton, the absence of herring shoals may simply be due to the scarcity of herring food.

The quality of the fish depends upon the abundance and nature of their food. The valuable oiliness of the herring reflects the fat-content of the

THE BIOLOGY OF SOME FOOD FISHES

plankton, which in turn depends to some extent upon climatic conditions. Generally, a warm, sunny summer produces a rich plankton and provides fish in excellent condition for the autumn fishery. A copious diet of copepods produces a specially oily flesh. If the food consists mainly of pteropods, the fat-content of the flesh is poorer.

Scale Markings

Annual growth rings are found in the scales of herring but are not often easy to detect. The rings are due to slight differences in refraction of different regions of the scale, and are best viewed under the microscope using a 1 in. or 2 in. objective with dark-ground illumination. The rings are usually most clearly seen in scales taken from the anterior part of the trunk region.

The rings indicate the interruption of growth of the scale that occurs during the winter months. Fish spawned in late summer or autumn probably fail to record their first winter as a scale ring, and in these the first ring relates to the second winter. Scale markings may be used to determine growth rates because the width of each zone between the rings is closely proportional to the growth in length of the fish during the period in which that zone was formed (*see Figure 8.33*, page 263). At sexual maturity the growth rate is reduced, and this is indicated on the scales by a narrowing of the growth zones. The pattern of scale rings therefore varies according to the age at which the fish mature. In shelf herring spawning in the southern North Sea, the first narrow zone is usually to be found between the second to fourth ring. In oceanic herring which mature later, the first narrow zone occurs after the fifth ring. In Baltic herring, most of which mature during their second

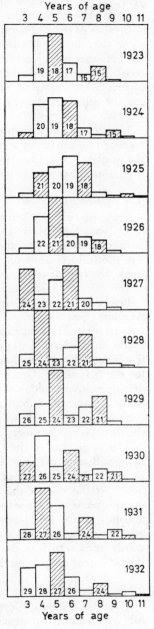

Figure 8.21. Age census of East Anglian herring shoals, 1923-32

(From Hodgson, W. C.[10] by courtesy of The Buckland Foundation)

243

year, there is usually only one wide zone, the first narrow zone being between the first and second ring.

Studies of the age-composition of herring shoals in the southern North Sea show that here the adult shoals contain fish from 3 to 11 years old, the maximum age normally reached by herring in this area. Off the Norwegian coast, where the fish mature later, the age range is from 4 to 15 years or older. Great variations occur from year to year in the numbers of fish entering the adult shoals, and this is reflected in the relative abundance of each year class (*Figure 8.21*).

Injuries to the fish may cause an interruption of growth, and may be recorded on the scales as a 'false' ring. After World War II, many herring in the southern North Sea were found to have unreadable scale markings with numerous false rings, probably caused by the effects of underwater explosions.

The East Anglian Shoals

Prior to 1950 an extremely rich fishery for herring, involving hundreds of drifters, took place annually along the coast of East Anglia during the period October–December. Since 1950 this fishery has sharply declined. The possible reasons for this are discussed by Dr. W. C. Hodgson in his book *The Herring and its Fishery*[11]. The following is a brief summary of some of the relevant information.

Before 1950, in a normal season, the shoals which appeared off the East Anglian coast in the autumn came in order of age, and usually comprised two groups. The first to arrive were the October shoals, or 'Recruit' Shoals, made up predominantly of Group IV fish, but with some Group III and a few older fish. These shoals consisted of virgin fish (matties) about to spawn for the first time. Gradually, these Recruit shoals were replaced by shoals of progressively older fish. These were the November shoals or 'Adult Shoals', fish which had already spawned in previous years. Before 1950 both groups of shoals were of great size, but in recent years the November shoals have progressively diminished and have now virtually disappeared.

The virgin fish of the October shoals are thought to have come from the Dogger Bank nursery areas. After spawning, the spent fish move northwards, probably following the drift of the water along the eastern part of the North Sea. At the end of the winter they form feeding shoals which appear in various parts of the northern North Sea during the summer months. In the autumn, these fish begin a southward migration back to the spawning grounds, congregating to form the November shoals off East Anglia. Afterwards, they return to their summer feeding grounds in the North Sea, and this cycle of migration is probably repeated annually.

Forecasting the Shoals

The East Anglian October Recruit shoals contain only recently-matured fish which have not previously spawned. Until 1951, it was always found that the majority of these fish were of Group IV, although some of Group III were also present. It appeared that, in each year-class, a minority matured early as Group III fish; but the greater part of that year-class did not join the October shoals until the following year as Group IV fish, and a few even later. The proportion of Group III fish therefore indicated the probable numbers of Group IV fish in next years' shoals. Whenever an unusually large number of Group III appeared (for example 1927, *Figure 8.21*), showing a specially good survival of that year-class, very large October shoals could be expected the following year when the majority of that year-class would reach maturity as Group IV fish.

Usually each year-class was at its maximum in the shoals as Group IV fish. Afterwards when the fish returned in the following years to spawn in the November shoals, the numbers of each year-class steadily declined, but remained in the same relative abundance to the older fish. Annual records of the age-composition of the shoals therefore provided a basis for prediction of the likely size, composition and time of arrival of the next year's shoals.

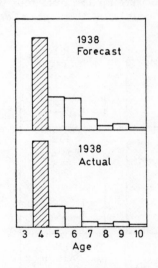

Figure 8.22. Forecast and actual composition of East Anglian herring shoals in 1938

(From Hodgson, W. C.[11] by courtesy of Routledge and Kegan Paul)

Such information could be useful to the fishery, enabling appropriate preparations to be made for catching, processing and marketing when the heaviest landings were expected.

The first forecast of the East Anglian shoals was made prior to the 1929 fishing season. The forecast, though not entirely accurate, proved to be sufficiently good to warrant further attempts in subsequent years. Methods were developed to ensure a more even sampling of the shoals, and the effects on their movements of such factors as wind, water circulation, plankton conditions and the moon were studied. By the late thirties, considerable success was achieved and the fishermen were beginning to take notice of the forecasts, and to regulate their activities accordingly. For example, analysis of the 1937 shoals showed a strong recruitment of Group III fish. A large influx of Group IV fish was therefore predicted for 1938, heavy landings being expected from the October shoals. The actual compositions of the 1938 shoals, and the accurate forecast, are shown in *Figure 8.22*. The proportion of Group III fish could not, of course, be predicted, as the success of the previous year's Group II's was not known.

The Decline of the East Anglian Fisheries

During World War II, fishing and forecasting ceased, and no further forecast was made until 1950 which proved to be very successful. A large number of Group III fish appeared in 1950, and was taken to presage the arrival of an exceptional number of Group IV fish in 1951, and the forecast shown in *Figure 8.23* was confidently made. In fact, however, the forecast proved to be accurate only with respect to the older fish. The shoals of Group IV fish were far less numerous than expected, their actual strength being 41 per cent below the forecast (*Figure 8.23*).

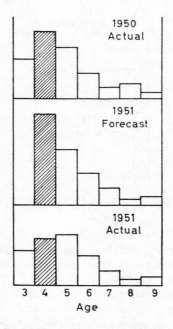

Figure 8.23. *1951 forecast based on observation of 1950 herring shoals, and actual 1951 shoal-strengths showing failure of 4-year-olds*
(From Hodgson, W. C.[11] by courtesy of Edward Arnold)

In 1952, a similar result was obtained. The forecast was accurate for the older year-classes, but Group IV fish again failed to appear in their expected strength, being 44 per cent below the forecast. Before 1951, the Group IV fish were always more numerous in the shoals than Group III fish, usually greatly so. In contrast to this, since 1951, the Group IV fish have been considerably fewer in number than those of Group III.

The period 1951–1952 can now be seen to have been a turning point in the fortunes of the East Anglian herring fishery. Although herring shoals are notoriously irregular in their behaviour, changing their movements from time to time for no obvious reasons with consequent failure of the fisheries in other areas, the East Anglian shoals had behaved with exceptional consistency over hundreds of years. There are no records of previous disasters to this fishery comparable with that which occurred in 1951. At first, the change affected only the Group IV fish in the Recruit shoals. However, as the years passed, the reduction in number of recruits gradually led to the depletion of the older age-groups until the November shoals have now virtually disappeared. The East Anglian fishing season, traditionally of 8 weeks duration, has been reduced to the four October weeks when the Recruit shoals, greatly depleted in size, make an appearance. Each year the landings have declined, and many drifters which regularly put in at Great Yarmouth and Lowestoft for the season no longer find it worth while to make the trip.

Various causes, such as changes in the habits of the fish, changes in water temperature or the condition of the plankton have been advanced to account for the decline of the East Anglian shoals. Dr. Hodgson found an explanation[11] in the changing character of the herring fishery rather than in any change in habits or environment of the fish. Drifters were gradually giving place to trawlers as the main method of capturing herring. Although trawling for herring has been practised for many years, it greatly increased after World War II. Since 1948 an intensive trawl fishery for immature herring has grown up on the Dogger Bank nursery grounds from which the East Anglian shoals are recruited. Landings were at first small, growing to 5,450 tons in 1950. In 1951, it leapt to 31,500 tons, and to 64,000 tons by 1954. This fishery is operated mainly by Danish, German and French trawlers but has recently been joined by British vessels in the view that, though probably having a destructive effect, this country might as well join the others in taking what can be had while there are still sufficient fish left to provide profit. Trawling continues throughout the greater part of the year, and in a single year over a thousand million immature fish have been taken from only three age-groups, I, II and III. The East Anglian drift net fishery, in one of its best post-war seasons, 1948, took only 650,000,000 fish from nine groups, III to XI. The trawled fish are too small for human consumption and are used for fishmeal for cattle food.

At the very time that the number of Group IV fish in the October shoals first fell so greatly, the fish of this age-group underwent another remarkable change. Before 1951, the average length of Group IV fish was always close to 24·5 cm. In 1951, the average length of this age-group increased suddenly to nearly 26 cm, and has continued so in subsequent years. In 1952, Group V fish (1951's Group IV fish) were also found to be larger; in 1953, the size increase was noted in Groups IV, V and VI; and so on. It was clear that the drastic reduction in number of recruits was accompanied by a simultaneous increase in their size, which has since remained constant.

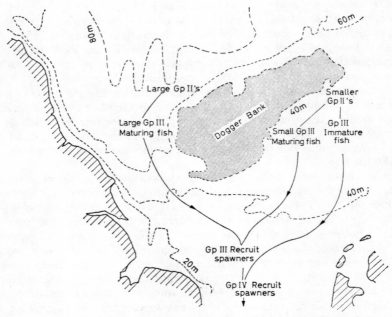

Figure 8.24. Theoretical distribution of young herring around the Dogger Bank, and migration routes of recruits to spawning shoals, according to Hodgson

(After Hodgson, W. C.[11] by courtesy of Routledge and Kegan Paul)

A possible explanation of this may lie in the distribution of the different size-groups of herring in the Dogger Bank nursery areas. Young herring are very sensitive to depth and, in each age-group, the larger fish are generally found in deeper water. Around the Dogger Bank, where Groups II and III fish occur in large numbers, there is a distinct length/depth relationship. At 40 m, the average length of Group II fish is 14 cm, at 70 m it is 16 cm. The contours of the Dogger Bank area and the distribution of size and age groups are shown in *Figure 8.24* Although trawling is widespread, the major part of the catch is taken from the shallower water south and east of the Bank, where the smaller fish occur. Consequently, such fish as are left to join the

East Anglian Recruit shoals would be mostly the larger ones from deeper water north and west of the Bank where the fishery is less intensive.

Dr. Hodgson's explanation has not found general acceptance among fishery scientists, and the disagreement demonstrates the complexity of this problem. Some of the grounds of objection can be summarized as follows[32]:

(1) The decline in number of Group IV fish and the increase in mean length of Group III recruits occurred before the intensity of trawling had increased sufficiently to produce such a sudden and marked effect on the shoals.

(2) Subsequent tagging experiments to measure the proportion of the immature stock removed by trawling indicated that this was insufficient to account for the observed reduction of Group IV fish.

(3) The Dogger trawl fishery took mainly herring of Groups I and II, and would therefore be expected to reduce the number of Group III recruits as well as Group IV. But on the contrary, when the Group IV recruits declined, the number of Group III showed some increase.

(4) In the same period changes in growth and recruitment similar to those of the East Anglian shoals were also observed among summer–autumn spawners of the middle and northern North Sea (*Figure 8.25*), but the total numbers of these fish increased rather than declined.

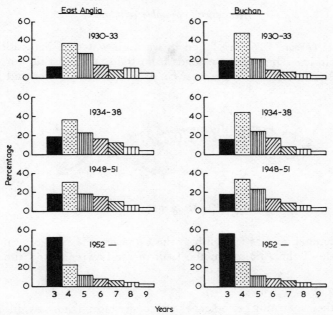

Figure 8.25. Age composition of catches in southern and northwestern North Sea herring fisheries before and after 1950

(From Parrish, B. B. and Saville, A.[32], by courtesy of Allen and Unwin)

(5) At the time of these changes there was evidence of a greater inflow of Atlantic water into the North Sea, and larger numbers of *Calanus*. Also there had been a gradual rise of water temperature over the previous 20 years, especially an elevation of the winter temperature. The observed changes in growth and earlier recruitment could therefore be attributed to a better supply of food and a longer feeding season in warmer water, and the decline of the East Anglian shoals to a change in distribution of the fish.

After 1955 there was a reduction in the intensity of herring trawling around the Dogger, but this did not result in any increase in the number of recruits to the East Anglian shoals. Instead the number of both recruits and adult fish declined even further, and a comparable decline was also observed in North Sea shoals further north. Possibly this overall decline might be due to the effects of trawl fishing for immature herring in the central and northern North Sea, where it increased when the Dogger trawling was reduced. Alternatively it could be the effects of changing water conditions altering the abundance of shelf herring in this area. Obviously there are many uncertainties about the causes of these changes in the stocks of North Sea herring. In view of the great importance of these shoals as a source of human food and fish meal, there is a need for greater understanding of the biology and ecology of these fish.

Mackerel—*Scomber scombrus*

Distribution

Mackerel (*Figure 8.26*) are found in warmish water on both sides of the North Atlantic. Their range extends from the south coast of Norway and northern North Sea, along the west coasts of the British Isles and into the

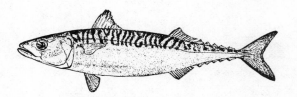

Figure 8.26. The Mackerel, Scomber scombrus

English Channel, and as far south as the Canaries. They also occur on the western side of the Atlantic in the Gulf of St. Lawrence and the Gulf of Maine.

Life History[40]

During their spawning period, mackerel congregate on the spawning grounds in vast schools. Off the west coasts of Britain, they spawn between March and July, the most intensive spawning occurring during April in a spawning area some 50–80 miles west of the Scilly Isles (*Figure 8.27*). As the season proceeds, spawning fish are found in progressively shallower water further to the east, until by July they are spawning in St. Georges Channel, the

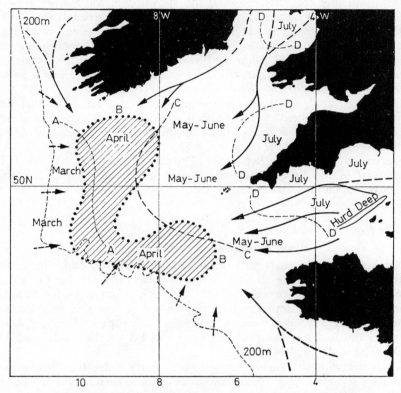

Figure 8.27. Generalized diagram indicating the migrations of mackerel in the English Channel and Celtic Sea in relation to the spawning ground. Arrows indicate movement of shoals towards spawning areas. Spawning begins in March to the west of line A near the 200 m contour. The greatest intensity of spawning is in April within the shaded area B. In May and June, all spawning has ceased to the west of line C. Residual spawning occurs in July to east of lines D

(From Steven, G. A.[40] published by Cambridge University Press)

Bristol Channel and the western part of the English Channel. Spawning also occurs in the north of the North Sea off the south coast of Norway.

The early stages of the life history of mackerel are not well known. The female is thought to produce 400,000 to 500,000 pelagic eggs. These are not all shed together but are produced in successive batches over a period. The eggs are about 1·2 mm in diameter and are probably shed into the water between the surface and middle depths. At first they float to the surface, but after 2 days they lose buoyancy slightly and sink to mid-depth water. At the time of hatching the larva is about 2·5 mm long. The yolk-sac is resorbed in about 9 days, and the next stages of the life history are not known with any certainty. Young fish spawned in the Scillies area are presumably carried eastwards by the drift of the water towards the English, Irish and Welsh coasts. In some years, sizable shoals of young mackerel of about 13–17 cm in length are found along the English south-west coast during the summer. In autumn and winter small mackerel have sometimes been brought up from the bottom in fine-mesh trawls, suggesting a pattern of behaviour similar to that of the adult schools (*see* below).

Growth is rapid, reaching approximately 23 cm during the first year. Mackerel mature when 2 years of age at approximately 29 cm, and grow to about 36 cm by their sixth year, but after this period there is no satisfactory method of determining their age (page 253).

Steven investigated the behaviour of the adult mackerel schools off the south-west of the British Isles, and found a seasonal cycle of migration which he subdivided as follows:

1. The Demersal Period

(*a*) A Compact phase, November–December.

Mackerel pass the early part of the winter on the sea-bed. In the late autumn they congregate on the bottom in dense concentrations in various localities selected for overwintering. The fish seem to prefer situations adjacent to banks or depressions on the sea-bed. Numerous, very localized concentrations of fish occur over a wide area in the English Channel, Bristol Channel and Irish Sea and probably extend well down the continental slope to the west of the Scillies.

These overwintering regions show no uniformity of depth or temperature. Probably, however, they are places where cascade currents occur (*see* page 147), carrying down food from the surface. Where so many fish are massed together, the food supply is unlikely to be very plentiful and throughout the winter a large proportion of the fish are found to have empty stomachs. But the fish will feed where food is available, taking mainly small demersal and benthic prey such as crustacea, polychaetes and small fish.

(*b*) A Deployment Phase, December–February. Towards the end of winter fish begin to leave their overwintering concentrations. At first they still keep to the bottom, but soon begin to perform diurnal vertical movements, ascending during darkness and eventually forming surface schools.

During the early months of the year the surface plankton is sparse and the majority of the fish are still without food, but they readily feed when suitable food is found, for example, on schools of small fish (*Maurolicus muelleri*).

2. The Pelagic Period

(a) A Schooling phase, January–July. During the late winter and spring the surface schools begin to move towards the spawning areas (*Figure 8.27*). Those which passed the compact phase in the English Channel move westwards, those from the Irish Sea move south-westwards, while those which wintered on the continental slope move towards the east, all tending to converge towards the west of the Scilly Isles.

The schools with the shortest distance to travel will be the first arrivals in the spawning area, and these are likely to be fish that have wintered on the continental slope. Probably these are also the first to have fully ripe gonads, having wintered in slightly warmer water than the fish on the continental shelf, and perhaps having had a rather better food supply derived from cascade currents. The first spawners are found near the continental edge as early as March, but the majority of schools arrive in the main spawning zone in April. After that, the fish with greater distances to travel gradually arrive, spawning further to the east and later in the season until July. After spawning, the shoals move mainly towards coastal regions. Throughout this shoaling phase, the food of mackerel consists largely of planktonic crustacea, mainly copepods, but small fish are also taken. Fishermen associate good catches of mackerel with an appearance of the sea which they call 'yellow water', and it has been shown that these areas are particularly well-filled with the copepods *Pseudocalanus* and *Calanus*. On the other hand, leaden-coloured water with a slightly unpleasant smell, known to fishermen as 'stinking water' and associated with the poorest catches, was found to be loaded with phytoplankton.

(b) A Dispersal phase, June–October. As the shoals approach the coast they tend to disperse, the fish ranging to-and-fro along the coastline seeking food. The diet undergoes a change, consisting now largely of other fish, for example, sand-eels and pilchards, and also various inshore crustacea such as mysids, shrimps and prawns.

3. The Transition Period

A Reconcentration phase, October–November. Towards the end of the autumn the fish disappear from coastal regions and return to their over-wintering concentrations on the sea-bed, i.e. the Compact phase.

Age Determination

Two otoliths are present on each side of the brain and annual rings are present in the larger of each pair. This provides the most satisfactory method of age determination. Annual rings are also present in certain scales, but the scales are so easily rubbed off and transferred from fish to fish that this method is only applicable where the fish are caught singly by line and kept isolated. Neither method is reliable in fish of over 6 years of age.

Fishery

During the winter months, mackerel may be caught by trawl. During the spring and summer months, there is a pelagic fishery based on Newlyn in Cornwall, when the fish are caught in and around the spawning grounds by means of drift nets. There are also small inshore fisheries for mackerel along the Cornish and Devon coasts during the summer. Significant numbers of mackerel are also caught by short lines.

THE OVERFISHING PROBLEM

Addressing the International Fishery Exhibition in London in 1883, T. H. Huxley said

> I believe that it may be affirmed with confidence that, in relation to our present modes of fishing, a number of the most important fisheries, such as the cod fishery, the herring fishery and the mackerel fishery, are inexhaustible. And I base this conviction on two grounds, first, that the multitude of these fishes is so inconceivably great that the number we catch is relatively insignificant; and, secondly, that the magnitude of the destructive agencies at work upon them is so prodigious, that the destruction effected by the fisherman cannot sensibly increase the death-rate.

Since then the landings of fish from the seas around the North Atlantic have increased to an extent that Huxley could hardly have envisaged. The old sailing vessels have been superseded by trawlers with powerful engines. The old beam trawl has been replaced by the far more effective otter trawl, and radically new techniques for detecting fish-shoals have been invented. As a result, the stock of some fish has been greatly reduced on the more intensively exploited fishing-grounds of the northern Atlantic area. Overall stocks of pelagic fish such as herring, smelt and mackerel are probably still more than adequate to meet present demands, but there is now much concern at the extent to which the favourite demersal species, such as cod, haddock, hake and plaice, are being fished in some areas. E. S. Russell[15] in 1942 wrote.

> A state of overfishing exists in many of the trawl fisheries in north-west European waters. Two things are wrong. First, there is too much fishing, resulting in catches below the possible steady maximum, and secondly, the incidence of fishing falls too early in the fishes' life resulting in a great destruction of undersized fish which ought to be left in the sea to grow.

In order to understand what is involved in 'over-fishing', we will briefly consider the effects that fishing is likely to have on the size and composition of fish populations. First, we will take the case of a stock of fish which is subjected only to very light fishing. This population can be regarded as having grown to the limits imposed by the food supply, which restricts both the number of fish surviving and the size to which the individual fish grow. Scarcity of food prevents all the fish from making as much growth as they would if they were better fed, and the slow-growing, older fish compete for food with the rapidly-growing younger specimens. The small catches taken by fishermen are likely to consist mainly of the larger, older fish; but due to age, undernourishment or disease, these may not be of good quality, and therefore fetch correspondingly poor prices on the market.

A stock of fish in this condition may be regarded as 'underfished'. The population is overcrowded. An undue proportion of the food is consumed by old fish of poor market quality at the expense of young fish, and none can realize their full growth potentials. This stock could support a larger, more profitable fishery of better-quality fish if more fish were caught. A reduction in the size of the population, particularly by the removal of older fish, would promote a better growth-rate throughout the remaining stock and improve the condition of the fish, the process being analogous to a gardener thinning-out his plants to encourage the best growth and quality of his specimens.

Considering now the opposite case, a stock of fish subject to extremely heavy fishing, this population is likely to consist mainly of young, small specimens because the fish are caught as soon as they reach catchable size. Landings of undersized fish are likely to fetch poor prices because these fish carry little edible meat, being mainly skin and bone.

This stock is obviously 'overfished'. Too many fish are caught too early in life. Young fish make rapid growth and, if left longer in the sea, would soon reach a more valuable size, providing heavier landings and better prices. Productivity and profits would eventually improve if the amount of fishing was reduced. However, in these conditions the fisherman is tempted to try and increase his profits by catching even more fish, but this is a vicious spiral which can only lead to a further reduction in the size of the stock, and a further dwindling of the fisherman's income.

Wars and Fish Stocks in British Waters

During the first half of the present century the fisheries of the north-east Atlantic have twice been interrupted by war. Conditions on some of the major fishing grounds visited by British vessels have fluctuated correspondingly, with symptoms of underfishing during war years and severe overfishing in peacetime. For example, the North Sea provides an important fishery for haddock. During the years prior to 1914, the total landings of haddock from this area showed a fairly steady decline. During the period of the 1914–18 war, fishing in the North Sea was greatly reduced. When normal fishing was resumed after the war, greatly increased yields were at first obtained, amounting to approximately double the pre-war landings, but during subsequent years the catches gradually diminished. By the late 30s' the annual landings of haddock had fallen below their pre-1914 level to about a quarter of their 1920 weight.

During the war of 1939–45, fishing in the North Sea virtually ceased.

When normal fishing was re-established, there was at first a marked increase in haddock landings, but subsequently the yield of the fishery has again fallen (*Figure 8.28*).

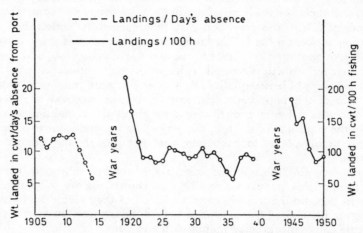

Figure 8.28.   *Haddock catch per unit of fishing effort by Scottish trawlers in the North Sea*

(From Graham, M.[6] by courtesy of Edward Arnold)

These diminishing peacetime yields have not been the result of any reduction of fishing effort, rather the reverse. Furthermore, as the yields decline the percentage of the catch consisting of small fish shows a great increase (*Figure 8.29*).

In the plaice fishery, comparable changes have occurred (*Figure 8.30*). The largest sizes of plaice in the North Sea have only been abundant in immediate post-war years, and since then the catch has consisted mainly of small fish (*Figure 8.31*). Although the total weight of plaice landed from the North Sea has remained fairly constant over the first half of this century, the effort and expense of catching them has greatly increased, i.e. the yield per unit of fishing effort has become less. During the '60s there was some reduction of fishing for plaice in the North Sea because of poor returns. Attention turned to other species and some old vessels were laid up and not replaced. Consequently there was some recovery of plaice stocks in the late '60s and landings considerably improved.

In normal peacetime operation, plaice and haddock fisheries of the North Sea have therefore shown major features of overfishing, namely, declining yields and a preponderance of small fish. Similar trends have been observed in many areas for such important species as cod, sole, herring, halibut and hake.

THE OVERFISHING PROBLEM

The war years, on the other hand, produced some symptoms of under-fishing. The stock of fish increased greatly, and this was reflected in the heavier catches obtained when fishing was resumed. These early post-war catches contained a high proportion of large fish of the older age-groups,

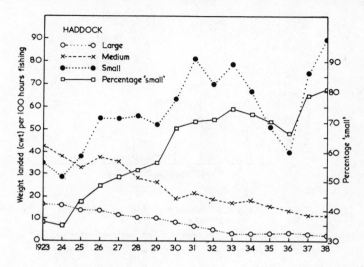

Figure 8.29. Catch, per unit of fishing effort, of North Sea haddock, and percentage of the 'small' category, 1923–38

(From Graham, M., *Sea Fisheries—Their Investigation in the U.K.*, (1956) by courtesy of Edward Arnold)

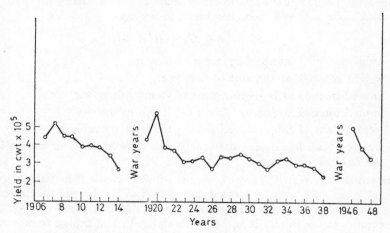

Figure 8.30. Catches of North Sea plaice by first class English steam trawlers

(From Wimpenny, R. S.[17] by courtesy of The Buckland Foundation)

but some were diseased and of poor quality. In the case of plaice, there is some evidence that the stock increased to such an extent that the average growth-rate of the fish was reduced, presumably by food shortage.

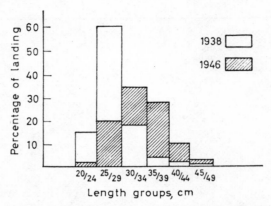

Figure 8.31. Size distribution in plaice landings at Lowestoft before and after World War II

(From Wimpenny, R. S.[17] by courtesy of The Buckland Foundation)

The Optimum Fishing-Rate

E. S. Russell[15] has pointed out that the weight of a stock of fish tends to increase as the fish grow and young fish join the stock. The effects of fishing and natural mortality operate against this tendency to natural increase. These factors can be combined in a simple expression

$$S_2 = S_1 + (A + G) - (C + M)$$

S_1 = weight of stock at the beginning of a year,

S_2 = weight of stock at the end of that year,

A = annual increment by recruitment of young fish to the stock,

G = annual increment due to growth of all fish in the stock,

C = total weight of fish removed during the year by fishing,

M = weight of fish lost during the year by death from all other causes, i.e. the natural mortality.

The amount by which the stock weight would increase if no fishing took place, $A + G - M$, can be termed the *natural yield*. In the special case where the total stock remains unchanged, $C + M$ must equal $A + G$. In these conditions the fishery is said to be 'stabilized', and fishing removes an 'equilibrium catch', i.e. a weight of fish that exactly corresponds with the natural yield of the stock. In theory, equilibrium conditions might be established for any level of stock weight, but the natural yield will vary for different weights and compositions of stock.

Underfishing and overfishing are extremes both of which result in low natural yields. Between the two there must be some intermediate intensity of fishing which, by matching the size of the stock to the environmental resources, especially the availability of food, would provide in equilibrium conditions the maximum possible sustained natural yield, i.e. an *optimum yield*. Such a fishing intensity could be termed the *biological optimum fishing rate*.

It is the major aim of fishery research to achieve a sufficient understanding of the variables in Russell's equation to be able to predict the conditions of optimum fishing for each species, and to achieve accordingly an international fisheries policy. The problems are extremely complex and incompletely understood because so many factors have to be taken into account. Fish stocks are subject to considerable fluctuations in rates of recruitment, mortality and growth due to natural causes such as variations in water movements, temperature, salinity, the condition of the plankton, food supplies, and the numbers of predators. The evaluation of Russell's equation therefore requires knowledge over a wide field of oceanography, meteorology, hydrography and biology.

There are also economic factors to be considered. Highest profits do not necessarily result from the heaviest landings. If a market is glutted, prices collapse. A biological optimum fishing rate, giving the heaviest possible sustained landings, may not be the same as an *economic optimum fishing rate*, i.e. one giving the greatest financial returns. If a fishing policy is to be acceptable to the industry, it must aim to ensure the maximum output consistent with a fair return to all those engaged in fishing. The theoretical effects of controlled fishing on fish populations and fishing industries are discussed in greater depth in references listed at the end of this chapter[2,13,27].

At the present time, control of natural factors influencing the yield of commercial sea fisheries cannot be envisaged, and it is therefore only in the regulation of fishing activity that there is a practical possibility of achieving optimum yields. But the relationships between fishing, stock, yields and profitability are by no means simple. The same weight of fish can be taken in innumerable ways; as a small number of large fish, a large number of small ones, or any combination of different sizes. Every variation in the composition of the catch will have a different effect upon the composition of the stock and its natural yield. To achieve anything approaching optimum yields it would therefore be necessary to control not simply the gross weight of fish landed, but also the numbers of each size of fish.

There are broadly two types of study upon which the formulation of fishery regulations may be based, the empirical and the analytical. Empirically, if sufficient observations are made of the general trends of a fishery, noting the nature and size of the catches and the extent to which they vary from year to year with different intensities of fishing, predictions may be made of the effects of varying the fishing rate. The analytical method attempts a fundamental elucidation of all factors that produce changes in size and yield of fish stocks, from both natural causes and the effects of fishing, an approach involving a very wide field of fishery research.

FISHERY RESEARCH

Many inter-related fields of study are included under the general heading of Fishery Research, including investigations of the distribution and natural history of fish, the size and composition of stocks, and the extent and causes of fluctuations in stocks. The following is a brief summary of some of the methods used.

Distribution

Although the general distribution of the most important commercial species has been known for many years from the observations of fishermen, detailed investigations sometimes reveal that the population is not biologically homogeneous, but comprises several more or less separate breeding groups. For the rational exploitation of a stock it is necessary to know of the existence of such subdivisions, the limits of the distribution of each, the extent to which interchanges may occur between them and the contribution that each makes to the fishery in any area.

In some cases, measurements of the sizes of fish may point to a heterogeneity of the population. Normally, the length-frequency curve for fish of equal age from the same stock is unimodal. A curve having an obviously bimodal or polymodal form for fish of the same age-group suggests that a mixed population is being sampled. Sometimes anatomical or physiological differences indicate subdivisions of the population. Precise studies of the anatomy of fish from different areas may reveal slight differences in structure; for example, in number of vertebrae, the number of fin-rays or the pattern of rings in scales or otoliths. There may also be physiological differences between different parts of the population; for example, in salinity- or temperature-tolerances, fecundity, breeding season or blood-cell counts. Recently, increasing attention has been given to biochemical and serological studies as evidence of genetic differences between subdivisions of fish stocks[23]. Fishery research attempts to discover if such differences are characteristic racial features, or are simply variations due to environmental causes.

Information on the movements of fish can be gained from marking and tagging experiments. Fish marking is done by making small marks or mutilations on the fish whereby they may be recognized on later occasions. Tagging involves the attachment to the fish of some form of label, usually a small metal or plastic disc (*see* page 235), and is a much more precise method than marking, enabling records to be kept of the movements of individual fish. Tagging experiments have been performed on most species of commercial importance, including plaice, lemon sole, halibut, cod, haddock, hake and herring. In addition to indicating the extent of migrations and interchanges of population between different areas, tagging also provides data for calculations of fishing intensity, stock size and mortality.

Natural History

Fishery biologists seek to accumulate knowledge covering the entire life of a species from egg to adult, including details of spawning areas and seasons, eggs and larval stages, fecundity, the factors that influence brood survival, nursery areas of young fish, their growth and maturation, feeding habits throughout life, changes in distribution at different stages of life and throughout the annual cycles of breeding and feeding, predators, diseases and all causes of mortality. The techniques of investigation vary according to the habits of particular fish, the nature of the area and the research facilities available.

If the fish spawn satisfactorily in aquaria, the early stages of life can usually be investigated in the laboratory. The eggs and larval stages are studied so that they may be readily identified when found at sea, and the effects on development of environmental factors such as salinity or temperature can be experimentally determined.

To discover areas and seasons of spawning, the landings of the commercial fishery may be examined. When the catch includes fish with fully ripe gonads, the regions from which such fish were obtained may be learned from the fishermen. This gives a general indication of the probable sites of spawning, but for precise information it is usually necessary for research vessels to investigate the area. If the eggs are demersal, an attempt may be made to find them by dredging. If they are pelagic, tow-net hauls are made at a large number of stations to determine precisely in which areas and at what depths the main concentration of eggs is found.

Standardized techniques of net-hauling and egg-counting have been devised for the calculation of the number of eggs in unit volume of water, or beneath unit surface area. The concentration of eggs at each sampling station can be plotted on a map, and lines drawn joining the points where equal numbers of eggs have been found. These 'egg-density contours' indicate the zones and limits of the spawning areas (*Figure 8.32*). As the spawning period proceeds, the areas of main spawning may shift, so these investigations need to be repeated at intervals.

Studies are also made of the fecundity of the fish of different ages, and in different areas. These may have a bearing on recruitment rates, and also provide some data which may be used for calculations of total population (*see* page 265).

During the pelagic phase of the life history, distribution and development of larvae and young fish can usually be studied from tow-net hauls, but once the fish leave the surface water it is often difficult to find them, especially when they occur in rocky areas where fine-mesh trawls cannot easily be used. The biology of some species is not at all well known during this intermediate period between the end of the pelagic phase and the time when they become large enough to be caught by commercial nets.

Some information on feeding habits may be gained from observation of fish in aquaria, but this may give no true picture of food preferences in natural conditions. The most reliable information on feeding habits comes from studies of the gut-contents of fish captured at sea, but new techniques of under-water observation, such as free-diving or under-water television, now open up new possibilities for the study of feeding and many other features of fish behaviour in natural surroundings.

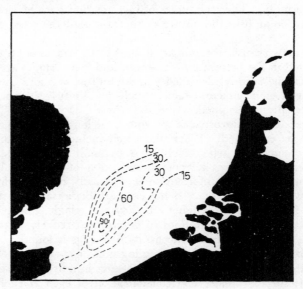

Figure 8.32. Egg density contours for plaice in southern North Sea. Numerals give number of eggs per m³ of water.
(From Wollaston, H. J. B.[42] by permission of the Controller H.M. Stationery Office)

Growth-rates can seldom be determined by direct observation. The growth of fish in aquaria is not a reliable guide to growth in natural surroundings. Where tagging is thought to have no detrimental effect on the fish, measurements of tagged fish before release and after recapture provide some information about growth in the sea. In some cases, the width of growth-zones between seasonal marks on meristic structures such as scales or otoliths is proportional to the overall growth of the fish. For example, the seasonal rings of herring scales give a close indication of the length of the fish in previous years (*Figure 8.33*). In most cases, however, average growth-rates are determined by correlating size with age in an analysis of the stock composition (*see* page 264).

It is evident that individual fish differ greatly in their rates of growth. Intrinsic differences may be partly responsible for this, but growth-rates also vary with locality due to environmental conditions, particularly the water temperature and the availability of food.

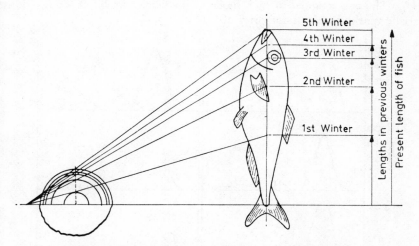

Figure 8.33. Correlation between scale growth and fish growth in herring

Stock Analysis

(*a*) *Sampling*—The first requirement for any analysis of the composition of a stock of fish is a reliable method of obtaining representative samples. Where there are no market or mesh restrictions leading to the elimination of small fish from the commercial landings, these may include the full range of sizes and provide good samples of the fishable stock. If the commercial catch is not representative of the stock, it may still supply useful data if the sampling error is known. In most cases, however, special fishing gear operated by fishery research vessels is needed to obtain good samples of all age-groups. The fishing grounds are explored using trawls with cod ends covered with fine-mesh net capable of retaining the smallest fish. For pelagic fish, drift nets may be used which include a range of mesh sizes suited to the different sizes of fish in the schools.

(*b*) *Age census*—Stock samples are examined to determine the number of fish in each age-group. In some cases, age determination is based on the periodic markings of scales, otoliths, or other structures showing meristic features. There are many difficulties of interpretation of these markings, and they tend to become less reliable as the age of a fish increases, but the method is useful and widely applied.

Where annual markings are absent, unreadable or of doubtful reliability, Petersen's method of age analysis is used. Measurements are made of the length of each fish in representative samples of stock, and length/frequency graphs are plotted. These polymodal curves can be broken down to their separate modes, each representing an age-group within the sample (*Figure 8.34*). The age assigned to each mode of the curve can be checked against meristic markings of the majority of that group.

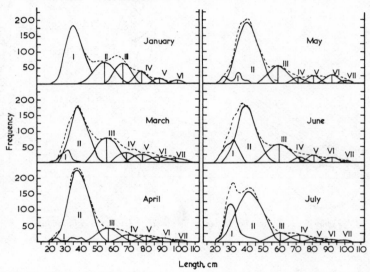

Figure 8.34. Dissection of length/frequency data for cod on Petersen's principle. Polymodal curves are pecked

(From Graham, M.[5], Buckland Lecture (1948), based on *Fisheries Investigation Ser. 2*, **XIII**, No. 4, p. 54, Fig. 18 by permission of the Controller H.M. Stationery Office)

(*c*) *Sizes and growth rates*—The modes of the length/frequency curve indicate the length distribution of each age-group. From these, average growth rates can be estimated.

(*d*) *Sex ratio*—From examination of stock samples, the sex ratio and proportion of mature and immature fish can be determined.

(*e*) *Stock size*—There are two principal ways in which estimates of total numbers in fish stocks may be attempted. One is the tagging and recapture method[20]. A known number of tagged fish are released into the sea, and time allowed for dispersal. Records are then kept of the numbers of tagged and untagged fish that are captured. There are several sources of error, but taking the simplest case where mortality can be ignored, where no tags are lost and the tagged fish are evenly distributed throughout a stock unaffected by migrations, the total population could be obtained from the relationship:

$$\text{Total population} = \frac{\text{No. of tagged fish released} \times \text{Total No. of fish caught}}{\text{No. of tagged fish recaptured}}$$

As an illustration, Graham[4] gives the following figures. Over a period of years before 1914 the results of tagging experiments on North Sea plaice indicated a fishing mortality of about 70 per cent of the stock. Average landings of plaice from the area being some 50,000 tons, the stock weight was presumably about 70,000 tons. Allowing 3,000 fish to the ton gives an average stock of 210,000,000 fish.

An alternative method is based on egg-counts. We have previously referred to the method of plotting egg-density contours. From these, an estimate can be made of the total number of eggs laid in a season within a spawning area. If the average number of eggs produced by spawning females is known, the total number of spawning females may be determined. If the sex ratio is known, the total number of spawning males can also be calculated. Taking account of the proportion of the population that are immature, and any fish that may spawn outside the main spawning area, a calculation may be made of total stock numbers.

This method was applied by Wollaston to estimate the plaice population of the southern North Sea. In 1914, an exceptionally good spawning year, the total egg production in the Flemish Bight spawning area was reckoned to be about 3.5×10^{12}. Taking a mean fecundity of 70,000 eggs per female plaice, the number of females spawning in this area would be

$$\frac{3.5 \times 10^{12}}{70,000} = 50,000,000.$$

This figure must be doubled to allow for the number of mature males, and should probably be about doubled again to allow for plaice spawning in other parts of the southern North Sea, making 200,000,000. As approximately half of the fishable stock are immature fish this gives a final figure for the total fishable population of 400,000,000 plaice. These figures[6] are a revision of Wollaston's original estimate, which is now regarded as considerably too low, his figure of 200,000 for the mean fecundity of plaice probably being too high by a factor of about three.

(e) *Stock growth*—The growth-potential of the stock may be estimated from data on the composition of the stock and the mean growth-rates of each age-group, due allowance being made for differences of growth-rate of the two sexes and in different areas over which the stock is distributed.

(f) *Recruitment*—Studies of fecundity indicate that this usually varies with size, large fish producing more eggs than small ones (*Figure 8.35*).

From such data, estimates of total egg production can be made for various sizes and compositions of stock. In most cases, however, the reproductive capacity of a stock does not seem to have any great effect on brood strength. The major factors controlling recruitment appear to be environmental, chiefly temperature, food supply and movements of the water. The prediction of recruitment rate is therefore very uncertain unless methods are available for sampling the younger age-groups before they join the fishable stock. As previously mentioned, this phase of the life history is often the least well known.

(g) *Total, natural and fishing mortality*—Estimates of total mortality (death from all causes), natural mortality (death from causes other than fishing) and fishing mortality may be made by relating an analysis of stock composition

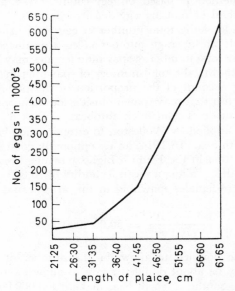

Figure 8.35. Relation between numbers of eggs and length for North Sea plaice
(From Wimpenny, R. S [17] by courtesy of The Buckland Foundation)

to an analysis of the composition of the catch of the fishery. Total mortality can be derived by noting over a period of years the numbers of fish in each age-group of the stock. Examination of the commercial catch reveals the numbers of fish removed each year from each age-group by fishing, i.e. the fishing mortality. From these, it is possible by subtraction to determine the rate at which the numbers of fish are reduced from causes other than fishing, i.e. the natural mortality.

Mortality rates may also be calculated from the data provided by tagging experiments. If the rate of recapture of tagged fish is known, an estimate may be made of the mortality directly attributable to fishing and the overall mortality from all causes, the natural mortality rate again being the difference between the two.

THE REGULATION OF FISHERIES

The primary purpose of regulating commercial sea fisheries is an economic one, being to ensure ample supplies of good quality fish for consumption and to safeguard and, if possible, increase the profits of fishing. Fishing is a risky enterprise, and profits must be reasonably assured to attract the large investments needed by an up-to-date industry. Conservation measures for fish stocks are therefore designed principally to preserve the fishermen's livelihood by ensuring that there are plenty of good-sized fish for him to catch, rather than simply to protect the fish. Where it is considered that overfishing is taking place, it may seem obvious that the cure is to catch fewer fish, but the fisherman may not see the matter in such simple terms. The immediate consequence of any reduction of fishing is to cut down his earnings. It may be difficult to convince a man whose income is derived from such a precarious occupation that, by reducing his catch, he will eventually be more than reimbursed by larger yields from a recuperated stock. He may feel less confidence in the predictions of fishery scientists than in his own ability to maintain his livelihood by increasing his efforts to catch fish so long as the stocks last. If he does not catch the fish himself, he may well suspect that someone else will; and, unless restrictions can be easily enforced on all fishermen, he is probably right. Regulations to control fisheries are therefore likely to be opposed by the very people they are designed primarily to benefit.

We may summarize the principal means of attempting the protection of fish stocks as follows:

(1) Restricting the total amount of fishing by

(a) limiting the total permitted annual catch,

(b) limiting the size of the fishing fleet,

(c) limiting the length of the fishing season,

(d) limiting the type and size of fishing gear,

(e) limiting the areas in which fish may be caught,

(2) Affording special protection to young fish so as to ensure that they reach a good size before capture by

(a) prohibiting the landing of undersized fish,

(b) closing nursery areas to fishing,

(c) prescribing minimum mesh or hook sizes.

The suitability of different methods of control vary with the nature of the fishery. Whatever methods are chosen, it is clearly necessary that a limit must be set to the total catch. No other protective measures can make overfishing impossible so long as the fishing intensity can be sufficiently increased to circumvent them. Limitations of total annual catch have been applied with success in the North Pacific halibut fishery. Attempts have also been made to limit total catches of whales in the Antarctic; but here the limits have either been too high or have been ignored with the result that depletion of whale stocks has progressed to the point at which Antarctic whaling has now greatly declined[29].

In the north-east Atlantic, where most British fishing vessels operate, the problems of devising satisfactory fishery regulations are particularly complex because of the number of nations involved in fishing, and the range of species sought. The United Kingdom, France, Belgium, Holland, Denmark, Germany, Norway, Sweden, Iceland, Poland and Russia all have fishing vessels operating in these waters. Obviously, complete international agreement is essential if regulations are to be fair and effective, and this has not been easy to obtain. Not until 1954 was any agreement reached, when an International Convention recommended minimum sizes of mesh for trawl cod ends and seines over a wide area, and also minimum sizes of fish that may be landed. This affords some protection to smaller fish, but there is as yet no agreement to limit total landings from any part of the area.

It was at one time argued that mesh regulations for trawlers must prove ineffective in protecting young fish because it was thought that, during trawling, the net becomes pulled out lengthways, almost completely closing the meshes so that any small fish entering the cod end must be retained. This was disproved in experiments by Davis and Goodchild[22], 1928–1933; they enclosed the cod end of a trawl with a long bag of fine-mesh netting capable of trapping any small fish that might escape through the cod end. The fine-mesh bag was encircled by a noose which after a period of trawling was automatically drawn tight, closing the bag. This demonstrated conclusively that large numbers of small fish passed through the cod end unharmed.

Placing a fine mesh bag around the cod end also provides a means of studying the extent to which different sizes of fish are retained by different meshes of trawl. By counting the number of each size of fish in the cod end and in the bag, this data can be used to construct a graph showing the percentage of each size of fish that escape through the mesh. This is termed a Percentage Release Curve or Selection Ogive, and usually has the form shown in *Figure 8.36*. We see that, for each mesh size, all fish above a certain size are retained in the net. There is also a lower limit of size beneath which virtually all the fish escape. Between these two limits, the proportion of fish which are caught or escape varies with their length, there being a particular size at which 50 per cent of the fish escape. This 50 per cent release length is a convenient index for the comparison of the selective action of different meshes.

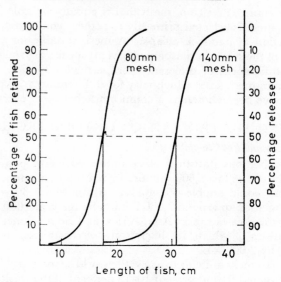

Figure 8.36. Selection Ogive for plaice in nets of 80 *and* 140 mm
mesh
(From Wimpenny, R. S.[17] by courtesy of The Buckland Foundation)

Experiments have demonstrated that drift nets are also to some extent selective, and that increasing the size of the mesh allows the smaller fish to escape (*Figure 8.37*).

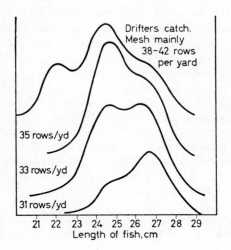

Figure 8.37. Measurements of herring retained by drift nets of various mesh sizes. Increasing the mesh aperture allows the smaller fish to escape

(From Hodgson, W. C.[10] by courtesy of The Buckland Foundation)

There is, of course, no size of mesh that is equally suitable for all species of fish. If the mesh is selected primarily to protect young cod, it will let an undue proportion of haddock escape. A mesh suitable for plaice will let through most of the sole. In most areas, several species are sought, and there are national differences of preference for particular species. The present agreed minimum mesh sizes, which vary for different areas, types of net and net material have been chosen as a compromise between several interests.

INCREASING THE PRODUCTION OF FOOD FROM THE SEA

World Population and Food Supplies

The United Nations statistical office estimated world population in the middle months of 1969 as 3,600 million, increasing at a rate which, short of global disasters, could double this number within about 30 years, bringing the world population to around 7,000 million by the year 2000. The highest population growth-rates are in Asia, Africa and South America, the combined populations of which are likely, on present trends, to reach nearly 5,500 million by 2,000 A.D.

While it is extremely difficult to assess the world's food supply and requirements, it is estimated that at present 10–15 per cent of the world's population are underfed, having insufficient food to provide even enough calories for a normally active life. In addition to these hungry millions, there are many more, possibly as many as 30 per cent of the total, whose calorie-intake is adequate, but who suffer from malnutrition because their diet lacks various substances essential for health.

The food situation differs greatly in different areas. In Western Europe, North America, Australia and New Zealand the consumption of food is, in general, fully adequate for daily needs, and health problems arise more from over-indulgence or improperly balanced diets than from any insufficiency of quantity. In Africa, South America and the Near East, however, food intake is precariously balanced with requirements, and often inadequate, while serious food shortages occur in parts of Asia. In all communities inadequate diet is closely associated with poverty.

These unsatisfactory features of the world's food budget are likely to get worse. Hunger and malnutrition are found mainly in under-developed areas where population is increasing most rapidly. Proper balance between food supply and demand must depend ultimately on a more equitable distribution of wealth and effective measures of population control, but there is also the immediate task for science and technology of finding means to increase total food production and improve food distribution to bring present diets up to adequate levels and keep pace with the expanding population.

Overall the organic production of the oceans cannot be greatly less than that of the land surface but, although 71% of the earth's surface is covered with sea-water, only about 1% of the total supply of human food comes directly from the sea. About 10% of the total human consumption of animal protein is derived from the sea, either directly or indirectly via fishmeal fed to livestock. Fish is a high-protein food of excellent nutritive quality, in some respects better than meat, and any increase in fish supplies would be a valuable supplement to the world's sources of protein. Many fish are also rich in edible oils, and fish livers are an important source of vitamins A and D. Whales, molluscs and crustacea are other marine products which are useful foods, and might in some cases be more fully exploited, especially the cephalopods. It is also possible that food might be obtained from the sea by unconventional methods, such as direct harvesting of the marine plankton. We will briefly consider some of the ways in which the seas might make a greater contribution to the world's pressing need for more food.

Increasing the Output of Fisheries

We have previously discussed the danger of attempting to raise the output of fisheries by simply increasing the intensity of fishing because this may lead to diminishing returns due to gradual depletion of fish stocks. Improved techniques for finding and catching fish must therefore be applied with caution if overfishing is to be avoided. Overfishing, however, is at present limited to particular areas and species. There is no doubt that more fish could be safely removed from the sea if a greater variety of species were exploited, and fishing operations were conducted over wider areas. It can be calculated that, *if carefully regulated*, world landings of fish might be more than doubled without overfishing (page 294), but it is doubtful to what extent the economic returns could justify the very large capital investment required. The pelagic fisheries in particular could provide considerably larger landings in many areas. Even on the intensively fished north-east Atlantic fishing grounds, more herring, mackerel, smelt and bluefish could be taken quite safely if a ready market existed. There are also good stocks of some less-popular demersal species, such as pollock and skates, which are more appreciated on the mainland of Europe than in the United States. Conservative habits of diet are partly to blame for the concentration of fishing on a few popular species, and new methods of processing, for example, 'fish-sticks', are now doing something to open the market to a wider variety of fish.

In many of the under-developed parts of the world, the productivity of local fisheries is severely limited by their primitive methods of fishing, preserving and marketing. Small craft can only make short voyages, with the result that the close inshore grounds are intensively overfished. Much better landings might be obtained if the fishermen could extend their operations over wider areas. Furthermore, the consumption of sea fish is often restricted to coastal regions because nothing is done to preserve the catch for distribution inland. The expansion of these primitive fisheries requires large capital investments for the mechanization of fishing, for the training of fishermen in new methods, and for the provision of proper facilities for preservation and distribution.

At present, most fishing operations are carried on in the highly-productive, shallow waters overlying continental shelves, amounting to only some 8 per cent of the total sea area. In deep waters beyond the continental edge, quantities of pelagic fish are virtually unexploited, offering great scope for the expansion of oceanic fisheries. For example, in recent years an off-shore fishery along the Peruvian coast has been rapidly developed. Here, the continental shelf is very narrow, but the upwelling of deep water along the extensive coastline (pages 10, 129) gives rise to prolific production of a great variety of marine life. For many years, fishing in this area was conducted only on a small scale. The main output of the region was guano, the fertilizer formed from the droppings of millions of sea birds which fed on the teeming life of the sea. Now an intensive deep-water fishery has grown up for the small pelagic fish, anchoveta (*Engraulis ringens*), which occur in prodigious numbers in the surface layers. Unfortunately, this is an un-appetising fish for which there is little market for direct human consumption, and so virtually the entire catch is converted into fishmeal for cattle or poultry food. Peru now takes a larger annual catch of fish than any other nation, about 10 million tons, approximately one-sixth of annual world landings. For comparison, annual landings of U.S. fisheries are below 3 million tons and declining.

Waste fish from many sources is now converted into fishmeal. This has become a popular animal feeding-stuff and the trend is for an increasing proportion of the catch to be processed for this purpose. Nearly 50% of world landings of marine fish are now used for animal food. This, however, inserts an additional link in the food chain and is therefore a relatively wasteful way of using valuable protein. Techniques are now being developed for the preparation of odourless and flavourless fishmeals suitable as additives for human food. Although these have not yet proved popular, they may eventually provide a better use for unpalatable fish products.

Apart from the possibilities of increasing fishing in underfished areas, and discovering new sources of fish, there are various ways in which the productivity of the more intensively fished areas might be maintained or increased. For example, yields might be increased by transplanting young fish to areas of more abundant food. Since the early years of this century, Professor Garstang repeatedly advocated the transplantation of young plaice, and demonstrated a threefold to fourfold increase in growth rate of young North Sea plaice moved from their crowded coastal nurseries to the Dogger Bank, where bivalves suitable as plaice food are plentiful. Calculations indicated that, on a large scale, this could prove a profitable enterprise, but such a project would require close international co-operation to ensure that the transplanted fish were left long enough in the sea to benefit, and that costs and profits were equitably shared. As long as present laissez-faire attitudes in fishing persist there can be no possibility of this.

There are now some hopes that it might become possible to increase the stocks of certain species of marine fish by artificial rearing during the early stages of life. The first few weeks are a period of special danger when extremely heavy fish mortality always occurs. The number of fish surviving in each year-brood seldom bears any close relationship to the number of eggs spawned—generally, the more eggs, the greater the number of casualties. Survival is mainly determined by environmental factors, for example, temperature, salinity, food supply, currents and predation. It might, however, be possible to increase the stocks of fish if eggs and larvae could be raised in very large numbers in protected conditions, and supplied with ample food until the danger period is passed before setting them free in the sea. The unusually heavy catches made on some of the north-east Atlantic fishing grounds immediately following the two World Wars, during which very little fishing took place, indicate that these areas are able to support many more fish than they ordinarily do when fishing is proceeding at normal peacetime rates.

In experiments proceeding since 1951, J. E. Shelbourne[37] and his colleagues of the Lowestoft Fisheries Laboratory have developed successful methods of rearing large numbers of young plaice by stocking the open-circulation sea-water tanks at the Marine Biological Station at Port Erin, Isle of Man, with plaice eggs spawned in captivity. Methods were devised for bulk preparation of suitable planktonic food for the developing larvae, mainly nauplii of *Artemia salina*. A measure of bacterial control was achieved by treating the water with antibiotics and ultra-violet light. In these conditions many thousands of young plaice have been reared to the completion of metamorphosis, and survival rates of over 30 per cent of the original egg stock have been achieved. Future development of this type of work might eventually lead to the establishment of large-scale fish-hatcheries, and the continual re-stocking of the sea with millions of artificially-reared young fish set free in areas of plentiful food. So far a good survival of artificially reared young fish after transplantation to the open sea has not been demonstrated. Even if the biological difficulties can be overcome, there remains the problem of ensuring that those who pay the cost reap the benefit, and at present there appear to be better prospects in using certain artificially reared species for stocking fish farms (*see* below).

We have previously referred to the competition for food that exists between bottom-feeding fish and the large numbers of predatory invertebrates, and to Thorson's calculations that 'only 1–2 per cent of the fish-food is actually eaten by fish, the rest is taken by invertebrates' (*see* page 168). An area could presumably support a greater number of food fish if competitors could be eliminated, or at least considerably reduced. Hardy has suggested that it may eventually become possible to weed the sea-bed of unwanted creatures so that a larger proportion of the food becomes available for fish. He writes[7]

> If Thorson's calculations are correct, and if man could eliminate just a quarter of the pests and so allow the fish to have some 20 per cent, instead of 2 per cent, of the potential food supply, then he could make a given area support ten times the quantity of fish. How are such pests as star fish to be eliminated? I believe that just as we harrow and roll the land in addition to reaping our crops, we shall in time systematically drag some combing or other devices over the sea-floor at intervals to weed out the creatures that take food from the more valuable fish; and the pests themselves may well be ground into meal to feed poultry ashore. It will require much more ecological research to determine just which of the animals we can do with in smaller numbers.

Fish Farming

From early times fish culture in fresh water has been practised in many parts of the world, particularly in the warmer areas. Usually, fast-growing vegetarian species are cultivated in shallow ponds. The growth of suitable pond weed for their food is encouraged by enriching the water with sewage or organic refuse. Where sunshine keeps the water temperature high, biological processes proceed very rapidly and remarkably high rates of food-production can be obtained from efficiently managed fish ponds. Capital and labour costs are low, and land unsuitable for ordinary agriculture can often be profitably farmed in this way.

Primitive forms of husbandry are applied to the culture of various marine creatures, notably molluscs such as oysters and mussels, and also certain crustacea. Some remarkably high rates of meat production are claimed. For example, annual yields of about 250 metric tons of flesh per hectare (2·5 acres approx.) have been quoted for certain areas around the Spanish coast where mussels are cultured by allowing them to settle and grow on strings suspended below rafts. But there are greater difficulties in applying husbandry techniques to the cultivation of marine fish. A primary requirement for animal farming is to keep one's stock within a protected area, where they are safe from predators and can grow under controlled conditions without severe competition for food from unwanted species. Oysters and mussels, once they have settled, remain virtually fixed; and the beds where they are grown can to some extent be protected from enemies and competitors. But fish roam about, often over considerable distances, and fish-pens cannot easily be constructed in the open sea.

A simple form of enclosure for sea fish is the coastal inlet where a body of sea-water can be isolated behind some sort of dam or screen. Some interesting experiments[24-26,30,34-36] which illustrate the sort of problems that arise in marine fish farming were conducted over the years 1942–1947 in Loch Craiglin and Kyle Scotnish, small sea-water lochs on the west coast of Scotland.

274

INCREASING THE PRODUCTION OF FOOD FROM THE SEA

Loch Craiglin has an area of about 18 acres. The water is quite shallow, about 6 m in the deepest part but most of it less than 2 m in depth. The bottom is mostly covered with soft mud, with some sand along the shore. During the investigations, the interchange of water with the sea was partially controlled by a small dam. From time to time the loch water was enriched by the addition of various types of nitrogenous and phosphatic fertilizers, with the object of promoting the growth of phytoplankton. In the course of the experiment, numbers of small fish, mostly flounders, but also plaice, dabs, soles and witches were transplanted from the sea to the loch. The main purpose of the investigations was to study the effects of fertilization on the flora and fauna of the loch, with special reference to the growth of fish.

It was found that, unless great quantities of salts were used, the addition of fertilizers to the water produced only a brief increase in the concentration of nitrate or phosphate, usually of 2–3 days duration. The effects of fertilization on the growth of phytoplankton were equally short-lived unless very large quantities of salts were added. The rapid disappearance of the nutrients was probably due partly to their adsorption on the surface of mud or detritus particles, but mainly to their uptake by the large attached algae growing in the shallow water. Following the addition of nutrients, microflagellates generally showed a transient increase, but the numbers of diatoms and dino-flagellates did not increase until sufficient fertilizer was added to meet fully the requirements of the attached algae and leave some remaining for the phytoplankton. Only if this condition was satisfied was it found possible to produce any large and long-lasting increase in the quantities of diatoms and dinoflagellates. The numbers of zooplankton showed no clear correlation with the addition of fertilizers.

As the experiments proceeded, the increased plant growth brought about by the use of large amounts of fertilizer was followed gradually by a great increase in the numbers of the bottom fauna. The growth of the benthic population was slow and only reached its maximum after 3 years of continual fertilization. The main increase occurred only in the shallowest water, 2 m or less in depth, around the edge of the loch. In the deeper water, the hydrographic conditions were unfavourable. Long closure of the dam to prevent the escape of fertilizer caused poor circulation, leading to gradual stagnation and deoxygenation of the deeper water, and a concomitant production of hydrogen sulphide in which few organisms could survive.

Of the various species of fish transplanted to the loch, flounders were the only ones to flourish; the conditions did not suit the other species. Dabs, soles and witches soon died. Small plaice did rather better, and some appeared to make as much growth in one year as they would ordinarily make in two. But the flounders showed a remarkable acceleration of growth, some increasing four times as much in length, and sixteen times as much in weight, as those left in the area from which the flounders were transplanted.

Overall, these flounders accomplished in 2 years the growth normally made in 5–6 years.

The success of flounders in these conditions may be attributed partly to their ability to tolerate the peculiar hydrographic conditions of the loch, and partly to the suitability of the food supply. The loch is brackish, being sea-water diluted with varying amounts of fresh water from two streams. Flounders are euryhaline fish, common in estuaries and well able to tolerate low and fluctuating salinities. In the bottom mud were large numbers of midge larvae, and these proved to be a popular food for flounders, on which the fish could feed and grow throughout the year. However, once the loch contained many sizable fish, it became known to herons and cormorants as a rich feeding ground, and their daily depredations quickly removed all the larger fish.

The experiments in Kyle Scotnish were conducted without the construction of any barrier to the sea. Kyle Scotnish is a much larger area than Loch Craiglin, being about 160 acres with a maximum depth of 20 m, mostly 6–10 m deep, and having a narrow connection with Loch Sween which was left entirely open. The results of fertilization here were generally similar to those of Loch Craiglin, although much greater increases of zooplankton were obtained in this unenclosed loch where the hydrographic conditions were more favourable. The growth of flatfish was extremely good, Group O plaice gaining weight some four to five times faster, and Group O flounders about three times faster, than is normal in natural conditions. However, as the fish reached larger size they tended to move out of the fertilized area into deeper water, a normal pattern of behaviour in flatfish. They therefore received benefit from the fertilizations only during the first year or two of life.

Such experiments indicate some of the difficulties of marine fish farming as a profitable enterprise. There are the problems of keeping the fish in the farmed area, and of preventing the loss of nutrients to the sea. If an isolated body of sea-water is selected, problems may arise from poor circulation and abnormal hydrographic conditions. Few marine species are well suited to enclosed shallow water, and even those that can thrive in such conditions while young may prefer deeper water as they grow larger. There are also the problems of weeding the area to ensure that added nutrients do not lead to the growth of unwanted organisms, but are taken up only by plants which lead to the production of fish food. The water must also be kept free from competing animals which take food needed for the fish, or predatory creatures which devour the fish it is intended to harvest. There is the further difficulty of finding fast-growing fish, suited to enclosed conditions, which are sufficiently palatable to fetch a good price

Despite these problems, a simple type of fish farming in salt water does take place in some parts of the world, mainly in the east, notably in India, Hong Kong, Formosa, Indonesia and the Philippines, where salt or brackish water ponds are stocked with species of grey mullet or with the milk fish,

Chanos chanos. These are both hardy species which can tolerate a wide range of salinity and temperature, and have the further advantage of being vegetarian feeders, mainly on detrital and microscopic plant material. The food chain is therefore short and correspondingly efficient, and the fish make rapid growth and are good eating. In some of the brackish fish-ponds in Indonesia, enriched with sewage, production estimates for *C. chanos* have been as high as 2 tons per acre.

There are also some marine fish-ponds in Western Europe; for example, in the great lagoon at Arcachon in France, famous for its oyster beds, there are brackish water ponds for the culture of grey mullet. These ponds are stocked during the spring by encouraging mullet fry to swim into the ponds from the lagoon. The sluices between the ponds and the lagoon are partially opened during low tide, allowing a small seaward flow of water from the ponds. In response to the movement of the water, the mullet fry from the lagoon swim upstream into the sluices, and thence enter the ponds.

Recently, because of uncertain profits and increasing capital costs of sea fishing, there has been growing interest in the U.K. in the possibilities of intensive fish farming. The work started by Sheldon and his team (page 273) on the rearing of young fish has been extended to investigate possibilities of growing fish to marketable size in enclosures. Under the auspices of the White Fish Authority there have been fish-farming experiments with artificially reared plaice and Dover soles in an enclosed part of a sea loch at Ardtoe on the coast of Argyle, and also in tanks constructed at the electricity generating station at Hunterston, Ayrshire. These tanks are supplied with warm sea-water from the cooling system of the power station.

In the loch at Ardtoe, problems were encountered at first from fluctuating hydrographic conditions, predation and difficulties of recapturing the fish. Recently the fish have been better controlled and better protected by keeping them in netting cages and pens. Some fast growth rates have been obtained, and these studies are continuing in this and other lochs with the co-operation of the Highlands and Islands Development Board. In the tanks at Hunterston the fish can be closely supervised, and in this warm water they are able to feed and grow continuously throughout the year. Plaice and soles have reached market sizes in less than half the time taken in the sea. Fed on minced mussel their weight increase can be as high as 25% of food consumed, and trials are proceeding to find less expensive food. Probably various waste foods could be used for feeding fish. Other species which have been successfully grown in these tanks are turbot, brill and lemon sole. It is also possible in tanks in this country to rear the prawn *Palaemon serratus* and various bivalves, notably oysters (*Crassostrea*) and American clams (*Mercenaria mercenaria*).

These experiments have demonstrated the technical feasibility of raising marine animals in dense cultures in small enclosures in Britain. The profitability of commercial applications of these techniques remains uncertain.

Obviously this would need the development of reliable supplies of fish food, and this may require special culture methods for feeding larval and young fish. However, for stocking the tanks it is not necessary to raise all species from eggs because in some cases, for example plaice, ample supplies of juveniles can readily be obtained from the sea.

If it proves commercially viable, fish farming offers many attractions as a means of supplementing natural fish supplies. It could provide a fully controllable source of fresh fish, unaffected by weather and natural fluctuations, with the desired species and sizes available as required. Rapid growth rates and good conversion efficiencies can be achieved with appropriate feeding and temperature, and flavour might be controlled through diet. There are possibilities of genetic selection for good growth rate, conversion efficiency, disease resistance and palatability. Probably the same tanks could be used simultaneously for several species; for example, flatfish on the bottom, mullet or prawns above and molluscs on the sides. There are good export prospects for high-priced species such as Dover sole and large prawns.

Increasing the Fertility of the Seas

If fish farming in enclosed bodies of sea-water presents difficulties, why not simply raise the productivity of the open sea by enriching the surface waters over wide areas by the addition of plant nutrients? We spread fertilizers on the land to promote the growth of crops; why not spread them on the sea?

To produce any appreciable increase in concentration of plant nutrients in the open sea would require enormous quantities of fertilizer, and the costs would be tremendous. It would be an extremely wasteful process because so small a proportion of the nutrients absorbed by phytoplankton eventually become incorporated in fish flesh. Any additional plant growth obtained as a result of fertilization would contribute very largely to the production of unwanted organisms. Calculations of the increase of yield from sea fisheries that might be obtained by large-scale fertilization of sea-water do not stand up to comparison with those known to be obtained from the use of equal quantities of fertilizer applied to the land.

None the less, there are unintentional processes of artificial enrichment of sea-water going on in some areas, though not with the deliberate aim of benefiting fisheries. Close to large centres of population, great quantities of sewage are discharged into the sea, which decompose to provide nutrients for marine plant growth. For example, the fertility of the southern North Sea is augmented by the outflow of London's sewage via the Thames. It may be doubted, however, whether this form of sewage disposal is really in our long-term interests. Dr. L. H. N. Cooper[19], chemist at the Plymouth Laboratory, writes:

As matters now stand, very large amounts of nutrients are being poured into the sea, the great sink, as sewage from coastal towns and by way of the rivers from inland towns and farms fertilized and unfertilized. Phosphorus is a very precious commodity which in not so many years will become very scarce. The scale on which phosphorus even now is being dissipated to the sea is more than the world can afford. In years to come the cry will be for more methods for recovering phosphorus from the sea, not for putting it in.

Other measures for raising the productivity of sea-water have been proposed which do not involve the addition of valuable fertilizers. The seas already contain vast reserves of plant nutrients in deep water, and it might be possible to devise means of bringing these to the surface. Any measures which increase the mixing of surface and deep water would be likely to lead to increased production. For example, it has been suggested that upwelling could be brought about artificially by sinking atomic reactors in deep water to generate heat and cause convection currents to carry the nutrient-rich deep water to the surface. Alternatively, in some areas it might be possible to alter the natural circulation in ways that could lead to a better supply of nutrients at the surface, or a higher water temperature. Digging out the Strait of Gibraltar to a greater depth has been advocated as a means of raising the productivity of the Mediterranean by allowing the entry of deeper levels of water from the Atlantic. It has also been predicted that a barrage across the Strait of Dover which allowed only a one-way flow of water from the Channel to the North Sea would reduce the entry of colder water into the northern part of the North Sea, and gradually raise the North Sea temperature, thereby promoting higher productivity.

Harvesting Plankton

The losses of organic material that occur at each stage of a food chain are thought generally to amount to some 80–90 per cent. On this reckoning, 1 kg of phytoplankton provides about 100 g of herbivorous zooplankton, which in turn yields 10 g of first-rank carnivore, 1 g of second-rank carnivore, and so on.

Most of the food that man takes from the sea comes from food chains involving several links, and therefore the harvest can be only a small fraction of the primary production. Some of the most plentiful pelagic fish are first-rank carnivores, but the majority of the most popular species for human food feed at later stages of the chain. Cod, for example, feed largely on other carnivorous fish or on carnivorous benthic animals. It is, therefore, apparent that far larger quantities of food could be obtained from the sea by collecting the earlier stages of food chains than can ever be provided by fishing. Instead of catching fish, why not directly harvest the plankton itself and process it to extract the food materials?

The practical difficulty of collection presents a major obstacle to obtaining large quantities of food in this way. Usually, plankton is dispersed in a very large volume of water, and even in the most productive areas enormous quantities of water would have to be filtered to obtain plankton in bulk. If the smaller organisms are to be retained, and particularly if the aim is to collect the phytoplankton, very fine filters would be required and the process of filtration could therefore proceed only very slowly. It seems unlikely that direct harvesting of the plankton from the open sea could be carried on economically, except perhaps in a few areas where there are very dense aggregations of the larger zooplankton. For example, the Southern Ocean at times contains enormous numbers of *Euphausia superba* (krill), providing the great whalebone whales with sufficient food to grow very quickly to a huge size simply by sieving this crustacean from the water. As whales can do this so efficiently, we might be able to devise ways of copying them. 'Artificial whales' have been suggested, perhaps constructed as atomic submarines with gaping bows opening to revolving filter drums, and provided with means for the continuous removal, processing and storing of the filtered zooplankton. It has been claimed that krill-harvesting could prove more profitable than whaling[33].

In recent years, there have been a number of experiments to investigate the possibilities of mass culture of marine phytoplankton. There seems little doubt that methods can be developed for culturing diatoms in large, shallow, sea-water tanks enriched with plant nutrients. In dense fast-growing cultures, availability of carbon dioxide becomes a limiting factor; but if the culture tanks are sited near industrial installations, washed flue-gases can be used to supply the carbon dioxide for photosynthesis, and waste heat to maintain the optimum water temperature. In this way, a rapid growth of phytoplankton can be maintained, and it might be possible to develop continuous culture methods similar to those now used in brewing or the preparation of antibiotics. However, although diatoms are rich in protein and oil, there are considerable difficulties in the separation of the plants from salt-water and the subsequent extraction of the food materials from the cells. These processes are fairly efficiently performed biologically, and it seems likely that mass cultures of phytoplankton will find their chief usefulness in association with the rearing of young fish or the culture of some of the popular species of bivalve molluscs.

Conclusions

Apart from the obvious technical difficulties, proposals for increasing the yield of human food from the sea by attempting to bring about wide-scale alterations of the marine environment must be considered with much caution. Our knowledge of most aspects of the working of marine ecosystems is inadequate for us to be able to make predictions with certainty. There are many risks of unforeseen, detrimental consequences from tampering on a large scale with a vast environment we do not well understand, however well-intentioned our actions. At present there seems more likelihood that we may poison the oceans by continually using them as a drain into which we can indiscriminately tip our products, rather than that we can improve their natural productivity by artificial means. First we must safeguard the oceans from pollution.

Marine fish farming and shellfish culture are already practised to some extent, and it seems probable that more intensive cultivation of selected marine species, for example flatfish and shellfish, may eventually become possible in closely controlled conditions in isolated bodies of water. High efficiencies of production should be achievable in artificial environments comparable with those already used in factory-farming of poultry and cattle.

The immediate prospect of obtaining greater quantities of food from marine sources seems to lie mainly in the possibility of wider, controlled exploitation of natural stocks. This requires concurrent developments along several lines, including the utilization of a greater variety of species— especially the pelagic stocks, improvements in pelagic fishing techniques, the extension of fishing over wider areas, the modernization of primitive fisheries and better methods of preserving, processing and marketing their catches. A prerequisite for major advance is international co-operation in fishing and fishery science, without which optimum yields cannot be estimated and fishing appropriately regulated. The possibilities are enormous but the resources of the sea are not limitless. Uncontrolled, competitive laissez-faire hunting inevitably leads eventually to declining yields from diminishing stocks. In fishing, as in most human affairs, progressive improvement depends upon intelligent control of human behaviour.

REFERENCES AND FURTHER READING

Books

[1] Balls, R. (1961). *Fish Capture*. Buckland Lecture. London; Arnold
[2] Cushing, D. H. (1968). *Fisheries Biology. A Study in Population Dynamics*. Madison, Milwaukee and London; University of Wisconsin Press
[3] Cushing, D. H. (1966). *The Arctic Cod*. Oxford; Pergamon
[4] Graham, M. (1943). *The Fish Gate*. London; Faber and Faber
[5] Graham, M. (1948). *The Rational Fishing of Cod in The North Sea*. Buckland Lecture. London; Arnold
[6] Graham, M. (1956). *Sea Fisheries*. London; Arnold
[7] Hardy, A. C. (1959). *The Open Sea*, 'Part II. Fish and Fisheries.' London; Collins
[8] Hickling, C. F. (1935). *The Hake and the Hake Fishery*. Buckland Lecture. London; Arnold
[9] Hickling, C. F. (1962). *Fish Culture*. London; Faber
[10] Hodgson, W. C. (1934). *The Natural History of the Herring of the Southern North Sea*. London; Arnold
[11] Hodgson, W. C. (1967). *The Herring and its Fishery*. London; Routledge and Kegan Paul
[12] Jenkins, J. T. (1936). *The Fishes of the British Isles*. London; Warne

[13] Nikolskii, G. V. (1969). *Theory of Fish Population Dynamics*. Edinburgh; Oliver and Boyd

[14] Rae, B. B. (1965). *The Lemon Sole*. London; Fishing News (Books) Ltd.

[15] Russell, E. S. (1942). *The Overfishing Problem*. Cambridge University Press

[16] Tait, J. B. (1952). *Hydrography in Relation to Fisheries*. Buckland Lecture. London; Arnold

[17] Wimpenny, R. S. (1953). *The Plaice*. Buckland Lecture. London; Arnold

Papers

[18] Blaxter, J. H. S. and Holliday, F. G. T. (1963). 'The Behaviour and Physiology of Herring and Other Clupeids.' *Adv. Mar. Biol.* **1**, 261

[19] Cooper, L. H. N. (1948). 'Phosphate and Fisheries.' *J. mar. biol. Ass. U.K.* **27**, 326

[20] Cormack, R. M. (1968). 'The Statistics of Capture–Recapture Methods.' *Oceanogr. Mar. Biol. Ann. Rev.* **6**, 455

[21] Cushing, D. H. (1967). 'The Grouping of Herring Populations.' *J. mar. biol. Ass. U.K.* **47**, 193

[22] Davis, F. M. (1934). 'Mesh Experiments with Trawls.' *Fish. Investig. Ser. II*, **14**, No. 1

[23] de Ligny, W. (1969). 'Serological and Biochemical Studies on Fish Populations.' *Oceanogr. Mar. Biol. Ann. Rev.* **7**, 411

[24] Gross, F. (1947). 'An Experiment in Marine Fish Cultivation. I. Introduction. V. Fish Growth in a Fertilised Sea Loch.' *Proc. R. Soc. Edinb.*, B. **63**, 1 and 56

[25] Gross, F. (1949). 'A Fish Cultivation Experiment in an Arm of a Sea Loch. V. Fish Growth in Kyle Scotnish.' *Proc. R. Soc. Edinb.*, B. **64**, 109

[26] Gross, F. (1949). 'Further Observations on Fish Growth in a Fertilised Sea Loch (Loch Craiglin).' *J. mar. biol. Ass. U.K.* **28**, 1

[27] Gulland, J. A. and Carroz, J. E. (1968). 'Management of Fishery Resources.' *Adv. Mar. Biol.* **6**, 1

[28] Hickling, C. F. (1970). 'Estuarine Fish Farming.' *Adv. Mar. Biol.* **8**, 119

[29] Lucas, J. (1966). 'Conserving Whales.' *Sci. J.* April

[30] Marshall, S. M. and Orr, A. P. (1948). 'Further Observations on the Fertilisation of a Sea Loch (Loch Craiglin).' *J. mar. biol. Ass. U.K.* **27**, 360

[31] Parrish, B. B. and Saville, A. (1965). 'The Biology of the North-East Atlantic Herring Population.' *Oceanogr. Mar. Biol. Ann. Rev.* **3**, 323

[32] Parrish, B. B. and Saville, A. (1967). 'Changes in the Fisheries of North Sea and Atlanto-Scandian Herring Stocks and Their Causes.' *Oceanogr. Mar. Biol. Ann. Rev.* **5**, 409

[33] Pequegnat, W. E. (1958). 'Whales, Plankton and Man.' *Scient. Am.* January

[34] Raymont, J. E. G. (1947). 'An Experiment in Marine Fish Cultivation. IV. Bottom Fauna and Food of Flatfish.' *Proc. R. Soc. Edinb.*, B. **63**, 34

[35] Raymont, J. E. G. (1949). 'A Fish Cultivation Experiment in an Arm of a Sea Loch. IV. The Bottom Fauna of Kyle Scotnish.' *Proc. R. Soc. Edinb.*, B. **64**, 65

[36] Raymont, J. E. G. (1949). 'Further Observations on Changes in Bottom Fauna of a Fertilised Sea Loch.' *J. Mar. biol. Ass. U.K.* **28**, 9

[37] Shelbourne, J. E. (1964). 'The Artificial Propagation of Marine Fish.' *Adv. mar. Biol.* **2**, 1

[38] Simpson, A. C. (1959). 'The Spawning of the Plaice in the North Sea.' *Fishery Invest., Lond., Ser. II*, **22**, No. 7

[39] Simpson, A. C. (1959). 'The Spawning of the Plaice in the Irish Sea.' *Fishery Invest., Lond., Ser. II*, **22**, No. 8

[40] Steven, G. A. (1948). 'Contributions to the Biology of the Mackerel, *Scomber scombrus*.'

REFERENCES AND FURTHER READING

1. Mackerel Migrations in the English Channel and Celtic Sea.' *J. mar. biol. Ass. U.K.* **27**, 517

(1949). 2. 'A Study of the Fishery in the South-West of England with Special Reference to Spawning, Feeding and Fishermens' Signs.' *J. mar. biol. Ass. U.K.* **28**, 555

(1952). 3. 'Age and Growth.' *J. mar. biol. Ass. U.K.* **30**, 549

[41] Wollaston, H. J. B. (1915). 'Report on Spawning Grounds of Plaice in the North Sea.' *Fishery Invest., Lond., Ser. II*, **2**, No. 4

[42] Wollaston, H. J. B. (1923). 'Spawning of Plaice in the Southern Part of the North Sea in 1913–14.' *Fishery Invest., London., Ser. II*, **5**, No. 4

Books

[43] Bigelow, H. B. and Schroeder, W. C. (1925). *Fishes of the Gulf of Maine*. Washington; U.S. Dept. of Interior

283

CHAPTER 9

ENERGY RELATIONSHIPS OF MARINE
PRODUCTION

Living organisms, unlike machines, cease to exist once they stop working. All biological activity depends upon continual transfers and transformations of energy, without which any natural living system almost immediately disintegrates irreversibly. In this concluding chapter we will draw together some of the information from preceding pages in an elementary consideration of certain energy relationships of marine life, with particular reference to shelf seas around the major continental land masses.

Directly or indirectly the source of all energy for life is the sun, which continually emits radiant energy into space. A tiny fraction of this radiation reaches the earth, where a considerable part is lost by reflection from the earth's atmosphere, clouds and surface. Probably a global average of about 40% of the incoming radiation is reflected. The remainder is absorbed by the atmosphere and the land and ocean surfaces, where its main effect is to cause the heating which generates the movements of atmosphere and ocean. However, despite the continuous absorption of solar radiation, the climate does not appear in the long term to become hotter. This indicates that there is overall an output of radiant energy from the earth equal to that received, and the total heat content of atmosphere, surface and oceans, remains virtually constant except for minor fluctuations due to the elliptical form of the earth's orbit round the sun and to changes in solar activity (for example solar flares or sunspots). The incoming energy is received largely at wavelengths within the visible spectrum. The balancing emission from the earth is low-frequency heat radiation which passes out in all directions of space.

The routes through which light energy can flow between penetrating the earth's atmosphere and re-radiation into space as heat are numerous and complex. A small amount, probably only about 1–2% of the light energy reaching the earth's surface, enters pathways beginning with the absorption of sunlight by plants in photosynthesis. In this process radiant energy is transformed to chemical energy by an energy-fixing reduction of carbon dioxide. For instance, the synthesis of 1 g-mol of glucose from carbon dioxide and water involves the intake of 673 kcal of light energy.

$$6\ CO_2 + 6\ H_2O + \underset{\text{(light energy)}}{673\ kcal} \xrightarrow{\text{chlorophyll}} C_6H_{12}O_6 + 6\ O_2$$

This energy is then available in biological processes, for when 1 g-mol of glucose is oxidized in respiration, 673 kcal of energy is released. It is by means of transfers and transformations of the energy of chemical compounds formed initially by photosynthesis that power is provided for the activity of living organisms. The movements of materials involved in nutrition occur almost entirely as means of effecting energy transfers. The global total of energy fixation by photosynthesis determines the total amount of biological activity which the earth can support. The intake of radiant energy into the living system by photosynthesis is balanced by a corresponding outflow of energy as heat through pathways of respiration and movement.

We have insufficient knowledge of the energy relationships of marine organisms to be able to trace with much certainty the passage of energy through marine ecosystems, but until we can do this our understanding of the food webs of the sea must remain at an elementary stage. As an indication of some of the processes involved we conclude this book by attempting a simple analysis of the energetics of production and feeding in the shelf waters of the northern Atlantic. No reliance must be placed on the figures given, which should only be regarded as reasoned guesswork based on a modicum of firm information. The exercise is illustrative rather than factual. Obviously, for simplicity, innumerable interactions within the food web have been ignored, and by averaging out all values over a period of a year no account has been taken of the highly fluctuating nature of the system. Nevertheless, something may be learned from critical examination of the figures and from comparison with various sources of data relating to these considerations. The analysis is summarized diagrammatically in *Figure 9.1*.

PRIMARY PRODUCTION

A mean figure for solar energy entering the earth's atmosphere has been given as $15\cdot3 \times 10^5$ kcal/m²/yr. Much of this is reflected or absorbed on passage through the atmosphere, and the amount reaching the surface varies greatly with locality. North Atlantic[3] areas receive about $2\cdot5 \times 10^5$ kcal/m²/yr, and we will take this as the energy at the sea surface. Here there are losses by reflection (page 96), and on penetrating the surface the light is rapidly absorbed by the water and causes heating. Estimates of the fraction of the incident radiation which is fixed by photosynthesis in aquatic environments generally fall within the range $0\cdot1$ to $0\cdot3\%$ of the energy at the surface. Taking the middle figure of $0\cdot2\%$, this gives a value of $0\cdot002 \times (2\cdot5 \times 10^5)$ $= 500$ kcal/m²/yr for Gross Primary Production (GPP) (*see* page 120), equivalent to the synthesis of $500/673 = $ approx. $0\cdot75$ g-mol of glucose by phytoplankton beneath a square metre of sea surface in one year. Part of the energy fixed in this way is subsequently lost in the respiration of the phytoplankton. The remainder, the Net Primary Production (NPP), is the energy content of new plant tissue. Estimates of energy loss in plant respiration fall between 10% and 50% of GPP and we will again take the middle

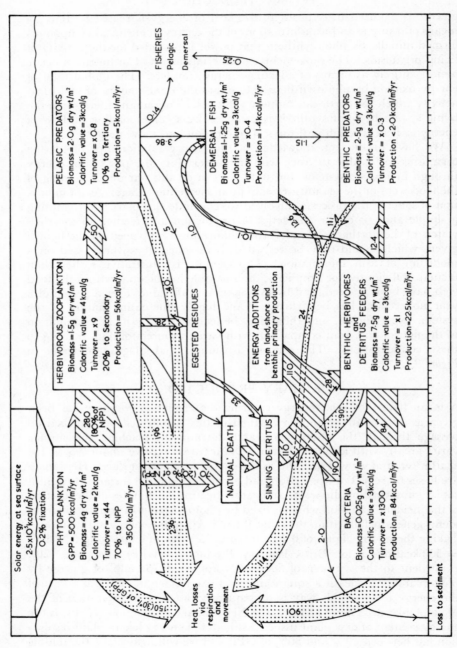

Figure 9.1. *Diagrammatic representation of hypothetical energy relationships in a marine ecosystem in coastal waters of the North Atlantic. Figures on the arrows have units of kcal/m²/yr*

286

value of 30%. The respiratory loss from the system at this stage will therefore be $0.3 \times 500 = 150$ kcal/m²/yr, leaving a figure of 350 kcal/m²/yr for
NPP.

Measurements of calorific values of a variety of vegetable tissues fall in the
range of 0.2 to nearly 9.0 kcal/g dry wt. Because of the relatively high
proportion of inorganic matter in diatoms, we will take a figure of only
2.0 kcal/g dry wt for the energy content of the phytoplankton. The energy
value for NPP is then equivalent to the formation of a dry weight of new
phytoplankton cells of $350/2 = 175$ g dry wt/m²/yr. As a generalization,
1 g dry wt of organic matter is approximately equivalent to 0.44 g of carbon,
so the NPP could be expressed as $175 \times 0.44 = 75$ gC/m²/yr. The values
of GPP calculated similarly are 250 g dry wt/m²/yr or 110 gC/m²/yr. These
quantities compare reasonably with values obtained from various production measurements in seas around the United States.

In Table 9.1 are given some estimates by Harvey of the mean annual

TABLE 9.1

Estimates of the Mean Annual Biomass
of Several Trophic Levels in the English
Channel
(From Harvey, H. W.[4], published by
Cambridge University Press)

Level	Dry wt of organic matter g/m²
Phytoplankton	~4·0
Zooplankton	1·5
Pelagic fish	1·8
Demersal fish	1–1·25

biomass of several trophic levels in the English Channel, and we will use
these figures in our analysis. The value for mean annual biomass of standing
stock of phytoplankton for the English Channel is 4.0 g dry wt/m². Comparing this figure with our value for annual production, it can be seen that
there is a high rate of turnover. The weight of new plant tissue produced in
the year is nearly 44 times (175/4) the mean weight of standing stock. The
production of new phytoplankton does not increase the standing stock from
one year to the next because it is balanced by a corresponding loss of plants
from the water. Death of plant cells occurs in two main ways, by sinking
and by consumption by herbivorous zooplankton. Sinking cells die through
lack of light and constitute a large part of the organic detritus reaching
the sea bottom. This provides a major energy source to support the
benthos. However, in contrast to most terrestrial ecosystems, it appears

certain that in the sea a greater proportion of the vegetation is consumed by animals than is lost by death and decomposition, the main energy transfer from NPP going to grazing zooplanktonts. We will assume that 20% of NPP contributes directly to sinking detritus, with 80% consumed by pelagic herbivores. The energy content of detritus from this source is therefore $0.2 \times 350 = 70$ kcal/m²/yr. This will be given further consideration below (page 290). The pelagic grazing population receives an energy inflow of $0.8 \times 350 = 280$ kcal/m²/yr.

THE GRAZING CHAIN

Secondary Pelagic Production

Some of the food ingested by the planktonic herbivores is not fully digested and absorbed but passes through the gut and is egested, contributing to the fall of organic debris to the sea-floor. We will assume that 10% of the energy content of the food consumed goes in this way to detritus. Of the assimilated food, most is used for respiration and the remainder forms new animal tissue. Herbivorous planktonts are relatively efficient converters of food to new tissue, with some early larval stages apparently using about 50% of food intake for growth. The general level of efficiency is certainly lower than this, and we will assume that 70% of food intake is used for respiration and 20% for Secondary Production. We will call this a Gross Conversion Efficiency (GCE) of 0.2, where

$$GCE = \frac{\text{calorific value of new tissue formed}}{\text{calorific value of ingested food}}.$$

We can summarize the energetics as follows:

10% egested unassimilated	= 28 kcal/m²/yr to detritus
70% utilized for respiration and movement	= 196 kcal/m²/yr lost from the system
20% utilized for secondary production	= 56 kcal/m²/yr available for predators
100%	= 280 kcal/m²/yr

Calorific values for copepod and euphausid species occurring around the North Atlantic are generally about 4·0 kcal/g dry wt. Using this figure, Secondary Production amounts to $56/4 = 14$ g dry wt/m²/yr, or (if 1 g dry wt = 0.44 g C) $14 \times 0.44 = 6.2$ g C/m²/yr.

General observation of plankton samples taken over the continental shelf suggests that a mean value for biomass of standing stock of herbivores must often be appreciably greater than the standing stock of phytoplankton, and some quantitative investigations indicate this. However, taking our value from Harvey's figures (Table 9.1), the mean annual biomass of zooplankton throughout the year in the English Channel is given as 1·5 g dry wt/m². Even at this apparently low value, secondary production amounts to less than ten times the weight of standing crop. This much lower rate of turnover compared with phytoplankton corresponds with the slower rates of growth and reproduction of zooplankton.

Some herbivorous planktonts die and reach the bottom uneaten by pelagic predators, but probably the great majority are consumed by carnivorous zooplanktonts or pelagic fish such as herring and mackerel. Assuming that dead pelagic herbivores sinking to the bottom amount to 10% of secondary pelagic production, this adds another $0·1 \times 56 = 5·6$ kcal/m²/yr to sinking detritus, leaving approximately 50 kcal/m²/yr for consumption by pelagic carnivores.

Tertiary Pelagic Production

Compared with herbivores, predators generally use a larger proportion of their food for respiration. A figure of 10% (GCE = 0·1) for conversion of food to new tissue by carnivores is generally quoted. This permits us to divide the energy transfers at this trophic level as follows:

10% egested unassimilated	= 5 kcal/m²/yr to detritus
80% utilized in respiration	= 40 kcal/m²/yr lost from the system
10% to tertiary production	= 5 kcal/m²/yr
100%	= 50 kcal/m²/yr

Taking 3 kcal/g dry wt as a mean calorific value for first-rank carnivores, tertiary production may be expressed as $5/3 = 1·66$ g dry wt/m²/yr or (if 1 g dry wt = 0·44 g C) approximately 0·7 g C/m²/yr.

The mean annual biomass of pelagic fish in the English Channel has been estimated at 1·8 g dry wt/m² (Table 9.1). If we assume that this is 90% of the weight of first-rank carnivores, the total biomass of pelagic predators is 2·0 g dry wt/m². We note that annual production at this level is only 83%, i.e. (1·66/2·0) of the mean weight of standing stock. Compared to plankton, fish are long lived but slow growing.

Pelagic Fisheries

Fishery statistics for the year 1969 give the total landings of pelagic species from the English Channel as approximately 23,000 metric tons wet weight, i.e. 23×10^9 g wet wt. Taking the area of the English Channel as about 82×10^9 m², this catch of pelagic fish amounts to $23/82 = 0·28$ g wet wt/m²/yr. Assuming the wet weight of fish to be six times the dry weight, pelagic fishing therefore took 0.047 g dry wt/m^2/yr. If the calorific value of the fish is 3.0 kcal/g dry wt, the energy content of the catch was

$$3 \times 0.047 = 0.14 \text{ kcal/m}^2/\text{yr},$$

or,

$$82 \times 0.14 \times 10^9 = 11.5 \times 10^9 \text{ kcal/yr}$$

for the whole area of the English Channel.

According to our previous calculation, tertiary production amounts to about 5 kcal/m^2/yr, so in 1969 pelagic fishing cropped only about $0.14 \times 100/5 = 2.8\%$ of production at this trophic level.

THE DETRITUS CHAIN

From the preceding sections we have derived several figures for energy content of various contributions to the organic detritus reaching the sea bottom. To these must be added a small amount to represent the bodies of dead pelagic predators sinking to the sea-floor, which we will take to be equivalent to 20% of tertiary pelagic production, giving a further $0.2 \times 5 = 1$ kcal/m^2/yr. These energy sources in detrital form can be summarized as follows:

		kcal/m^2/yr
From primary production	sinking, uneaten phytoplankton	70
	phytoplankton eaten but egested	28
From secondary production	dead, uneaten zooplankton	5.6
	zooplankton eaten but egested	5
From tertiary production:	dead pelagic predators	1
	Total =	109.6

This gives us a round figure of 110 kcal/m^2/yr for food energy reaching the shallow sea bottom from several trophic levels in the water above.

Examination of the composition of shallow-water sediments often reveals a significant additional amount of organic matter recognizably derived from the sea shore or the land, mainly fragments of large algae, wood or leaf. This is generally greater than the amount that can be attributed to pelagic sources. Also, where sufficient light reaches the bottom there will be primary production by plants, the compensation depth for benthic diatoms and other algae being generally lower than for phytoplankton. There will also be a slight contribution from chemosynthetic autotrophs. We must therefore make a considerable addition to the energy available for biological processes on the sea bottom to allow for these sources, which we will assume in total to be about equal to that derived from the debris of pelagic organisms, i.e. another 110 kcal/m^2/yr, making in all 220 kcal/m^2/yr. From this we must make small deductions to allow for organic matter lost from the living system by permanent inclusion within the sediment or by oxidation not effected by organisms, say 2 kcal/m²/yr, bringing us to a figure of 218 kcal/m²/yr for the food energy utilized by benthic organisms.

Some of this energy is directly available to the benthic fauna by digestion and assimilation of detritus, but many detritic materials reaching the bottom cannot be digested by animals. These materials are acted upon by bacteria, which utilize these energy sources to multiply rapidly. This production of bacterial protoplasm contributes a significant proportion of the food of benthic animals. We have insufficient knowledge of the feeding metabolism of the benthos to make a reasoned generalization about the proportions of energy derived from detritus directly by digestion and assimilation of organic debris or indirectly via consumption of bacterial protoplasm, but some experimental work suggests that bacteria contribute much the greater part. We shall assume that 25% of the energy intake of the benthic herbivores comes directly from detritus and 75% from the consumption of bacteria.

Benthic Fauna

For the mean annual biomass of the benthic fauna in shallow water around the continents we will take a value of 10 g dry wt/m²/ (*see* page 165). Part of this is benthic predators, some estimates indicating about 25%, leaving a standing stock of 7·5 g dry wt of benthic herbivores feeding on detritus and bacteria. With a calorific value of 3 kcal/g dry wt we have an energy content of mean standing stock of benthic herbivores of 22·5 kcal/m².

The conversion efficiency of these animals is likely to be as high as in herbivorous zooplanktonts, so we will take a value of 0·2 for GCE. Compared with zooplanktonts many of the benthic animals are much bigger and live longer—years rather than weeks—so the rate of turnover is low relative to the weight of standing stock. On the assumption that the stock produces its own weight of new tissue in a year, the energy intake necessary to produce this is 22·5/0·2 = 112·5 kcal/m²/yr. Because we have also assumed that 25% of this comes from direct assimilation of digested detritus and the remainder from ingestion of bacteria, we have detrital and bacterial contributions of 28 kcal/m²/yr and 84 kcal/m²/yr respectively.

Bacteria

From our assessment of 218 kcal/m²/yr of food energy available on the sea bottom, we must now deduct 28 kcal/m²/yr directly utilized by benthic fauna, leaving 190 kcal/m²/yr for bacteria. We have estimated the consumption of bacteria by benthic animals to be 84 kcal/m²/yr, and for the stock of bacteria to remain constant this must be the annual production. This implies a high conversion efficiency of 84/190 = 0·44 approximately, and a respiratory loss of 106 kcal/m²/yr.

The biomass of benthic bacteria is obviously very small compared with larger organisms. A mean value over all grades of shallow-water deposit may be in the order of 0·025 g dry wt/m². Assuming a calorific value of 3 kcal/g dry wt for bacteria, this gives an energy content for the stock of 3 × 0·025 = 0·075 kcal/m². For an annual production of 84 kcal/m²/yr this stock must reproduce itself over 1,100 times, (84/0·075), which is certainly not a very high rate of multiplication for bacteria.

Demersal Fish

Harvey estimated the mean annual biomass of demersal fish in the English Channel at 1·0–1·25 g dry wt/m² (Table 9.1). With a calorific value of 3 kcal/g dry wt the energy content of this biomass is 3·5 kcal/m² approximately. Harvey put the annual yield at 30–50% of stock (say 40%) so annual production can be calculated as 3·5 × 0·4 = 1·4 kcal/m²/yr. If the GCE is 0·1, this level of production requires an energy intake of 14 kcal/m²/yr.

Demersal fish feed partly on benthic and partly on pelagic prey. We have calculated above that from the annual production of 5 kcal/m²/yr of pelagic predators 0·14 kcal/m²/yr is removed by pelagic fisheries, and we have assumed that 1·0 kcal/m²/yr is lost by natural death, i.e. not eaten by other predators. This leaves 3·86 kcal/m²/yr available as food for demersal fish. If this is all consumed by demersal fish, there is a balance 14 − 3·9 ≃ 10·1 kcal/m²/yr to be made up by devouring benthic fauna. As a fraction of the annual production of benthic herbivores this is 10·1 × 100/22·5 ≃ 45%. Comparing this with Thorson's estimate quoted earlier (page 168) that demersal fish take only 1–2% of available fish food, with the rest consumed by invertebrate predators, it is clear that some assumptions are considerably astray. More information is necessary.

Demersal Fisheries

The total annual landings of demersal fish from the English Channel during 1969 were approximately 40,000 metric tons. This is equivalent to approximately 0·5 g wet wt/m²/yr, or 0·083 g dry wt/m²/yr. If the calorific value of the fish is taken as 3 kcal/g dry wt, the energy content of the catch was 0·25 kcal/m²/yr or 82 × 10⁹ × 0·25 = 20·5 × 10⁹ kcal/yr for the whole area. As a proportion of the annual production of demersal fish, the fishing yield amounted to 0·25 × 100/1·4 ≃ 18%.

Benthic Predators

From our figure of 10 g dry wt/m² for benthic fauna we attributed 7·5 g dry wt/m² to the herbivores, leaving a biomass of 2·5 g dry wt/m² for the predators. With a calorific value of 3 kcal/g dry wt the energy content of this biomass is 7·5 kcal/m².

These animals feed on the benthic herbivores and may also eat dead bodies of fish. From our assumed annual production of 22·5 kcal/m²/yr of

CONCLUSION

benthic herbivores, 10·1 kcal/m²/yr has been eaten by demersal fish, leaving
12·4 kcal/m²/yr available for benthic predators. There is also the balance
of production of demersal fish after the removal of 0·25 kcal/m²/yr by fisheries,
i.e. 1·4 − 0·25 = 1·15 kcal/m²/yr, making a total of about 13·5 kcal/m²/yr for
benthic predators. If they consume all this food energy and have a GCE of
0·15, the annual production would amount to about 2 kcal/m²/yr, the
respiratory loss about 11 kcal/m²/yr and the annual yield a little under 30%
of stock weight.

THE ENERGY BALANCE SHEET

From the foregoing figures we can now draw up a statement of the energy
input and output of the living system as outlined above (*see* Table 9.2).

TABLE 9.2
The Energy Balance Sheet

Energy input	Value kcal/m²/yr	Energy output	Value kcal/m²/yr
GPP by phytoplankton	500	Respiratory losses at each trophic level, i.e. heat loss:	
		Phytoplankton	150
Additions of organic		Pelagic herbivores	196
materials from shore and		Pelagic predators	40
land, plus some NPP by		Bacteria	106
benthic plants	110	Benthic herbivores	90
		Demersal fish	13
		Benthic predators	11
		Loss to permanent sediment and inorganic oxidations	2
		Balance to fisheries and benthic predators	2 approx.
Total	610	Total	610

CONCLUSION

In the preceding computations we may note that from a total energy fixation
of approximately 600 kcal/m²/yr, the energy content of the annual crop
taken by fisheries in the English Channel has been only about 0·4 kcal/m²/yr,
i.e. only about 1/1500 part. This figure is a reasonably conservative estimate
for sustainable yields of shelf fisheries. From the North Sea and Icelandic
shelf a slightly higher proportion of GPP is harvested. But for fisheries over
the deep ocean it seems unlikely that even 1/1500 part of GPP could be

gathered as fish because of greater difficulties of capture, even allowing for greater production at lower latitudes.

Despite many uncertainties there have been several attempts to compute a global quantity for total GPP over the whole ocean surface. A figure of the order of 10^{17} kcal/yr seems reasonable. Taking the optimistic view that world fisheries extended over the deep ocean could take 1/1500 part of this, we arrive at a maximum sustainable annual yield for world fisheries of about 7×10^{13} kcal/yr. Taking a calorific value of 3 kcal/g dry wt for fish, and a wet weight of six times dry weight, this converts to about 14×10^{13} g wet wt or 14×10^7 metric tons per annum of fresh fish. For an economic return on the effort of catching fish, it is probably more realistic to suppose that substantially less than this could be captured, and that the total realizable world catch may not be more than about 10×10^7 metric tons per annum. Considering that world fisheries are already taking over 6×10^7 metric tons per annum, it is clear that our calculations certainly do not support ideas that the future food needs of the rapidly rising world population can be greatly alleviated by a large extension of ocean fisheries. During recent years the world catch of fish has been increasing by about 7% per annum, with much of this increment coming from fishing beyond the shelf. Continued expansion at this rate may bring us to the maximum practicable yield within about 10 years, beyond which point lie dangers of diminishing returns through overcropping.

REFERENCES AND FURTHER READING

Books

[1] Odum, E. P. (1959). *Fundamentals of Ecology*. 2nd ed. Philadelphia; W. B. Saunders
[2] Odum, E. P. (1963). *Ecology*. New York; Holt, Rinehart and Winston
[3] Phillipson, J. (1966). *Ecological Energetics*. London; Arnold

Papers

[4] Harvey, H. W. (1950). 'On the Production of Living Matter in the Sea off Plymouth.' *J. mar. biol. Ass. U.K.* **29**, 97
[5] Corner, E. D. S., Cowey, C. B. and Marshall, S. M. (1967). 'On the Nutrition and Metabolism of Zooplankton. V. Feeding Efficiency of *Calanus finmarchicus*.' *J. mar. biol. Ass. U.K.* **47**, 259
[6] Corner, E. D. S. and Cowey, C. B. (1968). 'Biochemical Studies on the Production of Marine Zooplankton.' *Biol. Rev.* **43**, 393
[7] Mann, K. H. (1969). 'The Dynamics of Aquatic Ecosystems.' *Adv. Ecol. Res.* **6**, 1
[8] Reeve, M. R. (1969). 'Growth, Metamorphosis and Energy Conversion in the Larvae of the Prawn, *Palaemon serratus*.' *J. mar. biol. Ass. U.K.* **49**, 77
[9] Zhukova, A. I. (1963). 'On the Quantitative Significance of Microorganisms in Nutrition of Aquatic Invertebrates.' *Symposium on Marine Microbiology*. Ed. Oppenheimer, C. H., p. 699. Springfield, Illinois, U.S.A.; C. C. Thomas

APPENDIX 1

SUGGESTED TOPICS FOR FURTHER STUDY AND CLASS DISCUSSION OR WRITTEN WORK

1. Describe in outline the main ocean currents, at the surface and below. How are they set in motion, and what factors influence their courses? In what ways is this knowledge of interest to biologists?

2. In general terms describe the overall conditions of life in the marine environment. Giving your reasons, what do you consider to be the major subdivisions of the environment? In what respects do biological conditions in the sea differ from those of fresh water environments?

3. What are the chief processes which bring about vertical water-mixing in the seas? Where and when do they occur? Discuss the various effects of vertical water-mixing on marine organisms.

4. Discuss the problems involved in measuring the physical and chemical parameters of the oceans, and indicate why this data is important to biologists.

5. What do you understand by the term 'biomass' and how can this quantity be expressed? Review the methods and difficulties of quantitative sampling of marine plankton, nekton and benthos. Discuss the relevance of biomass estimates in the study of marine ecosystems.

6. Discuss the influence of temperature and salinity upon the distribution of marine species.

7. How do you account for vertical zonation in the sea? Discuss the phenomenon of diurnal vertical migrations, and the advantages and problems associated with this behaviour.

8. By what means do pelagic organisms keep afloat and adjust their depth? Review the mechanisms and problems of buoyancy control.

9. Describe the migrations of a named marine species, explaining how they were discovered. Discuss the ways in which pelagic animals may be able to navigate.

10. Review the adaptations of the abyssal fauna.

11. If you were the Creator designing a new species for the abyss, with what distinctive attributes would you endow it?

12. Discuss the factors influencing phytoplankton production. Review proposals for increasing the production of phytoplankton in the open sea and discuss their advisability.

13. Discuss the reasons why quantities of phytoplankton and zooplankton appear to be in inverse relationship.

14. How do you explain the patchy distribution of marine plankton?

15. Describe the nitrogen and phosphorus cycles of temperate seas and discuss the activities of bacteria in marine cycles.

16. Discuss the value of plankton studies as a means of investigating the movements of water.

17. Review the factors that bring about zonation of shore organisms. Discuss how intertidal zonation is established and maintained.

18. Discuss the effects of crowding on shore populations.

19. Discuss the causes of change in shore populations.

20. Give an account of marine wood-boring and rock-boring organisms, and discuss the problems they create for man.

21. All the discoverable specimens of *Gibbula cineraria* were collected within a quadrat of 3 m side at approximately MLWN level on a rocky shore. The maximum shell diameter of each specimen was measured to the nearest 0·5 mm, and the following data obtained:

Max. shell diameter	No. of specimens	Max. shell diameter	No. of specimens
<7·0 mm	0	13·5	2
7·0	1	14·0	5
7·5	3	14·5	9
8·0	5	15·0	11
8·5	2	15·5	6
9·0	2	16·0	4
9·5	9	16·5	7
10·0	5	17·0	4
10·5	1	17·5	4
11·0	0	18·0	2
11·5	0	18·5	0
12·0	1	19·0	2
12·5	4	19·5	1
13·0	0	>19·5	0

Discuss what inferences might be drawn from this information. What problems are posed, and what further investigations would you attempt?

22. Give outline accounts of the biology of any of the following, with special reference to their ecology; fucoids, littorinids, limpets, trochids, barnacles, *Arenicola marina*, *Nereis diversicolor*, *Calanus finmarchicus*, *Euphausia superba*, marine mammals of America's coastal waters, whales.

23. Review the ecological conditions and populations of any of the following: rocky shores, intertidal sands, estuaries, the bottom at shallow depths, fast-flowing tidal channels, the Atlantic Coast, the Pacific (both north and/or south).

24. Give an account of observations you have yourself made on any marine organism or group of organisms.

25. Discuss the role of pelagic larvae in the life-cycles of benthic marine animals.

26. Give a general account of selective settlement by marine larvae.

27. Give an account of food-webs in the sea.

28. In relation to the populations of the sea-shore or shallow sea bottom, discuss what meaning you attach to the term 'community'.

29. What do you understand by an 'ecosystem'? Discuss this concept in relation to (*a*) the surface layers of the open sea (*b*) the sea bottom at deep levels (*c*) the neritic province (*d*) the sea-shore.

30. Give an outline historical account of the development of oceanography and marine biology. What future advances in these fields of study do you foresee?

31. Give an account of camouflage by marine organisms.

32. Outline the chief methods of commercial fishing, explaining how these are related to the habits of the species sought.

33. Give an account of the biology of a named food fish of commercial importance, and indicate how this knowledge can be of value to the fishing industry.

34. Outline a programme of investigations to study the biology of a marine fish.

35. Discuss the problems of overfishing in Oceanic waters, and outline a policy for the regulation of sea fisheries.

36. Give some account of methods of marine fish-culture, and the associated problems.

37. Discuss proposals for obtaining more food from the sea.

APPENDIX 2

SUMMARY OF LABORATORY EXERCISES

Practical information on several of the experimental exercises can be found in *Laboratory Exercises in Invertebrate Physiology* by J. H. Welsh and R. I. Smith, Burgess Publishing Co., Minneapolis, and *The Invertebrates: Function and Form* by Sherman and Sherman, Macmillan Co., Toronto.

A. Salinity measurements by titration and conductimetry.

B. pH measurements in sea-water and determination of titration alkalinity.

C. Estimation of a minor constituent; for example phosphate[19].

D. Elementary studies on barnacles, for example *Balanus balanoides, B. perforatus, B. crenatus, B. eburneus, Chthamalus spp., Eliminius modestus.*

   Diagnostic characters. Measurement of rate of cirral activity over ranges of temperature, salinity and pH. Comparison of the activity ranges of different species in relation to distribution. Observation of the light reflex, and investigation of its sensitivity and fatigue.

E. Elementary studies on bivalves, for example *Mytilus, Pecten, Tellina, Ensis, Mya, Mercenaria, Crassostrea.*

   Diagnostic characters. Comparison of structure of siphon, shell, mantle, ctenidium, palps and foot in relation to habitat and mode of life. Use of suspensions or cultures to investigate filtering rates, and pathways of feeding, selection and rejection. Measurement of food-particle transport rates over ranges of temperature and salinity.

F. Studies on *Ligio* or *Palaeomonetes*

   Observation of melanophores. The rate of colour change associated with changes of illumination and background.

   The effects on colour change on covering part or whole of eyes.

   Study of the phototaxis, hydrotaxis and thigmotaxis of *Ligia*, and statistical treatment of results[8].

   Measurement of rate of water loss, and comparison with other shore forms, e.g. *Gammarus, Idothea.*

G. Studies on *Corophium*[15].

   Observation of swimming and burrowing behaviour. Substrate selection. Light reactions. Cuticle permeability.

H. Behaviour of *Hydrobia* or *Littorina*. Phototaxes and geotaxes.

I. Osmotic relationships.

Measurement of weight changes of various animals in relation to changes of salinity, for example *Nereis diversicolor, N. pelagica, N. viteris, Arenicola marina, Carcinus maenas.*

Measurement of ionic concentrations in body fluids of *Arenicola* and *Carcinus*, and the changes consequent on changing salinity. (Na, K and Ca by flame photometry, Cl by titration.) Freezing point of blood samples.

J. Examination of named species of planktonic plants and animals for diagnostic features.

Examination of plankton samples from various sources, with special reference to seasonal and geographical differences.

Observations on live plankton—flotation, swimming and filtering activity, phototaxis, etc.

K. Examination of representative collections of benthos from shallow bottoms, shores and estuaries with attention to adaptations, zonation, feeding relationships and community structure.

The use of shell measurements and size/frequency curves for population analysis and determination of mean growth rates.

L. Examination of commercial fish species with reference to recognition features, adaptations, gut contents, parasites, scale and otolith markings. The growth rate curve from herring scale rings.

(In addition to the foregoing exercises, laboratory time is also required following field-work, for sorting collected material, measuring, counting, tabulating and graphically representing results.)

APPENDIX 3

SYNOPSIS OF A FIELD COURSE

Students attending this course have already gained some familiarity with shore populations from previous fieldwork, and can recognize many of the common species. Each student is advised to bring to the course the following items:

Warm and waterproof clothing. Knee boots or waders. Anti-seasickness pills.

Knapsack, shoulderbag or canvas bucket.

For outdoor work, notebook with string-attached pencil. Strong, old penknife. Wide-mouth pipette with rubber teat. Magnifying glass. Plastic kitchen strainer. Assorted plastic specimen jars and tubes, and polythene bags. Tape measure. Chalk. Skewers. Ball of string. Flash light. Pocket compass.

Laboratory notebook and graphpaper. Dissecting instruments. Binocular microscope.

Also desirable—dividers and short steel rule, metric scale, part engraved to 0·5 mm divisions. Slide rule. Recommended books:

Lights et al., *Intertidal Invertebrates*[14]   Smith's *Keys to Marine*
Zim and Ingles', *Seashores*[2]   *Invertebrates etc.*[1]
Abbott's *Seashells of North America*[10]   Van Name's *Ascidians*[7]
Newells' *Marine Plankton*[9]   Dawson's *Seaweeds*[11]
Breder's *Fishes*[5]   Smith's *Corals*[13]

Any other convenient identification keys.

The projects summarized below are examples of exercises attempted by students during the course, which is of about a week's duration. These exercises require planning beforehand to avoid wasted time and effort. Transects are usually worked by teams of 4–6 students, and other exercises are allocated to students working in pairs. The value of these projects depends largely upon the provision of adequate class time for students to present accounts of their work for general discussion and criticism.

Rocky Shores

A. By plotting the occurrence of organisms on transects, and making reference to tide tables, investigate the relationships of the littoral fringe, eulittoral zone and sublittoral fringe[6] to the tidal levels of the shore. Compare the extent and levels of these zones on different shores, for example algal-dominated and barnacle-dominated shores.

B. Using the notation given below (page 304), plot the zonation of organisms on shores of various aspects, and attempt to relate to Ballantine's Exposure Scale[16].

C. Investigate the vital statistics of populations of selected shore molluscs, for example *Acmaea testudinalis*, *Mytilus edulis*, *Littorina littorea*, by plotting size/frequency curves from convenient shell measurements, for example height or maximum diameter. Compare the populations of different levels and different shores.

D. In appropriate quadrats on various shores and/or levels, investigate the following correlations; (*a*)/(*b*), and discuss your findings.

(*a*) Percentage cover or wet weight of macroflora.

(*b*) Numbers or wet weights of (*i*) *Acmaea spp.* (*ii*) *Littorina obtusata* (*iii*) *L. littorea*.

(*a*) Percentage cover, rough weights or numbers of barnacles.

(*b*) Rough weights or numbers of *Thais lapillus or Urosalpinx cinerea*.

(*a*) Ratio of shell height/length in *Acmaea* occupying surfaces of similar slope and aspect.

(*b*) **Tidal level.**

  (*a*) Mean size of *Acmaea* shells.

  (*b*) Numbers of *Acmaea* per unit area.

E. Marking experiments. For studying movements and homing tendencies, nail polish is a good marker for mollusc shells, for example *Acmaea*, *Thais*, littorinids and trochids. Felt-tipped pens can also be used. In the course of these observations, a visit to the shore during a night-time low tide should be made to compare the positions of specimens during day and night.

Sandy Shores

A. Plot zonation patterns for depositing shores by digging quadrats at points along line transects. For macrobenthos use coarse sieves. For meiobenthos use plastic kitchen strainers. Retain each collection separately for identification, counting and weighing. From each quadrat keep a sediment sample for estimation of grade and organic content, and an interstitial water sample for salinity determination. Note the depth of the unblackened layer. Measure the temperature gradient between the surface and deeper layers.

B. In the laboratory, determine the percentage weight of coarse sand, fine sand and silt in your sediment samples by sieving, and organic content by incineration. Relate your findings to your observations on the distribution of organisms.

C. Investigate the vital statistics of *Cardium* populations by sampling, and measuring growth zones on shells.

Time, weather and facilities permitting, the course also includes one or more of the following projects.

A. Visit to an estuary to observe the distribution of freshwater, estuarine and marine species. Where possible, methods are devised for measuring water depths and collecting water samples from the surface, middle depths and bottom at several stations along the estuary at intervals through the tidal cycle. Salinities, temperatures and oxygen-contents are measured. On diagrammatic sections of the estuary the isohalines, isotherms and percentage oxygen saturations are plotted at stages of the tide, and the zonation of organisms recorded.

B. Boatwork at sea for the following purposes.

  1. Demonstration of radio navigation equipment and sonic sounding apparatus.

  2. Demonstration of the use of insulated and reversing water bottles and thermometers.

3. Plankton sampling with various nets.

4. Collection of benthos and fish from various substrates using several types of collecting gear, for example Van-Veen grab, Agassiz trawl, Otter trawl. After sorting and identifying benthic material, the sampled communities are related to Jones' classification (page 158). Fish are examined for external and gill parasites, gut contents, scale and otolith markings and age/length relationship.

C. Visit to local fish auction and associated industries, for example fishing vessels, fish curing, processing and freezing installations, net factory and ice factory.

D. Day tour as far along the coast as practicable to observe biological, geographical and geological features.

E. Visit to a salmon and trout hatchery, and a fish ladder and counting station.

In applying the use of this text as a core for the presentation of a Marine Ecology course, it should be recognized that the classroom discussions and assignments which are generated by that course ought to comprise one-half the total curriculum. The practical application and demonstration of the principles of ecology defined in this text are best learned and reinforced through the use of field work as a means of learning by doing. The steadily growing interest which both teachers and students are showing for the experience of 'hands-on' learning in the sciences, is one of the best means we have of developing self-confidence in the student by making his education relevant to the real world and immediately productive in the class room.

The teaching of a Marine Ecology course may be based on an infinite variety of approaches since no two teachers, or physical locations, are alike. However, it is certainly possible to provide the broad parameters for such a course in order to guide the establishment of a new course, as provided in this Synopsis of a Field Course. Perhaps, the easiest curriculum to design is one which allows the class to spend part of its time in the field making the actual collections of data and specimens. Although the logistics of getting the class to the field station, and back, may be a major task, the excitement of that field work is contagious and works for the teacher to get, and keep, the students deeply absorbed in their learning experience. Once in the field, control of the class and collection of the necessary data may be facilitated by dividing the class into teams with pre-assigned duties for each member of those teams. Six general duties may be assigned to provide the following functions:

Recorder—This individual is in charge of a team and, like a pitcher in a baseball game, is responsible for the success or failure of that team. He is responsible for ensuring that all his team members understand their duties and have all their equipment. This is determined by using a master checklist. In the field, the Recorder keeps any notes of the data being collected and, therefore, serves as a focal point for the team.

Keeper—This individual is in charge of the collecting jars, trays, etc., and is responsible for keeping the specimens labelled and in good order. When numerous biological specimens are taken, this job is of particular importance since much of the identification and experimentation must be conducted back in the laboratory; and that requires the return of healthy and normal specimens to the laboratory.

Digger and Sifter—At least one pair of students should work together to locate the benthic organisms and sediment samples in the field station. Since many of the biological specimens are quite delicate, a fairly safe means of recovering a representative number of them is gently, but quickly, to scoop a shovel full of sediment and place it into a heavy duty screen (about 20 cm square) and then slowly raise and lower the screen just below the surface of the water. The sediments will fall through the screen and leave the biological specimens behind. The size of the screen determines the size of the sediments and specimens which will be retained.

Netters—One or more individuals may be assigned the duty of using various nets to collect nekton or plankton in the field station area and so their function will depend on both the equipment available and the geology of the field station.

Lookers—One or more individuals may be assigned the duty of using glass-bottomed buckets, styrene boxes, face plates, or the equivalent to look over the shallow water environment in the field station. Their duty is to record the epi-benthos, and to collect any other data which relates to the surface of the sediments and other substrate.

The object of making these assignments is not only to ensure some means of controlling a large class in the field, but also to help ensure that a field plan will have been thought out in advance and that the pertinent data will be collected without forgetting important areas. Perhaps of greater importance to a smoothly running programme, the assignment of these rotating duties ensures that all the class members have an opportunity to use different pieces of equipment and have a variety of responsibilities. Within a very short time, the teams work toward well-defined objectives and thus not only produce more, but they also learn the benefits of a division of labour. Such division does not dilute the rate of learning. On the contrary, it serves as a catalyst to instill self-confidence in both teacher and student to learn those ecological principles which serve to tie the overall investigation together.

If it is impossible to teach Marine Ecology in the field for any reason, a curriculum may be developed based on marine aquaria and materials shipped to the class from various supply houses throughout the country. Naturally, the specific nature of that course will be modified appropriately, but it may still reflect the excitement and productivity of using observations of living material as a prime source of laboratory work. Some of these sources of material are listed below.

Marine Science Equipment and Materials
(Key: B—biologicals, C—chemicals, E—equipment, K—kits, V—visuals)
Supply Department,
Marine Biological Laboratory,
Woods Hole, Mass. 02543.
(B)
Ward's Natural Science
  Establishment, Inc.,
PO Box 1712,
Rochester, N.Y. 14603
PO Box 1749,
Monterey, Calif. 93940.
(B, C, E, K, V)
Turtox Products,
General Biological Supply
  House, Inc.,
8200 So. Hoyne Avenue,
Chicago, Ill. 60620.
(B, C, E, K, V)
Carolina Biological Supply,
Burlington, N.C. 27215.
(B, C, E, K, V)
Maritime Biological Labs.,
PO Box 749,
St. Stephen, New Brunswick,
Canada.
(B)

Pacific Bio-Marine Supply Co.,
PO Box 536,
Venice, Calif. 90291.
(B)
Wildco,
Wildlife Supply Company,
2200 S. Hamilton Street,
Saginaw, Mich. 48602.
(E)
Marine World,
3452 West Devon Avenue,
Lincolnwood, Ill. 60646.
(B)
Western Biological Collectors,
PO Box 14675,
Phoenix, Ariz. 85001.
(B)
Hawaiian Marine Imports,
465 Town & Country Village,
Houston, Texas 77024.
(B)
Gulf Specimen Company, Inc.,
PO Box 237,
Panacea, Fla. 32346.
(B)
La Motte Chemical Products
  Company,
Chestertown, MD. 21620.
(C, E, K)

ABUNDANCE SCALE FOR INTERTIDAL ORGANISMS

Note. The following notation, devised by Crisp and Southward[17], Ballantine[16], and Moyse and Nelson-Smith[18], is useful for recording approximate numbers of certain littoral organisms.

A—abundant, *C*—common, *F*—frequent, *O*—occasional, *R*—rare.
N—not found

Anemones

A. Many in almost every pool and damp place.
C. Groups in pools and damp places.
F. Isolated specimens in few pools.
R. A small number, usually under 10, found after 30 min searching.

ABUNDANCE SCALE FOR INTERTIDAL ORGANISMS

Barnacles. Applicable to *B. balanoides, B. eburneus* and *B. crenatus*

    A. More than $1/cm^2$; rocks well covered.

    C. $10-100/dm^2$, up to one-third of rock space covered.

    F. $1-10/dm^2$; individuals never more than 10 cm apart.

    O. $1\cdot0-100/m^2$; few within 10 cm of each other.

    R. Less than $1/m^2$; only a few found in 30 min searching.

Limpets. Per square metre

    A. Over 50, or more than 50 per cent of limpets at certain levels.

    C. 10–50, or 10–50 per cent at certain levels.

    F. 1–10, or 1–10 per cent at certain levels.

    O. Less than $1/m^2$ on average, less than 1 per cent of population.

    R. Only a few found in 30 min searching.

Top shells, Urosalpiux cinerea and *Thais lapillus.* Per square metre

    A. Over 10 generally.

    C. 1–10, sometimes very locally over 10.

    F. Less than 1, locally sometimes more.

    O. Always less than 1.

    R. Only one or two found in 30 min searching.

Periwinkles

For small *Littorina saxatilis*

    A. Over $1/cm^2$ at HW, extending down the eulittoral zone.

    C. $10-100/dm^2$, mainly in littoral fringe.

    F. Less than $10/dm^2$, mainly in crevices.

    O. A few in most deep crevices.

    R. Only one or two found in 30 min searching.

For larger littorinids. Per square metre

    A. More than 50.

    C. 10–50.

    F. 1–10.

    O. Less than 1.

    R. Only one or two found in 30 min searching.

Mytilus edulis

    A. More than 20 per cent of cover at certain levels.

    C. Large patches at some levels.

    F. Many scattered individuals and small patches.

    O. Scattered individuals, no patches.

    R. Few seen in 30 min search.

APPENDIX 3

Spirorbis spp.
    A. 5 or more/cm² on 50 per cent of suitable surfaces.
    C. 5 or more/cm² on 5–50 per cent of suitable surfaces.
    F. 1–5/cm² or on 1–5 per cent of suitable surfaces.
    O. Less than 1/cm².
    R. Only a few found in 30 min search.

Fucoids, Laminaria spp, Alaria etc.
    A. More than 30 per cent cover.
    C. 5–30 per cent cover.
    F. Less than 5 per cent cover but zone still apparent.
    O. Scattered individuals, zone indistinct.
    R. A few found in 30 min search.

Lichens
    A. More than 20 per cent cover at some levels.
    C. 1–20 per cent cover, zone well-defined.
    F. Large scattered patches, zone ill-defined.
    O. Widely scattered patches, all small.
    R. A few small patches seen in 30 min search.

REFERENCES
Books
[1] Smith, R. I. (1964). *Keys to Marine Invertebrates of the Woods Hole region.* Woods Hole, Mass; Marine Biological Laboratory
[2] Zim, H. S. and Ingle, L. (1955). *Seashores.* New York; Golden Press
[3] Eales, N. B. (1967). *The Littoral Fauna of Great Britain.* 4th edn. Cambridge University Press
[4] Evans, S. M. and Hardy, J. M. (1970). *Seashore and Sand Dunes.* London; Heinemann
[5] Breder, C. M. (1948). *Field Book of Marine Fishes of the Atlantic Coast.* New York; G. P. Putnam
[6] Lewis, J. R. (1964). *The Ecology of Rocky Shores.* London; English University Press
[7] Van Name, W. G. (1945). *The North and South American Ascidians.* New York; American Museum of Natural History
[8] Moroney, M. J. (1965). *Facts from Figures.* Harmondsworth; Penguin
[9] Newell, G. E. and Newell, R. C. (1966). *Marine Plankton, A Practical Guide.* Revised edition. London; Hutchinson
[10] Abbott, R. T. (1968). *Seashells of North America.* New York; Golden Press
[11] Dawson, E. Y. (1956). *How to Know the Seaweeds.* Dubuque; W. C. Brown Co.
[12] Taylor, W. R. (1966). *Marine algae of the northeastern coast of North America.* Ann. Arbor; U. Michigan Press
[13] Smith, F. G. W. (1971). *Atlantic Reef Corals.* Coral Gables; U. Miami Press

[14] Light, S. F., Smith, R. I., Pitelka, F. A., Abbott, D. P. and Weesner, F. W. (1954). *Intertidal Invertebrates of the Central California Coast.* Berkley; U. California Press

Papers
[15] Barnes, W. J. P., Burn, J., Meadows, P. S. and McLusky, D. S. (1969). *'Corophium volutator*—An Intertidal Crustacean Useful for Teaching in Schools and Universities'. *J. Biol. Educ.* **3,** 283
[16] Ballantine, W. J. (1961). 'A Biologically Defined Exposure Scale for the Comparative Description of Rocky Shores'. *Fld Stud.* **1** (3), 1
[17] Crisp, D. J. and Southward, A. J. (1958). 'The Distribution of Intertidal Organisms along the Coasts of the English Channel'. *J. mar. biol. Ass. U.K.* **37,** 157
[18] Moyse, J. and Nelson-Smith, A. (1963). *Zonation of Animals and Plants on Rocky Shores around Dale, Pembrokeshire. Fld. Stud.* **1** (5), 1
[19] Murphy, J. and Riley, J. P. (1962). *Analytica chim. Acta.* **27,** 31

APPENDIX 4

SUPPLEMENTARY READINGS

One of the most generally available and useful journals which may serve to supplement this text is *Scientific American*, New York, published by Scientific American, Inc., 415 Madison Avenue. The following bibliography of articles which have appeared in that magazine are particularly useful as readings in a Marine Ecology course. It is followed by a listing of selected books similarly useful as supplements to a Marine Ecology course using the text as its core.

The title and author of the article are followed by the year and month of publication, and first page number. The subject arrangement of the articles follows that of the chapters of the text.

1. THE OCEANS
'Airborne Magnetomer, The', Homer Jensen, 1961, June, 151.
'Animal Sounds in the Sea', Marie Poland Fish, 1956, Apr., 93.
'Arctic Ocean, The', P. A. Gordienko, 1961, May, 88.
'Atlantic, The Anatomy of the', Henry Stommel, 1955, Jan., 30.
'Canyons, Submarine', Francis P. Shepard, 1949, Apr., 40.
'Canyons, The Origin of Submarine', Bruce C. Heezen, 1956, Aug., 36.
'Circulation, Models of Oceanic', D. J. Baker, Jr., 1970, Jan., 114.
'Circulation of the Oceans, The', Walter Munk, 1955, Sept., 96.
'Continental Drift', J. Tuzo Wilson, 1963, Apr., 86.
'Continental Drift and Evolution', Bjorn Kurteen, 1969, Mar., 54.
'Continental Drift, The Confirmation of', Patrick M. Hurley, 1968, Apr., 52.
'Continental Shelf, The', Henry C. Stetson, 1955, Mar., 82
'Continental Shelves, The', K. O. Emery, 1969, Sept., 106.

APPENDIX 4

'Continents, The Origin of', Marshall Kay, 1955, Sept., 62.
'Coriolis Effect, The', James E. McDonald, 1952, May, 72.
'Cromwell Current, The', John A. Knauss, 1961, Apr., 105.
'Current, The Peru', Gerald S. Posner, 1954, Mar., 66.
'East Pacific Rise, The', Henry W. Menard, 1961, Dec., 52.
'Fractures in the Pacific Floor', Henry W. Menard, 1955, July, 36.
'Geological Observatory, The Lamont', George W. Gray, 1956, Dec., 83.
'Geological Subsidence', S. S. Marsden, Jr., and S. N. Davis, 1967, June, 93.
'Islands in the Arctic, Ice', Kaare Rodahl, 1954, Dec., 40.
'Magnetism of The Ocean Floor, The', Arthur D. Raff, 1961, Oct., 1
'Mohole, The', Willard Bascom, 1959, Apr., 41.
'Ocean Floor, the Rift in the', Bruce C. Heezen, 1960, Oct., 98.
'Oceanic Ridges, The Origin of', E. Orowan, 1969, Nov., 102.
'Ocean, The', Roger Revelle, 1969, Sept., 54.
'Ocean, The Microstructure of the', Michael Gregg, 1973, Feb., 64.
'Oceans, The Atmosphere and the', R. W. Stewart, 1969, Sept., 76.
'Oceans, The Origin of the', Sir Edward Bullard, 1969, Sept., 66.
'Mediterranean Dried Up, When the', Kenneth J. Hsu, 1972, Dec., 26.
'Pacific Floor, Fractures in the', Henry W. Menard, 1955, July, 36.
'Pacific Rise, The East', Henry W. Menard, 1961, Dec., 52.
'Peru Current, The', Gerald S. Prosser, 1954, Mar., 66.
'Red Sea Hot Brines, The', E. T. Degens and D. A. Ross, 1970, Apr., 32.
'Sargasso Sea, The', John H. Ryther, 1956, Jan., 98.
'Sea-floor Spreading', J. R. Heirtzler, 1968, Dec., 60.
'Seismic Shooting at Sea', Maurice Ewing and Leonard Engel, 1962, May, 116.
'Submarine Canyons', Francis P. Shepard, 1949, Apr., 40.
'Tsunamic', Joseph Bernstein, 1954, Aug., 60.

2. MARINE PLANKTON
'Buoyancy of Marine Animals', Eric Denton, 1960, July, 118.
'Deep Scattering Layers, The Seas', Robert S. Dietz, 1962, Aug., 44.
'Man-of-War, The Portuguese', Charles E. Lane, 1960, Mar., 158.
'Man, Whales, Plankton and', Willis E. Pequegnat, 1958, Jan., 84.
'Salpa', N. J. Berrill, 1961, Jan., 150.

3. MEASURING AND SAMPLING
'Fossils, The Oldest', Elso S. Barghoorn, 1971, May, 30.
'Living Under the Sea', Joseph B. MacInnis, 1966, March, 24.
'Man, The Oceans and', Warren S. Wooster, 1969, Sept., 218.
'Oceanic Life, The Nature of', John D. Isaacs, 1969, Sept., 146.
'Sailing Yachts, The Study of', H. D. Herreshoff and J. N. Newman, 1966, Aug., 60.

'Technology and the Ocean', Willard Bascom, 1969, Sept., 198.
'Television, Underwater', W. R. Stamp, 1953, June, 32.

4. SOME PARAMETERS OF THE ENVIRONMENT
'Abyss, The Circulation of the', Henry Strommel, 1958, July, 85.
'Animals, The Buoyancy of Marine', Eric Denton, 1960, July, 118.
'Aquatic Life, Thermal Pollution and', John R. Clark, 1969, Mar., 18.
'Califaction of a River', D. Merriam, 1970, May, 42.
'Clouds, Trade-Wind', Joanne Starr Malkus, 1953, Nov., 31.
'Corals as Paleontological Clocks', S. K. Runcorn, 1966, Oct., 26.
'Desalting Water by Freezing', Asa E. Snyder, 1962, Dec., 41.
'Fresh Water from Salt', David S. Jenkins, 1957, Mar., 37.
'Hurricanes', R. H. Simpson, 1954, June, 32.
'Hurricanes, The Origin of', Joanne Starr Malkus, 1957, Aug., 33.
'Ocean Waves', Willard Bascom, 1959, Aug., 74.
'Mercury in the Environment', Leonard J. Goldwater, 1971, May, 15.
'Physical Resources of the Ocean, The', Edward Wenk, Jr., 1969, Sept., 166.
'Salt, Fresh Water From', David S. Jenkins, 1957, Mar., 37.
'Thermal Pollution and Aquatic Life', John R. Clark, 1969, Mar., 18.
'Water', Arthur M. Buswell and Worth H. Rodebush, 1956, Apr., 76.
'Water Cycle, The', P. Cloud and A. Gibor, 1970, Sept., 98.

5. ORGANIC PRODUCTION IN THE SEA
'Abyss, Animals of the', Anton F. Brunn, 1957, Nov., 50.
'Aging, The Great Lakes', Charles F. Powers and Andrew Robertson, 1966, Nov., 94.
'Algae as Food', Harold W. Milner, 1953, Oct., 31.
'Eelgrass Catastrophe, The', Lorus J. and Margery J. Milne, 1951, Jan., 52.
'Food From the Sea', Gordon A. Riley, 1949, Oct., 16.
'Food Resources of the Oceans, The', S. J. Holt, 1969, Sept., 178.
'Marine Farming', G. B. Pinchot, 1970, Dec., 14.
'Poisonous Tides', S. H. Hunter and John McLaughlin, 1958, Aug., 92

6. THE SEA BOTTOM
'Animals of the Bottom', Henry G. Vevers, 1952, July, 68.
'Antarctic Ocean, The', V. G. Kort, 1962, Sept., 113.
'Bathyscaph, The', Robert S. Dietz, Russel V. Lewis, and Andreas B. Rechnitzer, 1958, Apr., 27.
'Deep-ocean Floor, The', H. W. Menard, 1969, Sept., 126.
'Deep-sea Layer of Life, The', Lionel A. Walford, 1951, Aug., 24.
'Exploring the Ocean Floor', Hans Pettersson, 1950, Aug., 42.
'Minerals on the Ocean Floor', John L. Mero, 1960, Dec., 64.

'Ocean's Floor', 1949, Dec., 44.
'Pacific Floor, The', Robert S. Dietz, 1952, April, 19.
'Pacific, The Trenches of the', Robert L. Fisher and Roger Revelle, 1955, Nov., 36.

7. THE SEA-SHORE

'Beaches', Willard Bascom, 1960, Aug., 80.
'Behaviour in Gulls, The Evolution of', N. Tinbergen, 1960, Dec., 118.
'Biological Clocks and the Fiddler Crab', Frank A. Brown Jr., 1954, Apr., 34.
'Biological Luminescence', William D. McElroy and Howard H. Seliger, 1962, Dec., 76.
'Changing Level of the Sea, The', Rhodes W. Fairbridge, 1960, May, 70.
'Estuary, The Life of an', Robert M. Ingle, 1954, May, 64.
'Gulls, Visual Isolation in', Neal Griffith Smith, 1967, Oct., 94.
'Luminescence of Living Things, The', E. Newton Harvey, 1948, May, 46.
'Sand', Ph. H. Kuenen, 1960, April, 94.
'Tides and the Earth-Moon System', Peter Goldreich, 1972, May, 15.

8. SEA FISHERIES

'Behaviour of Sharks, The', Perry W. Gilbert, 1962, July, 60.
'Blue Whale, The', Jonhan T. Rudd, 1956, Dec., 46.
'Diving Women of Korea and Japan, The', Sukki Hong and Herman Rahn, 1967, May, 34.
'Electric Fishes', Harry Grundfest, 1960, Oct., 115.
'Electric Location by Fishes', H. W. Lissman, 1963, Mar., 50.
'Fish, The Ice', Johan T. Rudd, 1965, Nov., 108.
'Fishes, Air-Breathing', Kjell Johansen, 1968, Oct., 102.
'Fishes Swim, How', Sir James Gray, 1957, Aug., 48.
'Fishes, The Chemical Languages of', John H. Todd, 1971, May, 98.
'Fishes, The Schooling of', Evelyn Shaw, 1962, June, 128.
'Fishes with Warm Bodies', Francis G. Carey, 1973, Feb., 36.
'Fishes, Reflectors in', Eric Denton, 1971, Jan., 64.
'Hagfish, The', David Jansen, 1966, Feb., 82.
'Homing Salmon, The', Arthur D. Hasler and James A. Larsen, 1955, Aug., 72.
'Lamprey, The Sea', Vernon C. Applegate and James W. Moffett, 1955, Apr., 36.
'Last of the Great Whales, The', Scott McVay, 1966, Aug., 13.
'Penguins', William J. L. Sladen, 1957, Dec., 44.
'Penguins, The Navigation of', J. E. Emlen, and R. L. Penney, 1966, Oct., 104.
'Salmon, The Swimming Energetics of', J. R. Brett, 1965, Aug., 80.

'Seal, The Weddell', Gerald L. Kooyman, 1969, Aug., 100.
'Sharks, The Behaviour of', Perry W. Gilbert, 1962, July, 60.
'Sharks v. Men', George A. Llano 1957, June, 54.
'Squid, The', H. B. Steinback, 1951, Apr., 64.
'Whale Cardiogram', 1952, Oct., 68.
'Whale, The Return of the Gray', Raymond M. Gilmore, 1955, Jan., 62.
'Whales, The Last of the Great', Scott McVay, 1966, Aug., 13.
'Whales, The Physiology of', Cecil K. Drinker, 1949, July, 52.

9. ENERGY RELATIONSHIPS OF MARINE PRODUCTION
'Algae as Food', Harold W. Milner, 1953, Oct., 31.
'Algae, the Useful', Francis Joseph Weiss, 1952, Dec., 15.
'Coelacanth, The', Jacques Milot, 1955, Dec., 34.
'Discoveries in Nitrogen Fixation', Martin D. Kamen, 1953, Mar., 38.
'Energy in the Biosphere, The Flow of', David M. Gates, 1971, Sept., 88.
'Food from the Sea', Gordon A. Riley, 1949, Oct., 16.
'Food Resources of the Oceans, The', S. J. Holt, 1969, Sept., 178.
'Oysters', Pieter Korringa, 1953, Nov., 86.
'Salt-water Agriculture', Hugo Boyko, 1967, Mar., 89.

APPENDICES
'Brain Cells in Mollusks, Giant', A. O. D. Willows, 1971, Feb., 68.
'Cleaning Symbiosis', Conrad Limbaugh, 1961, Aug., 42.
'Electricity, Animal', H. B. Steinbach, 1950, Feb., 40.
'Learning in the Octopus', Brian B. Boycott, 1965, Mar., 42.
'Marine Invertebrates, Escape Responses', Howard M. Feder, 1972, July, 92.
'Phalarope, The', E. Otto Hohn, 1969, June, 104.
'Teredo, The', Charles E. Lane, 1961, Feb., 132.
'Water-breathing, Experiments in', Johannes A. Kylstra, 1968, Aug., 66.

Paperback Books

Paperback books useful as supplements to a course in Marine Ecology are listed below with an indication of the current retail price. They represent an easy means for the novice student and teacher to augment their learning of the principles of Marine Ecology and to acquire a personal library on the subject.

(Anon.) *University Curricula in the Marine Sciences and Related Fields*, Academic years 1960–70 and 1970–71, 273 pp. Washington, D.C.; Marine Sciences Affairs Staff, Office of the Oceanography of the Navy, U.S. Government Printing Office. ($2.00)
Abbott, R. T. *How to Know American Marine Shells*, New York KT375; Signet. ($.75)
Abbott, R. T. (1968). *A Guide to Field Identification, Seashells of North America*. New York; Golden Press. 280 pp. ($3.95)

APPENDIX 4

Bascom, W. (1964). *Waves and Beaches*, Anchor Science Study Series, N.Y.; Doubleday. ($1.45)

Behram, A. S. (1968). *Water is Everybody's Business*. Doubleday Anchor (AMC 4). 229 pp. ($1.45)

Berrill, N. J. *1001 Answers to Questions About the Seashore*, N.Y.; Grosset & Dunlap. ($1.75)

Blanchard, D. C. (1967). *From Raindrops to Volcanos, Adventures With Sea Surface Meterology*, Doubleday Anchor, (S50). 180 pp. ($1.25)

Boolootian, R. A. (1968). *Marine Biology, A Study of Life In The Sea*, N.Y.; Holt. 112 pp. ($2.75)

Burke, Wm. T. (1966). *Ocean Sciences, Technology and the Future International Law of the Sea*, Ohio State University Press. 91pp. ($1.50)

Carrington, R. A. *Guide to Earth History*, N.Y.; Signet, MT335. ($.75)

Carson, Rachel. *The Edge of the Sea*, N.Y., P2360; Signet. ($.60). *The Sea Around Us*, N.Y., P2361; Signet. ($.60). *Under the Sea Wind*, N.Y., P2339; Signet. ($.60). *Silent Spring*, N.Y.; Fawcett Publications. ($.60)

Carson, R. (1969). *Life Under the Sea*, N.Y.; Golden Press. 80 pp. ($.75)

Chapil, H. and Smith, F. G. W. *Ocean River*, N.Y., SL63; Scribners. ($1.65)

Chapman, V. J. *Coastal Vegetation*, Pergamon-MacMillan. ($3.75)

Clark, J. (1967). *Fish and Man, Conflict in the Atlantic Estuaries*, Special Pub. No. 5. American Littoral Society, N.J. 07732; Highlands. 78 pp. ($1.00)

Clarke, A. C. *The Challenge of the Sea* (Intro. by Werner von Braun), N.Y., #1159; Dell. 191 pp. ($.50)

Clarke, A. C. with Wilson, Mike. *The Treasure of the Great Reef*, Perennial Library, P25; Harper & Row. 209 pp. ($.85)

Coates, C. W. *Tropical Fish as Pets*, N.Y., AS 177; Collier Books. ($.95)

Coker, R. E. (1962). *This Great and Wide Sea*, Harper Torchbook, TB 551. 325 pp. ($2.25)

Coustea, J. Y. *The Silent World*, Pocket Books, GC119. ($.50)

Cowen, R. C. (1963). *Frontiers of the Sea*, N.Y.; Bantam Books HP29. ($.60)

Daniel, H. and Minot, F. *The Inexhaustible Sea*, N.Y., AS9K; Collier Books. ($.95)

Darwin, C. (1962). *The Structure and Distribution of Coral Reefs*, University of California Press. ($1.95)

Darwin, G. H. *The Tides*, San Francisco; Freeman & Co. ($2.75)

Dawson, E. Y. (1956). *How to Know the Seaweed*, Dubuque; W. C. Brown. ($2.25)

Defant, A. (1958). *Ebb and Flow: The Tides of Earth, Air and Water*, University Press, AAS506; Ann Arbor, Michigan. ($1.95)

Dibner, B. *The Atlantic Cable*, Blaisdel BP15. ($1.95)

Dowdeswell, W. *Animal Ecology*, Harper Torchbook, TB543. ($1.50)

Dubach, H. W. and Taber, R. W. *Questions about the Oceans*, Wash., D.C.; U.S. G.P.O. 121 pp. ($.55)

Eddy, S. (1957). *How to Know the Fresh-water Fishes*, Dubuque; W. C. Brown. ($2.75)

Forbes, L. Compiler. (1968). *Oceanography In Print*. Falmouth, Mass.; Oceanographic Education Center. 58 pp. ($.75)

Galstoff, P. S. (1964). *The American Oyster*, U.S.F.W.S. Fishery Bulletin Vd. 64. Wash. D.C. 20402; U.S. Government Printing Office. 480 pp. ($2.75)

Gordon, Bernard L. *Handbook for Advisers to Junior American Littoral Society*, Amer. Littoral Society. Highlands, N.J. 07732

Gross, M. G. (1967). *Oceanography*, Ohio; C. E. Merrill Books. 135 pp. ($2.75)

Guberlet, M. L. (1956). *Seaweeds at Ebb Tide*, University of Washington Press. ($2.95)

Gullion, E. A., Editor. (1969). *Uses of the Seas*, N.J. S-AA-24; Spectrum Prentice Hall, Englewood Cliffs. ($2.45)

SUPPLEMENTARY READINGS

Helm, I. (1963). *Shark, Shark!* N.Y.; Collier Books. ($.95)

Heyerdahl. *Kon Tiki*, Pocket Books. ($.50)

Hirsch, P. (Editor). (1966). *Underwater*, N.Y.; Pyramid Books. 128 pp. ($.50)

Keen, M. J. (1968). *An Introduction to Marine Geology*, Pergamon Press. 218 pp. ($4.50)

Kuenen, P. H. (1963). *Realms of Water*, N.Y.; Sci. Eds., Wiley. ($1.95)

Lane, F. W. (1963). *Kingdom of the Octopus*, N.Y., WS12; Pyramid Pub. ($.75)

Lawrence, L. George. (1967). *Electronics in Oceanography*, Indianapolis; H. W. Sams, Bobbs Merrill Co. 288 pp. ($4.95)

Lewis, W. M. (1963). *Man and Dolphins*, N.Y. WS11; Pyramid Pub. ($.75)

Lilly, J. C., M.D. (1969). *The Mind of The Dolphin*, N.Y. NS 38; Avon—Discus Book. 286 pp. ($.95)

Long, E. J. (1965). *New Worlds of Oceanography*, N.Y.; Pyramid Pub. ($.75)

Mager, N. H. *A Guide to Tropical Fish*, N.Y., W629; Wash. Sq. Press. ($.60)

Morgan, A. *Aquarium Book*, N.Y., SL101; Scribners. ($1.25)

Morton, J. E. *Mollusca*, N.Y., TB524; Harper Torchbook, ($1.40)

North, W. J. (1968). *Golden Guide to Scuba Diving, Handbook of Underwater Activities*, N.Y.; Golden Press. 160 pp. ($1.25)

Peterson, M. (1965). *History Under the Sea: A Handbook of Underwater Exploration*, Wash., D.C.; Smithsonian Institute. 108 pp., 56 plates. ($3.00)

Prescott, G. W. *How to Know the Fresh-water Algae*, Wm. Brown Co. ($2.25)

Rapport, S. *The Crust of the Earth*, Signet, P2083. ($.60)

Reid, G. K. et al. (1967). *Pond Life*, A Guide to Common Plants and Animals of N. American Ponds and Lakes. A Golden Nature Guide. N.Y.; Golden Press. 160 pp. ($1.25)

Scott, W. B. *Freshwater Fishes of Eastern Canada*, University of Toronto Press. 137 pp. ($2.25)

Shepard, F. P. (1964). *The Earth Beneath the Sea*, N.Y.; Antheneum. ($1.65)

Smith, R. F., Swartz, A. H. and Massmon, W. H., Editors. (1966). *A Symposium on Estuarine Fisheries*, Washington, D.C. 20005; American Fisheries Society Special Publ. No. 3. 154 pp. ($2.00)

Spar, J. *Earth, Sea and Air*, Reading, Mass.; Addison-Wesley. ($1.95)

Stewart, H. (1963). *The Global Sea*, N.J.; Van Nostrand. 126 pp. ($1.95)

Taber, R. W., LaPorte, L. R. and Smith, E. C. (1968). *An Oceanographic Curriculum For High Schools Outline*, Wash. 20402; U.S. Navy Oceanographic Office. For Sale by U.S. Government Printing Offices. 30 pp. ($.35)

Turekian, K. K. (1968). *The Oceans*, N.J.; Prentice Hall. 120 pp. ($2.75)

Wickstead, J. H. (1965). *An Introduction To the Study of Tropical Plankton*, London; Hutchinson & Co. 160 pp. (£1.25)

Yasso, W. E. (1965). *Oceanography: A Study of Inner Space*, N.Y.; Holt. ($1.95)

Younge, C. M. (1963). *The Seashore*, N.Y.; Atheneum. ($1.95)

Zim, H. S. *Fishes*, N.Y.; Golden Press. ($1.25). *Seashore*, N.Y.; Golden Press. ($1.25). *Seashells of the World*, N.Y.; Golden Press. ($1.25)

(list compiled by B. L. Gordon)

APPENDIX 5

METRIC UNITS OF MEASUREMENT AND THEIR CONVERSION

On 19 March 1791, a committee of five scientists presented a report to the Academie des Sciences in France proposing a new unit of measure equal to one ten-millionth the distance between the earth's equator and its pole. This unit of measure, the metre, also formed the basis for their definition of a standard of mass, the kilogram. That unit was defined as the mass of a cubic decimeter ($1/10$ metre^3) of water at its maximum density.

During the ensuing years, the metre has been redefined with increasing precision and in October, 1960, the Eleventh General (International) Conference on Weights and Measures defined it again as being equivalent to $1,650,763 \cdot 73$ wavelengths of the orange–red radiation in vacuum of krypton 86 corresponding to the unperturbed transition between the $2p_{10}$ and $5d_5$ levels.

The kilogram is now defined as the mass of a particular platinum–iridium standard called the International Prototype Kilogram, which is preserved at the International Bureau of Weights and Measures in Sèvres, France.

Units of Measurement—Conversion Factors*
UNITS OF LENGTH

To Convert from **Centimetres**			To Convert from **Metres**	
To	Multiply by		To	Multiply by
Inches	0·393 700 8		Inches	39·370 08
Feet	0·032 808 40		Feet	3·280 840
Yards	0·010 936 13		Yards	1·093 613
Metres	**0·01**		Miles	0·000 621 37
			Millimetres	**1 000**
			Centimetres	**100**
			Kilometres	**0·001**

* All boldface figures are exact; the other generally are given to seven significant figure.

In using conversion factors, it is possible to perform division as well as the multiplication process shown here. Division may be particularly advantageous where more than the significant figures published here are required. Division may be performed in lieu of multiplication by using the reciprocal of any indicated multiplier as divisor. For example, to convert from centimetres to inches by division, refer to the table headed 'To Convert from *Inches*' and use the factor listed at 'centimetres' (*2.54*) as divisor.

METRIC UNITS OF MEASUREMENT AND THEIR CONVERSION

To Convert from **Inches**	To Convert from **Feet**
To Multiply by	To Multiply by

To Convert from
Inches

To	Multiply by
Feet	0·083 333 33
Yards	0·027 777 78
Centimetres	2.54
Metres	0.025 4

To Convert from
Feet

To	Multiply by
Inches	12
Yards	0·333 333 3
Miles	0·000 189 39
Centimetres	30·48
Metres	0·304 8
Kilometres	0·000 304 8

To Convert from
Yards

To	Multiply by
Inches	36
Feet	3
Miles	0·000 568 18
Centimetres	91·44
Metres	0·914 4

To Convert from
Miles

To	Multiply by
Inches	63 360
Feet	5 280
Yards	1 760
Centimetres	160 934·4
Metres	1 609·344
Kilometres	1·609 344

UNITS OF MASS

To Convert from
Grams

To	Multiply by
Grains	15·432 36
Avoirdupois Drams	0·564 383 4
Avoirdupois Ounces	0·035 273 96
Troy Ounces	0·032 150 75
Troy Pounds	0·002 679 23
Avoirdupois Pounds	0·002 204 62
Milligrams	1 000
Kilograms	0·001

To Convert from
Kilograms

To	Multiply by
Grains	15 432·36
Avoirdupois Drams	564·383 4
Avoirdupois Ounces	35·273 96
Troy Ounces	32·150 75
Troy Pounds	2·679 229
Avoirdupois Pounds	2·204 623
Grams	1 000
Short Hundredweights	0·022 046 23
Short Tons	0·001 102 31
Long Tons	0·000 984 2
Metric Tons	0·001

APPENDIX 5

To Convert from **Millilitres**		To Convert from **Litres**	
To	Multiply by	To	Multiply by
Minims	16·230 73	Liquid Ounces	33·814 02
Liquid Ounces	0·033 814 12	Gills	8·453 506
Gills	0·008 453 5	Liquid Pints	2·113 376
Liquid Pints	0·002 113 4	Liquid Quarts	1·056 688
		Gallons	0·264 172 05
Liquid Quarts	0·001 056 7		
Gallons	0·000 264 17	Cubic Inches	61·023 74
Cubic Inches	0·061 023 74	Cubic Feet	0·035 314 67
Litres	**0·001**	Millitres	**1 000**
		Cubic Metres	**0·001**
		Cubic Yards	0·001 307 95

Units of Capacity, or Volume, Dry Measure

To Convert from **Litres**		To Convert from **Cubic Feet**	
To	Multiply by	To	Multiply by
Dry Pints	1·816 166	Liquid Ounces	957·506 5
Dry Quarts	0·908 082 98	Gills	239·376 6
Pecks	0·113 510 4	Liquid Pints	59·844 16
Bushels	0·028 377 59	Liquid Quarts	29·922 08
Decalitres	**0·1**	Gallons	7·480 519
		Cubic Inches	**1 728**
		Litres	**28·316 846 592**
		Cubic Metres	**0·028 316 846 592**
		Cubic Yards	0·037 037 04

Units of Area

To Convert from **Square Centimetres**		To Convert from **Square Metres**	
To	Multiply by	To	Multiply by
Square Inches	0·155 000 3	Square Inches	1 550·003
Square Feet	0·001 076 39	Square Feet	10·763 91
Square Yards	0·000 119 599	Square Yards	1·195 990
Square Metres	**0·000 1**	Acres	0·000 247 105
		Square Centimetres	**10 000**
		Hectares	**0·000 1**

METRIC UNITS OF MEASUREMENT AND THEIR CONVERSION

To Convert from
Square Inches

To	Multiply by
Square Feet	0·000 944 44
Square Yards	0·000 771 605
Square Centimetres	6·451 6
Square Metres	0·000 645 16

To Convert from
Square Feet

To	Multiply by
Square Inches	144
Square Yards	0·111 111 1
Acres	0·000 022 957
Square Centimetres	929·030 4
Square Metres	0·092 903 04

To Convert from
Square Yards

To	Multiply by
Sq. Inches	1 296
Sq. Feet	9
Acres	0·000 206 611 6
Sq. Miles	0·000 000 322 830 6
Sq. Cms	8 361·273 6
Sq. Metres	0·836 127 36
Hectares	0·000 083 612 736

To Convert from
Square Miles

To	Multiply by
Sq. Feet	27 878 400
Sq. Yards	3 097 600
Acres	640
Sq. Metres	2 589 988·110 336
Hectares	258·998 811 033 6

INDEX

Abra = *Syndosmya*, 157, 159, 164
Abyssalbenthic zone, 16
    food sources, 165–6
Abyssal plain, 3
Acartia, 28, 141
Acmaea, 69, 187, 196, 198, 203
Actinia, 186
Adiabatic compression, 86
Agassiz trawl, 54
Age-groups of fish, 228
Aglantha, 31, 111, 113
Alca, 198
Ammonium, 79, 122
Ammophila, 199
Amphidromic point, 173
Amphipoda, 29, 69, 70, 97, 113ff.
Amphisolenia, 22
Amphitrite, 200
Anaerobic conditions, 77, 78, 200, 201
Anchor dredge, 54
Anchovy, 63, 69, 272
Anemonia, 69
Animal exclusion (*see* Exclusion)
Antarctic, 9, 11, 13, 129, 146, 280
    Convergence, 12, 66
Antiboreal zone, 68
Aphotic zone, 97
Appendicularia, 32
Archibenthic zone, 16
Arctic, 3, 9, 11, 66, 68, 75, 99, 114, 115,
    147
Ardea, 205
Arenicola, 76, 158, 162, 164, 189, 200, 201
Argyropelecus, 106, 108
Artificial rearing of fish, 273, 277
Artificial sea-water, 79, 83
Ascophyllum, 194, 196, 198
Asterina, 69
Asterionella, 20, 21
Astronesthes, 107
Astropectin, 200
Aurelia, 30
Auxospore, 21

Bacteria, 49, 78–9, 122, 166, 200, 291
Balanus, 68, 69, 72, 77, 102, 141, 143, 158–
    159, 163, 164, 186, 195, 196, 203, 204
Baltic Sea, 75, 77
Bathymicrops, 106
Bathypelagic zone, 15–16
Bathyplankton, 18
Bathyscaphe, 59
Bathysphere, 59
Bathythermograph, 41
Beach,
    drifting, 178
    formation of, 177
    rocky, 193
    sandy, 199
Bends, 58
Benguela current, 7, 9
Benthic division, 15–16, 150ff., 290ff.
Benthos, 17
    communities, 155ff.
    energy flow, 290ff.
    food sources, 163ff.
    sampling, 52ff.
Benthoscope, 59
Bicarbonate, 78
Biddulphia, 20, 21, 113, 135, 141, 143
Biogeographic areas, 68
Bioluminescence, 106ff.
Bipolarity, 69
Birds, 183, 198, 205
Black Sea, 64, 75, 77
Bluefin tuna, 209
Blue-green algae (*see* Cyanophyceae)
Body fluids, 76
Boreal zone, 68
    communities of, 157ff.
Bossanyi hypoplankton net, 54
Botryllus, 68
Brissopsis communities, 157
Brotulids, 91
Bubbles, 84
Buffer action, 78
Buoyancy problems, 89ff.

319